Alternative Energy and Hybrid Fuels

Alternative Energy and Hybrid Fuels

Edited by **Craig Zodikoff**

SYRAWOOD
PUBLISHING HOUSE

New York

Published by Syrawood Publishing House,
750 Third Avenue, 9th Floor,
New York, NY 10017, USA
www.syrawoodpublishinghouse.com

Alternative Energy and Hybrid Fuels
Edited by Craig Zodikoff

International Standard Book Number: 978-1-68286-072-4 (Hardback)

Printed in the United States of America.

Contents

Preface

Hybrid fuels are considered as alternate options to replace fossil fuels because they have low carbon emission. They are evolving as alternative energy resources with least negative consequences. In recent years, this field has undergone extensive research and emerged as a field of great interest and importance. The book emphasizes on alternative energy services, energy efficiency and power engineering. It also deals with the various aspects and applications of energy resources, renewable energy and energy storage. The book elucidates the significant concepts and techniques of various related fields. Students and academicians related to this field will find this book helpful.

After months of intensive research and writing, this book is the end result of all who devoted their time and efforts in the initiation and progress of this book. It will surely be a source of reference in enhancing the required knowledge of the new developments in the area. During the course of developing this book, certain measures such as accuracy, authenticity and research focused analytical studies were given preference in order to produce a comprehensive book in the area of study.

This book would not have been possible without the efforts of the authors and the publisher. I extend my sincere thanks to them. Secondly, I express my gratitude to my family and well-wishers. And most importantly, I thank my students for constantly expressing their willingness and curiosity in enhancing their knowledge in the field, which encourages me to take up further research projects for the advancement of the area.

Editor

Preface

Next generation biorefineries will solve the food, biofuels, and environmental trilemma in the energy–food–water nexus

Y.-H Percival Zhang[1,2,3,4]

[1]Biological Systems Engineering Department, Virginia Tech, 304 Seitz Hall, Blacksburg, Virginia, 24061
[2]Institute for Critical Technology and Applied Science (ICTAS), Virginia Tech, Blacksburg, Virginia, 24061
[3]Gate Fuels Inc., 2200 Kraft Drive, Suites 1200B, Blacksburg, Virginia, 24060
[4]Cell-Free Bioinnovations Inc., Blacksburg, Virginia, 24060

Keywords

Artificial photosynthesis, biofuels, biomass role, biorefineries, energy–food–water nexus, food– biofuels–environmental trilemma, food security.

Abstract

The future roles of biomass and carbohydrate for meeting needs of food/feed, renewable materials, and transportation fuels (biofuels) remain controversial due to numerous issues, such as increasing food and feed needs, constraints of natural resources (land, water, phosphate, biomass, etc.), and limitations of natural photosynthesis, as well as competing energy conversion pathways and technologies. The goal of this opinion article is to clarify the future roles of biomass and biorefineries using quantitative data other than adjective words. In most scenarios, human beings could have enough biomass resource from plant photosynthesis for meeting the three goals at the same time: feeding 9 billion people, providing renewable materials, and producing transportation biofuels that could replace nearly all fossil fuel-based liquid fuels used in the land transportation in 2050. Land transport means will pass through transitions from internal combustion engines plus liquid fuels, to hybrid systems, to hydrogen fuel cell vehicles (FCVs), while battery electric vehicles (BEVs) could play a minor role. Next generation biorefineries based on artificial photosynthesis featuring ultra-high energy efficiency and low-water consumption could produce a large amount of carbohydrate and/or other biocommodities from hydrogen/electricity and CO_2. In conclusion, it is time to develop next generation biorefineries, which will efficiently utilize nonfood biomass for the coproduction of multiple products from biofuels, biochemicals, to food/feed, and even store electricity/hydrogen by fixing CO_2 to carbon-containing chemicals and biofuels. Next generation biorefineries will address the food, biofuels, and environment trilemma at the same time.

Introduction

Modern civilization is the product of incessant utilization of natural resources on large scales: fossil fuels (e.g., oil, gas, and coal), renewable energy (e.g., biomass, wind, and solar), water, and land [1–4]. Among finite fossil fuels, cheap crude oil will run out first within next several decades [5, 6]. Therefore, it is a great scientific and engineering challenge to replace cheap oil with something that can be produced from renewable resources [7]. Feeding the world population from 7 billion now to 9 billion in 2050 [8] poses another challenge by considering constraints of natural resources – limited farming land supplies and emerging water crisis [9–11]. In addition, food security is closely related to issues of food distribution, geopolitical stability, cost volatility, and functional nutrition [12], which are not discussed here. Although water is renewable, the collective fresh water demand of human beings could exceed foreseen supply by ca. 40% in 2030 [13]. This water shortage could escalate food prices, disrupt energy production, constrict trade, create refugees, and undermine authority [13].

Biomass is defined as biological materials. Nearly, all biomass (i.e., plant, animal, and microbial) originates from CO_2 fixation by natural photosynthesis. It has played important roles in human societies: (i) cereals from cultivated grains and grass from managed pastures are food and feed sources, respectively, accounting for approximately 2.0% of terrestrial net primary production (NPP); (ii) approximately, 2.3% of terrestrial biomass is directly burned for cooking and heating, especially in developing countries, or eventually converted to biogas as a secondary energy carrier; (iii) wood and other cellulosic materials accounting for approximately 1% of terrestrial NPP is used as construction materials and to make paper and renewable polymers (e.g., cellophane, rayon); and (iv) approximately, 0.2% of terrestrial NPP (e.g., corn kernels, sugarcane, and vegetable oil) is converted to liquid transportation biofuels in first generation biorefineries. It is important to retain biomass' irreplaceable roles as food/feed, construction materials, papers, and renewable polymers and then investigate whether there will be enough extra biomass resource to meet other needs.

Biofuels are defined as a secondary energy used in the transport sector, which is derived mainly from biomass [14–16]. First-generation biofuels include ethanol produced from sugarcane and starch-rich biomass (e.g., corn kernels, wheat, and aged cereals) and biodiesel produced from vegetable oils and animal fats. First-generation biofuels produced from food source receive severe criticisms because their impacts on the transportation fuels are minimal mainly due to limited feedstock supplies and their production has a minimum effect on a reduction of net greenhouse gas emissions [17, 18]. For example, it is estimated that replacing 5% of energy consumption through first generation biofuels could double water withdrawals for agriculture [13]. Clearly, the global production of first generation biofuels is not sustainable and is endangering current agricultural systems. Second-generation biofuels are produced mainly from nonfood biomass, such as cellulosic ethanol, butanol, fatty acid ethyl esters, methane, hydrogen, methanol, dimethylether, Fischer-Tropsch diesel, and bioelectricity [4, 19]. Because there are so many different energy conversion pathways (i.e., biological, thermochemical, and their hybrids) to converting nonfood biomass to a large variety of potential biofuels, which biofuels will become short-, middle- and long-term transportation fuels is a matter of vigorous debate [4, 19, 20]. Additionally, the future role of biomass in the nexus of energy, water, and food is not clear.

This article provides much-needed clarity on the desirability and feasibility of next generation biorefineries that will utilize nonfood biomass resource and/or even fix CO_2 through artificial photosynthesis. Such biorefineries will meet needs of food/feed and biofuels while not endangering water security and maintaining biodiversity.

Appraisal facts

It is necessary to provide some quantitative data pertaining to energy production and consumption, resource availability, and constraints of natural resources and biosystems before the potential impacts of next generation biorefineries are predicted.

Energy status quo

Generally speaking, energy demands determine energy production and conversion [4]. Typical energy systems are comprised of three basic components: primary (natural) energy sources, their conversion to secondary energies (i.e., #1 and #2 intermediates), and end applications from food/feed to energy to materials. In the past human societies, the simplest systems utilized one or two of energy sources (e.g., biomass) through few kinds of inefficient energy conversion for meeting basic needs – food/feed and cooking/heating. In contrast, modern societies can utilize numerous primary sources (e.g., fossil fuels, insolation, nuclear, and wind energy), convert them to a few energy carriers (e.g., electricity, hydrogen, and liquid fuels) with enhanced energy conversion efficiencies, and apply them in a myriad of ways to power complex high-energy societies [4, 21]. Figure 1 presents future pathways between basic needs and renewable primary energy without fossil fuels. In it, the needs of food/feed and renewable materials (i.e., paper, timber, and polymers) will have to be met by biomass and/or carbon-containing compounds made from artificial photosynthesis, while the energy needs (e.g., transport, heat/cooling) could be met through a variety of energy intermediates from numerous primary energy sources.

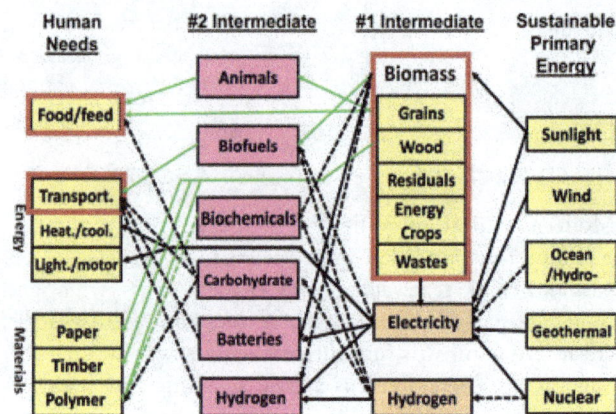

Figure 1. Human needs are and will be met from sustainable primary energies through numerous intermediates. Solid lines mean practical conversions; dash lines mean hypothetical conversions in the future.

Table 1. The world energy production and some major applications.

Name	Power (TW)	Percentage	References
Fossil fuels	13.10	72.02%	[22]
Oil	5.22	28.70%	[22]
Gasoline	1.20	6.60%	[22]
Middle distillates (diesel)	1.79	9.84%	[22]
Jet fuel	0.32	1.76%	[22]
Gas	3.61	19.85%	[22]
Coal	4.27	23.47%	[22]
Nuclear	0.29	1.61%	[22]
Renewables	4.80	26.39%	
Biomass - heating	1.50	8.25%	[29]
Food/feed[1]	1.33	7.31%	[11]
Wood[2]	1.28	7.04%	[100]
Hydroelectricity	0.39	2.15%	[22]
New renewables[3]	0.30	1.64%	[22, 32]
Total	18.19	100.01%	

[1]2.5 billion tons of cereals and grass [11].
[2]3.5 billion Cubic meters [100].
[3]Including solar cells, biofuels, wind, and geothermal energy.

Table 1 presents status quo of the world's energy production, where food/feed and wood consumption are included, because they are major energy consumption sections and their production greatly competes with the production of other energy for requiring water and land. Fossil fuels including oil, gas, and coal account for approximately 72% of the world's energy consumption. Crude oil is the largest primary energy and its major usage is transportation fuels – gasoline, middle distillates (e.g., diesel), jet fuels, and fuel oil, accounting for more than 60% of oil consumption [22]. Renewable energy resources accounts for approximately 26% of the world's energy consumption. Biomass is the largest utilized renewable energy source: heating fuel (i.e., 1.50 TW), 2.5 billion tons of food (i.e., 1.33 TW), and 3.5 billion cubic meters of wood (i.e., 1.28 TW). In all, the world's energy consumption is estimated to be 18.2 TW. This value is approximately 20% higher than the widely used value of 15 TW in the literature [23] because the smaller value does include neither food/feed nor wood consumption for materials.

Transportation fuels

Mobility reflects the level of civilization [4, 24, 25]. Human societies have passed through two transportation revolutions: from animal forces to external combustion engines to internal combustion engines (ICEs) [2, 16, 19, 26]. Affluent countries consume more transportation energy per capita than developing countries. For example, the global transport sector consumes approximately 20% of the energy produced (Table 1), and the transport sector

in the United States consumes approximately 28% of the total energy [16].

Vehicles running on land constitute the largest type of transportation energy consumption. They have some special requirements, such as high energy storage capacity in a small container, high power output, affordable fuel, affordable vehicle, low costs for rebuilding the relevant infrastructure, fast charging or refilling of the fuel, and high safety [16]. Such strict requirements lead to the outcome – that ICEs along with high-energy density liquid fuels are the dominant transport means [4]. However, the depletion of crude oil, rising prices of crude oil, the accumulation of greenhouse gases, and concerns of national energy security are motivating the development of new sustainable transport means.

Food and feed

Food is fundamental to human well-being and development [27]. Henry Kissinger, a former US Secretary of State, said "control oil and you control nations; control food and you control the people." In 5000 years of Chinese history, a lack of food supplies frequently resulted in dynasty shifts. Increasing food production is believed to effectively alleviate global food insecurity and stabilize societies.

The global energy market in terms of calories is approximately 13 times the food and feed market (Table 1). The ratios of the overall energy to food/feed are higher than 20 or even 40 in affluent countries and lower than 10 in developing countries [1, 28]. Therefore, the production of food/feed could be not as important in developed countries as in developing countries from a perspective of energy production and consumption.

Food and feed production is water-intensive. A simple rule of thumb is that it takes a half to one L of water to grow one calorie of cereals, depending on cultivation conditions and cereal types [13]. For example, the production of one kilogram of wheat requires the use of 1300 kg of water on average. Meat production, on average, requires about ten times the water per calorie than that of plants. For example, 10,000–20,000 kg of water are required to produce 1 kg of beef [13].

Currently, human beings consume approximately 2.5 billion tons of dry weight of harvested crops include approximately 2.3 billion tons of cereals (e.g., rice, wheat, and corn kernels) and grass from managed pastures [28]. Cultivated plants used for food and feed account for approximately 1.5% of the world's NPP, which is calculated from the data of Tables 1 and 2.

Growing population and continuous consumption growth per capita mean that the global demand for food/feed will increase by 50–100% in 2050 [9, 11]. Food

Table 2. The renewable energy sources and their potentials.

Renewable energy	Resource (TW) [1, 24, 29, 32]	Resource potential (TW) [31]	Utilization percentage
Surface insolation	87,000	50.7	0.06%
Wind	870	19.1	2.2%
Wave/Tide	63.7	1.6	2.5%
Geothermal energy	32	15.9	49.6%
Hydroelectric energy	7.2	1.0	13.9%
Photosynthesis	90		
Land	65	7.99	12.3%
Ocean	25		

security is inextricably linked to growing pressure on land, water, and energy resources [10]. Recent events of drought, large-scale land investments, and high energy prices underscore the world's food security. In addition, issues of food distribution, geopolitical stability, cost volatility, and functional nutrition could lead to hunger in some areas [12].

Natural resources

Renewable energy

Three major types of renewable energies are solar radiation, geothermal energy, and tidal energy. The six transformations of solar radiation are wind, wind-generated ocean waves, ocean currents, hydro energy, thermal difference between the ocean's surface and deep water, and biomass [29, 30]. Not all renewable energy sources can be utilized. For example, very low energy concentration (nonpoint) energies in terms of W/m^2, such as ocean thermal differences, currents, and biomass in ocean, are difficult to collect and utilize economically [4]. Additionally, some fraction of energy resources cannot be utilized economically. For example, it is estimated that only 2.2% of wind energy resource could be utilized in the future (Table 2). Similarly, most biomass on lands cannot be economically collected and utilized due to high collection and transportation costs and/or environmental concerns. Approximately, 12.3% of biomass could be utilized, nearly double to current biomass consumption (i.e., 4.11 TW) [19, 31]. This data suggest that biomass resource may not be as large as expected.

Solar radiation is the largest renewable energy source (Table 2). Approximately, 170 petawatt (PW, 10^{15} W) radiation reaches Earth and approximately 30% is immediately reflected and scattered in the upper atmosphere [24, 32]. Once the radiation enters the atmosphere, a complex series of reflections and absorptions take place. Thirty-one petawatt insolation is converted to thermal

energy in the atmosphere and the remaining solar radiation at the surface is approximately 87 PW [24, 32], approximately 5000 times of the world's energy consumption. Of 87 PW surface radiation, 38 PW becomes thermal energy in the land and ocean, 41 PW contributes to evaporating water, and 5 PW diffuse radiation is reflected off the surface and escapes into space, and a very small fraction goes to photosynthesis [32]. Earth's land surface, ocean, and atmosphere absorb solar radiation, and this raises their temperature and evaporates water, causing atmospheric circulation or convection. When the wet air reaches a high altitude where the temperature is low, water vapor condenses into clouds and then to rain/snow onto the Earth's surface, completing the water cycle. The latent heat of water condensation amplifies convection, producing atmospheric phenomena, such as wind, hurricanes, and cyclones [24, 32].

Water

Although it is renewable, water has no substitutes or alternative. Agriculture consumes approximately 3100 billion tons of water, accounting for 71% of fresh water withdrawals today for the production of approximately 2.5 billion tons of food [13]. Industrial withdrawals and domestic withdrawals account for 16% and 14%, respectively [13]. The changes in population growth from 7 billion to 8 billion in the next two decades, economic growth and urbanization, accompanied with increased food demand per capita will intensify global water consumption. It is expected that the collective demand of the humans for water will exceed foreseen supply by about 40% in 2030 [13]. Compared to availability of land and energy consumed, water is the biggest limiting factor in the world's ability to feed a growing population [13].

Land

The total arable land on Earth is 4.2 billion hectares [28]. Approximately, a third of arable land is being cultivated [28]. In reality, the potential to convert the remaining land is limited because most uncultivated land plays vital ecological roles [28]. Half of potential arable land is available only in seven countries (i.e., Brazil, Democratic Republic of the Congo, Angola, Sudan, Argentina, Columbia, and Bolivia) [28]. On the other extreme, South Asia and the Near East/North Africa have no spare land [13]. Overall, the world's net amount of arable land could expand an additional seventy million hectares, being 5% [13]. Also, aggressive expansion of agricultural lands from forest and grassland will impair biodiversity and release a large amount of new CO_2 emissions [18, 33].

The issues of energy, water, and land used for the food production have been interwoven ranging from ensuring access to services, to environmental impacts, to price volatility [34]. Systematic analysis and paradigm-shifting solutions are highly required to address challenges of the energy–food–water nexus.

Natural photosynthesis

Natural photosynthesis comprises a set of photochemical and redox reactions, called the "light reactions" and a sequence of enzymatic synthesis reactions, called "light-independent reactions" [17, 35–37]. In the light reactions, photosynthetic pigments (e.g., chlorophyll molecules) absorb approximately 47% of the light of the sun called "photosynthetic active radiation," but do not include green light, UV, and IR irradiation [37, 38]. The adsorbed energy is transferred to the reaction centers where the primary charge separation and transmembrane transport of electrons occurs. Subsequent electron- and proton-transfer reactions lead to the synthesis of ATP from ADP and inorganic phosphate and NADPH synthesis from $NADP^+$. In theory, eight photons are required to reduce two molecules of $NADP^+$ to NADPH. In reality, approximately 9.4 photons are consumed, that is, 11.8% of the energy of sunlight can be converted to the form of NADPH, which is close to the efficiency limit of the photosynthetic production of biohydrogen under optimal insolation [17, 39]. Light reactions have the highest photosynthesis efficiency at relatively low light intensities. The efficiency is saturated at 20% of full sunlight and decreases greatly at high light intensities. In addition, high light intensities lead to photo damage of a central protein subunit of the photosynthetic apparatus. The energy efficiency of light-independent reactions are limited by (i) low chemical synthesis efficiency of the enzyme RuBisCO for taking up low-concentration CO_2 from air and removing 2-phosphoglycolate; (ii) availability of sufficient amounts of water that is not met during much of the day and of fertilizers, and (iii) respiration of living organisms [17, 38, 40]. Light reactions operate on very short time scales from femtoseconds to milliseconds, while light-independent reactions operate over a timespan of seconds to hours [35, 36]. As a result, natural plant photosynthesis has low theoretical energy efficiencies from solar energy to chemical energy of 4.6 and 6.0% for C3 and C4 plants, respectively [38]. Although global efficiency of plant photosynthesis is 0.2%, the global primary biomass production is approximately five times the world's energy consumption (Tables 1 and 2).

Best energy efficiencies for well-fertilized and well-watered crops are between 2% and 3% [28]. In the past decades mainly due to the green revolution, yields of crops have increased by approximately three times [11]. Now global means of corn, wheat, and rice are 3.5, 2.0, and 2.5 ton/ha, respectively [28]. The highest corn, wheat, and rice harvest records are 22, 15.2, and 15.2 ton/ha, but such high crop productivities are achieved at the costs of high energy inputs, such as fertilizers, insecticides, and water [28]. As crop yields increases, the ratio of photosynthetic energy captured to energy spent on crop cultivation has decreased [41]. For example, ca. 50% fertilizers or even 70% used for cultivating high-yield crops in the United States and China cannot be utilized, resulting in serious nonpoint water pollution from farmland [42]. Therefore, it raises a challenge: how to increase crop yields while simultaneously decreasing energy consumption and utilizing natural resources, such as water and phosphate, more efficiently.

Key questions to clarify

The following addresses four key questions pertaining to the energy–food–water nexus and clarifies the roles of next generation biorefineries in the sustainability revolution.

Could we have enough biomass to feed the world?

There is no doubt that the production of food is more important than the production of energy and materials. Prior to the green revolution, the production of food was the first priority for human beings for several thousand years. For example, the former Soviet Union and United States investigated the production of single-cell proteins from crude oil. When the food supplies are abundant, the prices of food decrease greatly and the prices of crude oil soars, the production of liquid transportation fuels (i.e., ethanol and biodiesel) from food sources is in practice, especially the United States and Brazil. However, it is discouraged or even prohibited to expand the production capacity of first generation biofuels in most countries, such as China and the European Union, mainly due to the concern of food security.

How to meet increasing food needs is becoming a global challenge [10, 11, 43, 44]. Because the production of 2.5 billion tons of food has utilized ~30% arable lands and ~70% freshwater withdrawals, it is difficult to greatly increase agricultural lands and increase water withdrawals. Therefore, a group of scientists [9] suggests a variety of solutions to address food security: (i) closing yield gaps on underperforming lands, (ii) increasing agricultural resource efficiency, and (iii) increasing food delivery by shifting diets and reducing food waste, while halting agricultural land expansion. For example, several studies find that about one-third to one half of food is never consumed [45, 46]. For example, developing countries usually lose more than 40% of food postharvest or during processing, while industrialized countries often lose more than 40% of food at the retail or consumer levels [45]. On the other hand, some plant biologists, big plant companies, and policy makers promoted the genetically modified (GM) crops as a future solution [47]. However, long-term impacts of GM cereals on human health are not clear and their wide application is in heated debates [47–51].

Here, a paradigm-shifting solution is proposed – enzymatic biotransformation of cellulose to synthetic starch in next generation cellulosic biorefineries [52]. Via biomass fractionating [53], a variety of multiple products could be produced from major lignocellulosic components: cellulose, hemicellulose, and lignin (Fig. 2). I demonstrate simultaneous enzymatic biotransformation and fermentation (SEBF) that can transform cellulosic materials to starch, ethanol, and single-cell protein in one vessel in the presence of cascade enzymes isolated from bacterium, fungus, and plant sources, and a typical ethanol-producing yeast. Our data showed that up to 30% of the anhydroglucose units in cellulose were converted to synthetic starch; the remaining units were hydrolyzed to glucose suitable for yeast fermentation that can produce ethanol. This cellulose to starch biotransformation could be scaled up by increasing the stability of two key enzymes – cellobiose phosphorylase and starch phosphorylase because this process does not involve any labile coenzymes (e.g., CoA and NAD[P]); no glucose is wasted; neither energy nor costly reagents is added. The stability of both cellobiose phosphorylase [54] and starch phosphorylase [55] can be enhanced greatly by protein engineering. Also, starch

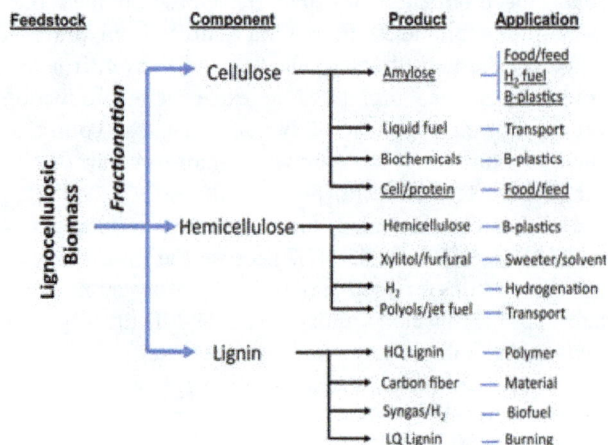

Figure 2. Next generation biorefineries based on fractionated lignocellulosic components for the production of multiple products for meeting different needs from biofuels to biochemicals to food/feed.

production from cellulose mediated by enzymes rather than GM organisms may avoid potential negative impacts of GM cereals and prevent bioethics debate.

Cellulose resource is approximately 40 times the starch produced by cultivated crops. Every ton of cereals harvested is usually accompanied by the production of at least two tons of cellulose-rich crop residues, most of which are not utilized [56]. In addition to the use of agricultural and forest residues (e.g., straws, corn stover, and wood dust), growing dedicated bioenergy crops could greatly increase biomass availability. Dedicated bioenergy crops usually have much higher productivities (e.g., approximately 40–80 ton/ha/y [57–59]), have much higher water utilization efficiency, require less energy-related inputs, such as fertilizers, insecticides, and herbicides, tolerate harsher environments, and could not require annual seedling, compared to cultivated starch-rich crops. Dedicated bioenergy crops can grow on low-quality arable land.

The Department of Energy (DOE) of the United States has summarized three distinct goals associated with potential bioenergy crops: (i) maximizing the total amount of biomass produced per hectare per year, (ii) producing sustainable biomass with minimal inputs (e.g., pesticides, fertilizers, seeds, and harvesting), and (iii) maximizing the amount of biofuels that can be produced per unit of biomass [60]. A yield of ca. 50 dry tons per hectare per year may be considered as a reasonable target in an area with adequate rainfall and good soil [60], which is about 15–25 times average yields of cultivated cereals. In addition to well-studies bioenergy crops, such as switchgrass, poplar, and Miscathanus [59, 61, 62], this study recommends two new promising bioenergy plants – bamboo and common reed. Although both of them have been cultivated and harvested in some areas, they are often ignored by most. Bamboos are giant woody, tree-like, perennial evergreen grasses [58]. They have been cultivated in East Asia and South East Asia [63]. *Phyllostachys pubescens* (Moso bamboo) grows in a subtropical monsoon climate but it can withstand as low as −20°C in winter. It can be cultivated in marginal lands, such as mountain valley, foot of mountain, and gentle slope. The bamboo productivity is highly dependent on soil, water, and climate conditions. The highest average yearly biomass productivity during 10-year plantation is approximately 76 tons of dry culms/ha/y, which can be easily collected [64]. *Phragmites australis* (common reed) is a widespread perennial grass that grows in wetlands or near inland water ways [57]. Although it is harvested for thatched roofs, ropes, baskets, and pulping feedstock, the common reed is more typically considered an invasive weed due to its vigorous growth and difficulty of eradication. Common reed could be used as a bioenergy crept due to three unique features: (i) high biomass productivity (e.g., ca. 45–71 tons/ha/y), (ii) low inputs needed for planting, such as water, fertilizers, and pesticides, and (iii) removal of phosphorus- and nitrogen-containing pollutants in water ways [57].

Intensive irrigation for cultivating dedicated bioenergy crops could not be recommended. Since it consumes approximately three and one orders of magnitude water based on energy content more than the production of oil from traditional oil drilling and advanced oil recovery, respectively [13], the production of biomass is believed to increase usage of freshwater [65, 66]. This issue has raised concerns about the increase in water stress, particularly in countries that are already facing water shortage [67]. Therefore, cultivating future dedicated bioenergy crops must take in account water consumption.

In a word, the cost-effective transformation of nonfood cellulose to starch could not only revolutionize agriculture by promoting the cultivation of plants chosen for rapid growth rather than those optimized for starch production [68–70] but also could maintain biodiversity and minimize agriculture's environmental footprint [71]. Also, wide implementation of cellulosic biorefineries would decrease postharvest food loss, especially for developing countries, so to increase overall food/feed availability [72].

What powertrain and fuel will become the dominant transport means in the future?

A number of scenarios (Fig. 1) can and could bridge between renewable primary energy and transportation energy demand through four powertrain systems: (i) ICEs and/or hybrid electric vehicles (HEVs) that burn liquid biofuels and compressed methane [19, 73], (ii) BEVs that run on electricity stored in rechargeable batteries, where electricity can be generated from sun radiation, tide, geothermal, wind, and nuclear energy [74], (iii) hydrogen FCVs that run on stored hydrogen through proton exchange membrane (PEM) fuel cells and electric motor [73], and (iv) sugar fuel cell vehicles (SFCVs) that run on stored sugar as a high-density hydrogen carrier based on FCVs [25]. Powertrain systems for vehicles must meet all of the following criteria: high energy storage capacity in a small container, high power output, economically competitive fuel, affordable vehicle, fast charging or refilling of the fuel, and high safety [16].

Table 3 compares the gravimetric energy densities of liquid fuels, stored hydrogen, rechargeable batteries, and capacitors, as well as kinetic energy output densities on wheels through different powertrain systems. The energy storage densities in a decreasing order are diesel, gasoline, butanol, ethanol, methanol, sugar, stored hydrogen,

Table 3. Gravimetric energy densities of stored energies and kinetic energy released through different powertrain systems [19].

Name	Gravimetric energy density (MJ/kg)	Kinetic energy output (MJ/kg)	Powertrain (efficiency, %)
H_2 without container	143	NA	NA
Diesel	46.2	8.32	ICE-diesel (18%)
		17.09	HEV-diesel (37%)
Gasoline	46.4	6.50	ICE-gas (14%)
		14.38	HEV-gas (31%)
Butanol	36.6	5.12	ICE-gas (14%)
		11.35	HEV-gas (31%)
Ethanol	30	4.20	ICE-gas (14%)
		9.30	HEV-gas (31%)
		11.10	HEV-diesel (37%)
Methanol	19.7	6.90	DMFC (35%)
Starch/Cellulose	17.0	8.16	Sugar-H_2-PEMFC/Motor (48%)
8% H_2 mass including container	11.4	5.13	PEMFC/Motor (45%)
Cryo-compressed H_2 including container	9.3	4.19	PEMFC/Motor (45%)
Compressed H_2 (700 bars) including container	6.0	2.70	PEMFC/Motor (45%)
4% H_2 mass including container	5.7	2.57	PEMFC/Motor (45%)
Compressed H_2 (350 bars) including container	5.0	2.25	PEMFC/Motor (45%)
Lithium ion rechargeable battery	0.56	0.381	BEV (68%)
NiMnH rechargeable battery	0.36	0.245	BEV (68%)
Lead acid rechargeable battery	0.14	0.095	BEV (68%)
Ultra-capacitor	0.02	0.016	Motor (80%)
Super-capacitor	0.01	0.008	Motor (80%)

DMFC: direct methanol fuel cell; PEMFC: proton exchange membrane fuel cell.

rechargeable batteries, and capacitors. Liquid gasoline and diesel plus their respective ICEs have kinetic energy output densities of 6.50 and 8.32 MJ/kg, respectively. When ICE's energy efficiencies are increased through hybrid electric systems, HEV-gas, and HEV-diesel can drive farther. Conventional hydrogen storage means have lower energy storage densities from 5.0 to 9.3 MJ/kg or even lower, resulting in shorter driving distance of FCVs compared to vehicles based on ICEs if the same weight fuel tank is used. Therefore, the DOE strongly encourages to develop novel high-density hydrogen storage means and provides the H-prize cash award [16]. Rechargeable batteries have at least one order magnitude lower energy storage densities than liquid fuels and stored hydrogen (Table 3). As a result, BEVs have very short driving distances. The energy densities of capacitors are very low, limiting its application in the transport sector.

Battery electric vehicles will not be a dominant future transport means. For example, the International Energy Agency and several studies predict that BEVs will play a minor role in the future [74, 75]. Rechargeable lithium (Li) batteries have energy densities of approximately 150 Wh/kg (i.e., 0.56 MJ/kg), resulting in very short driving distances for BEVs [76, 77]. If the energy densities of lithium batteries were increased by 5–10-fold [78, 79], other issues, such as safety, recharging time, and lifetime, could still prohibit their wide use in personal vehicles. In reality, future energy densities of rechargeable lithium batteries are expected to increase by twofold in next decades [76, 77] rather than 5–10 times by considering the configuration of Li batteries and its combustion energy (i.e., 43.1 MJ/kg lithium) [4]. Although developing lithium-air batteries are expected to have very high energy densities but the regeneration of lithium oxidize to lithium by electricity is energy intensive. Therefore, metal-air batteries are not suitable in the transport sector.

In addition to low energy densities of Li batteries, BEVs have other weaknesses. First, the recharging cycles and lifetime of high-density lithium batteries is approximately 1000 time and 2–3 years, respectively. Both are much shorter than requirement of the major car components lasting at least 10 years. (Think of lithium ion batteries in cellphones and laptops.) Second, lithium ion batteries are still costly for vehicles although its production costs could be decreased by several-fold. It is not realistic to believe that battery costs would be drastically decreased following Moore's Law because it is impossible to exponentially both decrease material consumption in batteries and increase battery performance according to the basic physical limits of materials. Third, Li batteries require a long recharging time. Although ultra-fast charging batteries have been developed [80], these capacitor-like batteries are made at the cost of decreasing energy storage densities [81]. Fourth, a huge infrastructure investment could be needed to upgrade the electrical grid, install sockets for fast recharge, and build power stations [21]. Fifth, disposing and recycling a large number of used rechargeable batteries could be another environmental challenge [21]. Sixth, the energy density loss rates of rechargeable batteries depend on temperature; for example, standard loss

rates per year are 6% at 0°C, 20% at 25°C, and 35% at 40°C [21]. Seventh, whether there is enough low-cost lithium for BEVs is not a certain thing. Goodenough, a pioneer of lithium batteries, pointed out that the principal challenges facing the development of rechargeable batteries for BEVs are cost, safety, energy density (voltage × capacity), rate of charge/discharge, and service life [82]. Due to BEVs' unique features such as cleanness and quietness, BEVs will still be popular in some special markets, for example, in golf courts. In a word, a complete switch to all battery electric cars is utterly unrealistic [21] by considering the above problems and the likelihood that better competing technologies will appear and mature.

This study suggests another paradigm-shifting solution for the future vehicles – SFCVs. Based on FCVs, carbohydrate (shorthand, CH_2O) is suggested to be a high-density hydrogen carrier so that its use could address hydrogen storage, distribution, and safety issues [40, 83–85]. In the hypothetical SFCV, an on-board biotransformer containing numerous thermoenzymes and (biomimetic) coenzymes that can achieve the reaction of $CH_2O + H_2O \rightarrow 2H_2 + CO_2$ [86, 87]. Because enzymes are 100% selective, work under moderate reaction conditions, and generate highly pure hydrogen, carbohydrates have a gravimetric density of 8.33 H_2 mass% for the carbohydrate/water slurry [16, 25]. During the past several years, we have increased enzymatic hydrogen generation rates to approximately 160 mmole H_2/L/h by nearly 800-fold (in preparation for publication). We anticipate to increase reaction rates by another 30-fold within next several years so that the on-board biotransformer will be small enough to store in a SFCV [16, 40].

In a word, HEVs based on ICEs are believed to be a short- and middle-term solution before FCVs [73]. SFCVs could be a good solution to address the problems of FCVs from hydrogen production, storage, distribution, infrastructure, and safety. SFCVs could have several advantages over BEVs: much higher energy storage densities, faster refilling rates, better safety, and less environmental burdens [19, 40].

Could we have enough extra biomass source to drive vehicles and feed the world?

As shown in Table 1, two irreplaceable applications of biomass resource are food/feed (1.33 TW) and wood for materials (1.28 TW). Compared to all terrestrial biomass resource (65 TW), the current biomass utilization efficiency is 6.32% and it is expected that biomass utilization efficiency will be increased to up to 12.3% [31]. This value is also partially supported by the DOE and USUA's a billion ton report [88].Two liquid fuels used for land transportation are gasoline (1.2 TW) and middle distillates (1.79 TW). Since the global average ICE-gas and ICE-diesel have fuel-to-wheel efficiencies of approximately 14% and 23%, respectively [19], the global kinetic energy output on wheels is 0.58 TW.

When we increase biomass utilization efficiency from 6.32% now to 12.3% in 2050, this study provides quantitative predictions for the worst, best, and most likely scenarios for the year 2050 based on different assumptions. In the worst scenarios, food/feed needs, wood consumption, and biomass for burning could increase by 100%, 50%, and 50%, respectively. At the same time, total biomass resource could be constant. Therefore, the remaining biomass source that could be collected and utilized will be 1.17 TW. The land transportation energy in terms of kinetic energy could increase to 0.85 TW from 0.58 TW based on an annual growth rate of 1%.

In the best scenarios, food/feed needs and wood consumption could increase by 50% and 20%, respectively. Slow growth in wood consumption could be attributed to less use of papers in affluent countries and better recycling. Biomass for burning could be decreased to half due to an increase in burning efficiency in developing countries [24]. At the same time, total biomass resource could increase to 94.9 TW at an annual growth rate of 1% due to (i) rising CO_2 levels in the atmosphere that fertilizes plant productivity [19, 38] and (ii) dedicated high-yield bioenergy crops [88]. Therefore, the biomass resource will be 7.52 TW. The land transportation energy in terms of kinetic energy could

increase to 0.70 TW based on an annual growth rate of 0.5%.

In the most likely scenarios, food/feed needs, wood consumption, and biomass for burning could increase by 70%, 35%, and 0%, respectively. Food/feed production from cultivated cereals could increase to 1.66 TW; the remaining food/feed need (0.60 TW) could be supplemented with synthetic starch made from biorefineries. At the same time, total biomass resource could increase to 78.6 TW at an annual growth rate of 0.5%. Therefore, the remaining biomass resource will be 4.84 TW. The land transportation energy in terms of kinetic energy could increase to 0.76 TW based on an annual growth rate of 0.7%.

The last uncertainty is the biomass-to-wheel (BTW) efficiency of future land transport means. The worst scenario is based on current ICE-gas (ethanol) system (BTW = 7%), while the best could be SFCVs (BTW = 27%). Several transitional powertrains could be HEV-gas (BTW = 20.7%), HEV-diesel (BTW = 24.8%), and FCV (BTW = 22%). In the 2050 market, it is likely that the transport sector could constitute different transportation means so that an average BTW efficiencies could range from 11% to 20%.

Table 4 presents the analysis for the future biomass and biofuels roles. In the worst scenarios, biomass could play a significant role in replacing approximately 10–25% transportation fuel need. On the contrast, in the best scenarios, biomass could be sufficient to meet all land trans-

Table 4. Scenarios of the roles of biomass for the production of food/feed, wood, heating, and land transportation fuels in 2050.

Worst 2050			Best 2050			Highly possible 2050		
Name	Power (TW)	Assumption	Name	Power (TW)	Assumption	Name	Power (TW)	Assumption
Food/Feed	2.66	100% gain	Food/Feed	2.00	50% gain	Food/Feed	2.26	70% gain
Food/Feed crops	2.66	100% gain	Food/Feed crops	1.86	40% gain	Food/Feed crops	1.66	35% gain in crop
New food/Feed	0.00	NA	New food/Feed	0.13		New food/Feed	0.60	
Wood	1.92	50% gain	Wood	1.54	20% gain	Wood	1.66	30% gain
Burning	2.25	50% gain	Burning	0.75	50% decrease	Burning	1.50	No change
Total land biomass	65	No change	Total land biomass	94.87	1% gain/year	Total biomass resource	78.56	0.5% gain/year
Available biomass	1.17	12.3% biomass use	Available biomass	7.52	12.3% biomass use	Available biomass	4.84	12.3% biomass use
Land kinetic energy	0.85	1% gain/year	Land kinetic energy	0.70	0.5% gain/year	Land transportation use	0.76	0.7% gain/year
Scenario	Land fuel replacement		Scenario	Land fuel replacement		Scenario	Land fuel replacement	
S1: ICE-gas (ethanol)	9.6%	BTW = 7%	S4: HEV-gas (ethanol)	156%	BTW = 15%	S7: ICE/HEV-gas (ethanol)	54.6%	BTW = 11%
S2: HEV-gas (ethanol)	20.7%	BTW = 15%	S5: FCV	229%	BTW = 22%	S8: HEV-gas (ethanol)	74.4%	BTW = 15%
S3: HEV-diesel (ethanol)	24.8%	BTW = 18%	S6: SFCV	280%	BTW = 27%	S9: SFCV/SFC/HEV	99.2%	BTW = 20%

portation energy need plus a large surplus. In the most likely scenarios, biofuels made from biomass could replace at least 50% to nearly 100% land transportation fuel need. The above analysis suggests that (i) we must increase powertrain system efficiency so to decrease biomass consumption, (ii) we must develop next generation biorefineries because it not only produce biofuels but also could produce food/feed and biochemicals, and (iii) we must utilize agricultural and forest residuals and then grow dedicated water-saving bioenergy crops by spatial segregation of food/feed and energy-producing areas by continuing producing food on established and productive agricultural land while growing dedicated energy crops on marginal land [89].

Could we surpass natural photosynthesis?

This study suggests developing next generation biorefineries by integrating high-efficiency solar cells or other electricity-generating systems, water electrolysis, with biological CO_2 fixation mediated by cell-free synthetic cascade enzymes (Fig. 3). This cell-free biosystem is believed to work based on the design principles of synthetic biology, knowledge in the literature, and thermodynamics analysis [40, 90]. This hypothetical system could have numerous advantages. First, solar cells have much broader light adsorption spectrum and higher efficiencies than plant pigments. Also, the efficiency of solar cells, unlike plants, does not change in response to insolation variation. Also, it is easy to concentrate nonpoint insolation to a point energy – electricity. Second, hydrogen generated by water electrolysis at daytime can be stored for a few hours so that it can be consumed at a constant synthesis rate for the biological CO_2 fixation process at night. Therefore, it is easy to regulate

and match changed-rate electricity generation and constant-rate biosynthesis process. Third, the products of artificial photosynthesis are carefully chosen: water-insoluble amylose, volatile alcohols, or water-insoluble fatty alcohols. So the product separation costs could be minimal. Fourth, ultra-high energy efficiency from hydrogen or electricity and CO_2 to chemical energy could be achieved, much better than natural processes mediated by living organisms that dissipate energy by respiration [91–93]. Table 5 presents the comparison between natural photosynthesis and artificial photosynthesis. Validation experiments and practical application of these systems will require worldwide collaborative efforts from biologists, chemists, electrochemists, and engineers [90] (Note: It is important to fix high concentration CO_2 generated from power stations rather than to capture atmospheric CO_2 because the latter requires extremely high energy inputs, resulting in economical infeasibility [94]).

In a word, next generation biorefineries based on artificial photosynthesis would not only bridge the current and future primary energy utilization systems aimed at facilitating electricity and hydrogen storage but also address such sustainability challenges such as renewable biofuel and chemical production, CO_2 utilization, and fresh water conservation [90]. Its large-scale implementation would foster the switch from fossil fuel-based resources to renewable bioresources.

Recommendations

First, the development of next generation biorefineries based on nonfood biomass is a must rather than an option because they will produce a variety of products that cannot be substituted by other renewable resources,

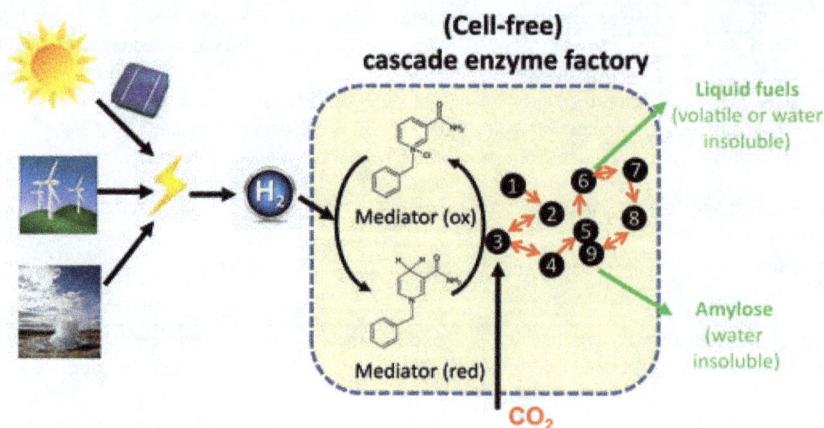

Figure 3. Next generation biorefineries based on artificial photosynthesis that can fix CO_2 and hydrogen to starch or other compounds. The enzymes involved in the synthetic enzymatic pathway responsible for CO_2 fixation are suggested in the reference [40], which are different from all natural CO_2 fixation pathways [99]. Also, the enzymes responsible for product formation are subject to change.

Table 5. Comparison matrix between photoautotrophic organisms and artificial photosynthesis based on cell-free cascade enzyme factories plus photovoltaic. Modified from Ref. [37].

	Natural photosynthesis	Artificial photosynthesis
Solar to chemical efficiency	Theoretical ~4–10%	Theoretical, 51%
	Practical: ~0.2–2%	Practical: >10%
Sunlight spectrum (e.g., 350–2350 nm)	Only 48% (only 400–700 nm)	Whole spectrum adsorption by solar cells (63%, theoretical; 42% highest; 18%, commercial)
Light-harvesting efficiency	Low under nonoptimal conditions	Nearly constant independent of insolation
Chemical synthesis pathway	Unmatched reaction rates between light reactions and dark reactions	Constant synthesis rate from stored hydrogen
Chemical synthesis efficiency	12–18 ATP + 6NAD(P)H equivalent consumed per hexose synthesis for utilizing low level CO_2	ATP-neutral synthesis pathway by using high levels CO_2 from power stations
Respiration	~50% loss	Not applicable
Complicated regulation between primary and secondary metabolisms	A (small) fraction of chemical energy could flux to product	99+% of energy could flux to product because of insulation of biocatalyst synthesis (e.g., cell growth) from product formation
Product separation costs	Separate intracellular product from aqueous cells	Generate water-insoluble product (e.g., amylose or fatty alcohols) or volatile products
Large water consumption	500 + kg of water needed for 1 kg of carbohydrate produced	0.6 kg water needed for 1 kg of carbohydrate synthesized
Contamination	Use of weedicides	Not applicable
Operation time	Daytime only	24/7
Temperature	Modest	Well-controlled bioreactors
Land resource	Limited due to the combined requirements of temperature, water and insolation	Nearly everywhere by separating solar harvesting systems from product synthesis systems
Waste generated	Nonpoint pollutants	Point pollutants from fermenters

such as transportation fuels, biochemicals, and food/feed. With respect to biomass fractionating and biorefining technologies, the production of multiple products in next generation biorefineries will be of importance to their economic viability because natural biomass feedstock contains multiple components (Fig. 2). With respect to feedstock, we need start utilizing the ready agricultural and forest residues before we grow dedicated bioenergy crops on a large scale. Also, it is strongly recommended not to change current agricultural lands used for food/feed production to the production of bioenergy crops, which could lead to food shortage. With respect to biofuels, biofuels must be produced from sugars through anaerobic fermentation because a fraction of sugar in aerobic fermentation is wasted, resulting in low energy efficiencies [20, 95]. The failure of Amyris's and LS9's efforts on biofuels production is a good example – hopeless aerobic fermentation.

Second, it is extremely important to develop more energy efficiency powertrain systems from ICEs to HEVs to FCVs to SFCVs. Increasing energy utilization efficiency is a megatrend for human societies [2, 4, 24]. Higher energy efficiency means less primary energy consumption and lower environmental footprints.

Third, it is important to develop next generation biorefineries based on artificial photosynthesis that can produce carbon-containing compounds from CO_2 and H_2/

electricity. Large-scale implementation of artificial photosynthesis would address such sustainability challenges as electricity and hydrogen storage, CO_2 utilization, fresh water conservation, and maintenance of a small closed ecosystem for human survival in emergency situations [90].

Fourth, to address food security, it is recommended (i) not to increase the production capacity of first generation biofuels, and (ii) not to grow GM cereals as the future food source. Human beings will have enough food/feed without GMs cereals by increasing traditional crop productivity, decreasing food waste, enhancing food distribution, and producing synthetic starch from nonfood cellulosic resource, and even producing amylose through artificial photosynthesis. Additionally, potential negative impacts of GM cereals on human health should not be underestimated because systematic long-term studies are not available and may not be conductive. For example, negative effects of saturated fat are recently realized after its long utilization [96, 97]. Cotton seed oil was once used to replace vegetable oil as food. After years, it was found that the use of cottonseed oil resulted in low fertility in males [98]. Similarly, chronic negative effects of tobacco were realized in 1960s after its use for several thousand years. Therefore, it is not necessary to take risk in consuming GM cereals but its benefits to food security are not irreplaceable.

In a word, biomass sugar isolated from nonfood biomass and/or produced from artificial photosynthesis could play an irreplaceable role in the sustainability revolution by providing food/feed, renewable materials, and transportation biofuels in the future.

Acknowledgments

This study was supported by the Biological Systems Engineering Department of Virginia Tech, the CALS Biodesign and Bioprocessing Research Center, and Shell Game-Changer Program.

Conflict of Interest

The authors declare competing financial interests. The enzymatic sugar-to-hydrogen technology is protected by the US patent 8211681. The enzymatic transformation of non-food biomass to edible starch is under protection of provisional patent disclosure filed by Virginia Tech. PZ has a financial interest in CFB9 and Gate Fuels.

References

1. MacKay, D. J. C. 2009. Sustainable energy – without the hot air. UIT Cambridge Ltd., Cambridge, U.K.
2. Smil, V. 2008. Energy in nature and society. MIT Press, Cambridge, MA.
3. Wei, J. 2012. Great inventions that changed the world. John Wiley & Sons, Inc., Hoboken, NJ.
4. Zhang, Y.-H. P. 2011. What is vital (and not vital) to advance economically-competitive biofuels production. Process Biochem. 46:2091–2110.
5. Smil, V. 2008. Oil: a beginner's guide. Oneworld Publications, Oxford, U.K.
6. Sorrell, S., J. Speirs, R. Bentley, A. Brandt, and R. Miller. 2010. Global oil depletion: a review of the evidence. Energy Policy 38:5290–5295.
7. Kerr, R. A., and R. F. Service. 2005. What can replace cheap oil – and when. Science 309:101.
8. Sibly, R. M., and J. Hone. 2002. Population growth rate and its determinants: an overview. Phil. Trans. R. Soc. Lond. B Biol. Sci. 357:1153–1170.
9. Foley, J. A., N. Ramankutty, K. A. Brauman, E. S. Cassidy, J. S. Gerber, M. Johnston, et al. 2011. Solutions for a cultivated planet. Nature 478:337–342.
10. Godfray, H. C. J. 2011. Food and biodiversity. Science 333:1231–1232.
11. Godfray, H. C. J., J. R. Beddington, I. R. Crute, L. Haddad, D. Lawrence, J. F. Muir, et al. 2010. Food security: the challenge of feeding 9 billion people. Science 327:812–818.
12. Lappé, F. M., J. Collins, P. Rosset, and L. Esparza. 1998. World hunger: twelve myths. Grove Press, New York, NY.
13. The World Economic Forum Water Initiative. 2011. Water security: the water-food-energy-climate nexus. Island Press, Washington, DC.
14. Lynd, L. R. 2010. Bioenergy: in search of clarity. Energy Environ. Sci. 3:1150–1152.
15. Vertès, A. A., M. Inui, and H. Yukawa. 2006. Implementing biofuels on a global scale. Nat. Biotechnol. 24:761–764.
16. Zhang, Y.-H. P. 2010. Renewable carbohydrates are a potential high density hydrogen carrier. Int. J. Hydrogen Energy 35:10334–10342.
17. Michel, H. 2012. Editorial: The Nonsense of Biofuels. Angew. Chem. Int. Ed. 51:2516–2518.
18. Searchinger, T., R. Heimlich, R. A. Houghton, F. Dong, A. Elobeid, J. Fabiosa, et al. 2008. Use of U.S. croplands for biofuels increases greenhouse gases through emissions from land-use change. Science 319:1238–1240.
19. Huang, W. D., and Y.-H. P. Zhang. 2011. Energy efficiency analysis: biomass-to-wheel efficiency related with biofuels production, fuel distribution, and powertrain systems. PLoS ONE 6:e22113.
20. Huang, W. D., and Y.-H. P. Zhang. 2011. Analysis of biofuels production from sugar based on three criteria: thermodynamics, bioenergetics, and product separation. Energy Environ. Sci. 4:784–792.
21. Smil, V. 2010. Energy myths and realities: bringing science to the energy policy debate. The AEI Press, Washington, DC.
22. International Energy Agency (IEA). 2012. Key World Energy Statistics 2012. Available at http://www.iea.org/publications/freepublications/publication/kwes.pdf (accessed 2012).
23. Lewis, N. S., and D. G. Nocera. 2006. Powering the planet: chemical challenges in solar energy utilization. Proc. Natl Acad. Sci. USA 103:15729–15735.
24. Smil, V. 1999. Energies: an illustrated guide to the biosphere and civilization. The MIT Press, Cambridge, MA.
25. Zhang, Y.-H. P. 2009. A sweet out-of-the-box solution to the hydrogen economy: is the sugar-powered car science fiction? Energy Environ. Sci. 2:272–282.
26. Zhang, Y.-H. P. 2008. Reviving the carbohydrate economy via multi-product biorefineries. J. Ind. Microbiol. Biotechnol. 35:367–375.
27. Misselhorn, A., P. Aggarwal, P. Ericksen, P. Gregory, L. Horn-Phathanothai, J. Ingram, et al. 2012. A vision for attaining food security. Curr. Opin. Environ. Sustainability 4:7–17.
28. Smil, V. 2000. Feeding the world: a challenge for the twenty-first century. MIT Press, Cambridge, MA.
29. Hoffert, M. I., K. Caldeira, G. Benford, D. R. Criswell, C. Green, H. Herzog, et al. 2002. Advanced technology paths to global climate stability: energy for a greenhouse planet. Science 298:981–987.

30. Smil, V. 2010. Energy Transitions: History, Requirements. Prospects, ABC-CLIO, LLC, Santa Barbara, CA.

31. Worldwatch Institute. 2009. State of the world 2009: into a warming world. Available at http://www.worldwatch.org/node/5984 (accessed 2009).

32. Hermann, W. A. 2006. Quantifying global exergy resources. Energy 31:1685–1702.

33. Fargione, J., J. Hill, D. Tilman, S. Polasky, and P. Hawthorne. 2008. Land clearing and the biofuel carbon debt. Science 319:1235–1238.

34. Bazilian, M., H. Rogner, M. Howells, S. Hermann, D. Arent, D. Gielen, et al. 2011. Considering the energy, water and food nexus: Towards an integrated modelling approach. Energy Policy 39:7896–7906.

35. Blankenship, R. E., D. M. Tiede, J. Barber, G. W. Brudvig, G. Fleming, M. Ghirardi, et al. 2011. Comparing photosynthetic and photovoltaic efficiencies and recognizing the potential for improvement. Science 332:805–809.

36. Williams, P. J. L., and L. M. L. Laurens. 2010. Microalgae as biodiesel & biomass feedstocks: Review & analysis of the biochemistry, energetics & economics. Energy Environ. Sci. 3:554–590.

37. Zhang, Y.-H. P., C. You, H. Chen, and R. Feng. 2012. Surpassing Photosynthesis: High-Efficiency and Scalable CO_2 Utilization through Artificial Photosynthesis. *ACS Sym. Ser. Recent Advances in Post-Combustion CO_2 Capture chemistry.* 1097:275–292.

38. Zhu, X.-G., S. P. Long, and D. R. Ort. 2008. What is the maximum efficiency with which photosynthesis can convert solar energy into biomass? Curr. Opin. Biotechnol. 19:153–159.

39. Weyer, K., D. Bush, A. Darzins, and B. Willson. 2009. Theoretical maximum algal oil production. Bioenergy Res. 3:204–213.

40. Zhang, Y.-H. P. 2011. Simpler is better: high-yield and potential low-cost biofuels production through cell-free synthetic pathway biotransformation (SyPaB). ACS Catal. 1:998–1009.

41. Gregory, P. J., and T. S. George. 2011. Feeding nine billion: the challenge to sustainable crop production. J. Exp. Bot. 62:5233–5239.

42. Braskerud, B. C. 2002. Factors affecting phosphorus retention in small constructed wetlands treating agricultural non-point source pollution. Ecol. Eng. 19:41–61.

43. Godfray, H. C. J. 2011. Food for thought. Proc. Natl Acad. Sci. USA 108:19845–19846.

44. Tilman, D., C. Balzer, J. Hill, and B. L. Befort. 2011. Global food demand and the sustainable intensification of agriculture. Proc. Natl Acad. Sci. USA 108:20260–20264.

45. Gustavsson, J., C. Cederberg, U. Sonesson, R. van Otterdijk, and A. Meybeck. 2011. Global food losses and food waste *in* Section 3.2 of international congress "Save Food" (FAO, rural infrastructure and agro-industries division). Available at: http://www.fao.org/fileadmin/user_upload/suistainability/pdf/Global_Food_Losses_and_Food_Waste.pdf (accessed May 2011).

46. Parfitt, J., M. Barthel, and S. Macnaughton. 2010. Food waste within food supply chains: quantification and potential for change to 2050. Philos. Trans. R. Soc. Lond. B Biol. Sci. 365:3065–3081.

47. Bruce, T. J. A. 2012. GM as a route for delivery of sustainable crop protection. J. Exp. Bot. 63:537–541.

48. Bagla, P. 2012. Negative Report on GM Crops Shakes Government's Food Agenda. Science 337:789.

49. Balmford, A., R. Green, and B. Phalan. 2012. What conservationists need to know about farming. Proc. R. Soc. B: Biological Sciences 279:2714–2724.

50. Beatty, P. H., and A. G. Good. 2011. Future Prospects for Cereals That Fix Nitrogen. Science 333:416–417.

51. Fedoroff, N. V. 2013. Will common sense prevail? Trends Genet. doi: 10.1016/j.tig.2012.09.002

52. Zhang, Y.-H. P., H. G. Chen, C. You, and R. L. Feng. 2012. Feeding the world: two out-of-the-box solutions. *in* The 243rd ACS national meeting, division of AGFD 258, Available at: http://abstracts.acs.org/chem/243nm/program/view.php. (accessed March 29, 2012).

53. Sathitsuksanoh, N., A. George, and Y.-H. P. Zhang. 2013. New lignocellulose pretreatments by using cellulose solvents: a review. J. Chem. Technol. Biotechnol. 88:169–180.

54. Ye, X., C. Zhang, and Y.-H. P. Zhang. 2012. Engineering a large protein by combined rational and random approaches: stabilizing the *Clostridium thermocellum* cellobiose phosphorylase. Mol. BioSyst. 8:1815–1823.

55. Yanase, M., H. Takata, K. Fujii, T. Takaha, and T. Kuriki. 2005. Cumulative Effect of Amino Acid Replacements Results in Enhanced Thermostability of Potato Type L α-Glucan Phosphorylase. Appl. Environ. Microbiol. 71:5433–5439.

56. Tuck, C. O., E. Pérez, I. T. Horváth, R. A. Sheldon, and M. Poliakoff. 2012. Valorization of Biomass: Deriving More Value from Waste. Science 337:695–699.

57. Sathitsuksanoh, N., Z. Zhu, N. Templeton, J. Rollin, S. Harvey, and Y.-H. P. Zhang. 2009. Saccharification of a potential bioenergy crop, *Phragmites australis* (common reed), by lignocellulose fractionation followed by enzymatic hydrolysis at decreased cellulase loadings. Ind. Eng. Chem. Res. 48:6441–6447.

58. Sathitsuksanoh, N., Z. Zhu, T.-J. Ho, M.-D. Bai, and Y.-H. P. Zhang. 2010. Bamboo saccharification through cellulose solvent-based biomass pretreatment followed by enzymatic hydrolysis at ultra-low cellulase loadings. Bioresour. Technol. 101:4926–4929.

59. Sathitsuksanoh, N., Z. Zhu, and Y.-H. P. Zhang. 2012. Cellulose solvent- and organic solvent-based lignocellulose fractionation enabled efficient sugar release from a variety of lignocellulosic feedstocks. Bioresour. Technol. 117:228–233.

60. DOE. 2006. Office of energy efficiency and renewable energy, office of science: breaking the biological barriers to cellulosic ethanol: a joint research agenda. a research roadmap resulting from the biomass to biofuels workshop. Available at http://www.doegenomestolife.org/biofuels/ (accessed June, 2006).

61. Clifton-Brown, J. C., B. Neilson, I. Lewandowski, and M. B. Jones. 2000. The modelled productivity of *Miscanthus×giganteus* (GREEF et DEU) in Ireland. Ind. Crops Prod. 12:97–109.

62. Murnen, H. K., V. Balan, S. P. S. Chundawat, and B. Bals. 2007. daCostaSousa L, Dale BE: Optimization of ammonia fiber expansion (AFEX) pretreatment and enzymatic hydrolysis of *Miscanthus* x giganteus to fermentable sugars. Biotechnol. Prog. 23:846–850.

63. Gratani, L., M. F. Crescente, L. Varone, G. Fabrini, and E. Digiulio. 2008. Growth pattern and photosynthetic activity of different bamboo species growing in the Botanical Garden of Rome. Flora 203:77–84.

64. Shanmughavel, P., and K. Francis. 2001. Physiology of bamboo. Scientific Publishers, India.

65. Batidzirai, B., E. M. W. Smeets, and A. P. C. Faaij. 2012. Harmonising bioenergy resource potentials—Methodological lessons from review of state of the art bioenergy potential assessments. Renew. Sustain. Energy Rev. 16:6598–6630.

66. Bernardi, A., S. Giarola, and F. Bezzo. 2012. Optimizing the economics and the carbon and water footprints of bioethanol supply chains. Biofuels, Bioprod. Biorefin. 6:656–672.

67. Gheewala, S. H., G. Berndes, and G. Jewitt. 2011. The bioenergy and water nexus. Biofuels, Bioprod. Biorefin. 5:353–360.

68. Casillas, C. E., and D. M. Kammen. 2010. The Energy-Poverty-Climate Nexus. Science 330:1181–1182.

69. French, C. E. 2009. Synthetic biology and biomass conversion: a match made in heaven? J. Roy. Soc. Interface 6:S547–S558.

70. Sheppard, A. W., I. Gillespie, M. Hirsch, and C. Begley. 2011. Biosecurity and sustainability within the growing global bioeconomy. Curr. Opin. Environ. Sustainability 3:4–10.

71. Somerville, C., H. Youngs, C. Taylor, S. C. Davis, and S. P. Long. 2010. Feedstocks for Lignocellulosic Biofuels. Science 329:790–792.

72. Lynd, L. R., and J. Woods. 2011. Perspective: A new hope for Africa. Nature 474:S20–S21.

73. Demirdoven, N., and J. Deutch. 2004. Hybrid cars now, fuel cell cars later. Science 305:974–976.

74. Thomas, C. E. 2009. Fuel cell and battery electric vehicles compared. Int. J. Hydrogen Energy 34:6005–6020.

75. Melamu, R., and H. von Blottnitz. 2009. A comparison of environmental benefits of transport and electricity applications of carbohydrate derived ethanol and hydrogen. Int. J. Hydrogen Energy 34:1126–1134.

76. Armand, M., and J. M. Tarascon. 2008. Building better batteries. Nature 451:652–657.

77. Tarascon, J. M., and M. Armand. 2001. Issues and challenges facing rechargeable lithium batteries. Nature 414:359–367.

78. Chan, C. K., H. Peng, G. Liu, K. McIlwrath, X. F. Zhang, R. A. Huggins, et al. 2008. High-performance lithium battery anodes using silicon nanowires. Nat. Nanotechnol. 3:31–35.

79. Hassoun, J., and B. Scrosati. 2010. A High-Performance Polymer Tin Sulfur Lithium Ion Battery. Angew. Chem. Int. Ed. 49:2371–2374.

80. Kang, K., Y. S. Meng, J. Breger, C. P. Grey, and G. Ceder. 2006. Electrodes with high power and high capacity for rechargeable lithium batteries. Science 311:977–980.

81. Zaghib, K., J. B. Goodenough, A. Mauger, and C. Julien. 2009. Unsupported claims of ultrafast charging of LiFePO4 Li-ion batteries. J. Power Sources 194:1021–1023.

82. Goodenough, J. B., and Y. Kim. 2010. Challenges for Rechargeable Li Batteries. Chem. Mater. 22:587–603.

83. Zhang, Y.-H. P. 2010. Production of biocommodities and bioelectricity by cell-free synthetic enzymatic pathway biotransformations: Challenges and opportunities. Biotechnol. Bioeng. 105:663–677.

84. Zhang, Y.-H. P., J.-B. Sun, and J.-J. Zhong. 2010. Biofuel production by *in vitro* synthetic pathway transformation. Curr. Opin. Biotechnol. 21:663–669.

85. Zhang, Y.-H. P., S. Myung, C. You, Z. G. Zhu, and J. Rollin. 2011. Toward low-cost biomanufacturing through cell-free synthetic biology: bottom-up design. J. Mater. Chem. 21:18877–18886.

86. Ye, X., Y. Wang, R. C. Hopkins, M. W. W. Adams, B. R. Evans, J. R. Mielenz, et al. 2009. Spontaneous high-yield production of hydrogen from cellulosic materials and water catalyzed by enzyme cocktails. ChemSusChem 2:149–152.

87. Zhang, Y.-H. P., B. R. Evans, J. R. Mielenz, R. C. Hopkins, and M. W. W. Adams. 2007. High-yield hydrogen production from starch and water by a synthetic enzymatic pathway. PLoS ONE 2:e456.

88. Perlack, R. D., L. L. Wright, R. L. Graham, B. J. Stokes, and D. C. Erbach. 2005. Biomass as feedstock for a bioenergy and bioproducts industries: The technical feasibility of a billion-ton annual supply. Oak Ridge National Laboratory, Oak Ridge, TN.

89. Dauber, J., C. Brown, A. L. Fernando, J. Finnan, E. Krasuska, J. Ponitka, et al. 2012. Bioenergy from "surplus" land: environmental and socio-economic implications. BioRisk 7:5–50.

90. Zhang, Y.-H. P., and W.-D. Huang. 2012. Constructing the electricity-carbohydrate-hydrogen cycle for a sustainability revolution. Trends Biotechnol. 30:301–306.

91. Li, H., P. H. Opgenorth, D. G. Wernick, S. Rogers, T.-Y. Wu, W. Higashide, et al. 2012. Integrated

Electromicrobial Conversion of CO_2 to Higher Alcohols. Science 335:1596.

92. Magnuson, A., M. Anderlund, O. Johansson, P. Lindblad, R. Lomoth, T. Polivka, et al. 2009. Biomimetic and Microbial Approaches to Solar Fuel Generation. Acc. Chem. Res. 42:1899–1909.

93. Nevin, K. P., T. L. Woodard, A. E. Franks, Z. M. Summers, and D. R. Lovley. 2010. Microbial electrosynthesis: feeding microbes electricity to convert carbon dioxide and water to multicarbon extracellular organic compounds. mBio 1:e00103–00110.

94. House, K. Z., A. C. Baclig, M. Ranjan, E. A. van Nierop, J. Wilcox, and H. J. Herzog. 2011. Economic and energetic analysis of capturing CO_2 from ambient air. Proc. Natl Acad. Sci. USA 51:20428–20433.

95. Bastian, S., X. Liu, J. T. Meyerowitz, C. D. Snow, M. M. Y. Chen, and F. H. Arnold. 2011. Engineered ketol-acid reductoisomerase and alcohol dehydrogenase enable anaerobic 2-methylpropan-1-ol production at theoretical yield in *Escherichia coli*. Metab. Eng. 13:345–352.

96. Mozaffarian, D., R. Micha, and S. Wallace. 2010. Effects on Coronary Heart Disease of Increasing Polyunsaturated Fat in Place of Saturated Fat: A Systematic Review and Meta-Analysis of Randomized Controlled Trials. PLoS Med. 7:e1000252.

97. Siri-Tarino, P. W., Q. Sun, F. B. Hu, and R. M. Krauss. 2010. Meta-analysis of prospective cohort studies evaluating the association of saturated fat with cardiovascular disease. Am. J. Clin. Nutr. 91:535–546.

98. Randel, R. D., C. C. Chase, and S. J. Wyse. 1992. Effects of gossypol and cottonseed products on reproduction of mammals. J. Anim. Sci. 70:1628–1638.

99. Berg, I. A., D. Kockelkorn, W. H. Ramos-Vera, R. F. Say, J. Zarzycki, M. Hügler, et al. 2010. Autotrophic carbon fixation in archaea. Nat. Rev. Microbiol. 8: 447–460.

100. The Global Institute of Sustainable Forestry at Yale University. 2009. Available at http://environment.yale.edu/gisf/programs/landscape-management/global-and-regional-forest-conditions/ (accessed 2009).

Numerical investigation of the effects of dwell time duration in a two-stage injection scheme on exergy terms in an IDI diesel engine by three-dimensional modeling

Samad Jafarmadar[1] & Alborz Zehni[2]

[1]Mechanical Engineering Department, University of Urmia, Urmia, West Azerbaijan 57561-15311, Iran
[2]Mechanical Engineering Department, University of Sahand, Sahand, Iran

Keywords
Dwell duration, exergy, exergy efficiency, IDI engine, irreversibility, two-step injection

Abstract

Dwell duration of multiple-injection scheme is an important parameter, which makes it possible to shift the tradeoff curve between soot and NOx closer to the origin. In this investigation, therefore, energy and exergy analyses are carried out for various two-step injection schemes in which 25% of the total fuel is injected during the second pulse and the dwell time is increased from 5°CA (crank angle) to 30°CA by 5°CA increments. The calculations are performed for a Lister 8.1 indirect injection (IDI) diesel engine at full load operation. The energy analysis for these schemes is performed during a closed cycle by using a three-dimensional CFD code. The cylinder pressure results for the baseline engine are compared with the corresponding experimental data and they show good agreement. For the exergy analysis, an in-house computational code has been developed, which uses the results of energy analysis in various cases. With crank angle positions and dwell durations set for different injection schemes, various rates of exergy are calculated and the cumulative exergy components are identified separately. The results show that the values of work exergy and exergy efficiency decrease when the dwell duration is changed from 5°CA to 30°CA. Also, there is a sharp change in the exergy parameters when the dwell time reaches 25°CA.

Introduction

The major pollutants from diesel engines are NOx and soot. NOx and soot emissions are of concern to the international community. Stringent exhaust emission standards require the simultaneous reduction in soot and NOx for diesel engines; however, it seems to be very difficult to reduce NOx emission without increasing soot emission. The reason is that there always is a contradiction between NOx and soot emissions whenever the injection timing is retarded or advanced. Split injection has been shown to be a powerful tool to simultaneously reduce soot and NOx emissions for direct injection (DI) and indirect injection (IDI) diesel engines when the injection timing is optimized. Split injection is defined as splitting the main single injection profile into two or more injection pulses with definite delay dwell between the injections. However, an optimum injection scheme of split injection for DI and IDI diesel engines has been always under investigation. Generally, the exhaust of IDI diesel engines is less smoky when compared to DI diesel engines [1]. Hence, investigation of the effect of split injection on combustion process and pollution reduction of IDI diesel engines can be quite valuable.

In recent years, some researchers have already spent significant effort on the effect of the split injection on the combustion process and pollution of DI and IDI diesel engines [2–11].

In a very competitive world, the improvement of engine performance has become an important issue for automotive manufacturers. In order to improve engine performance, the combustion and emission processes are studied more thoroughly these days by simultaneously applying the first and second laws of thermodynamics. Exergy is the key concept in the second law analysis; it is a special case of the more fundamental concept, the available energy, which has been introduced in [12]. For analyzing the performance of engine subsystems, exergy analysis can be a useful alternative to energy analysis, because it is able to reveal more information about engine processes [13–15]. Over the years, many reports have been published on the detailed use of the second law of thermodynamics with respect to internal combustion engines [16–19]. A summary of other studies on the subject has been provided below.

Rakopoulos et al. [20] developed a code for studying the effects of cylinder wall temperature on the second-law transient performance of an IDI turbocharged multi-cylinder diesel engine following a load increase, which had a special emphasis on the case of low heat rejection. The result of this work shows that the transient first-law properties are almost unaffected by the applied wall temperature scheme, while the second-law terms of engine and turbocharger are greatly affected, especially when a low heat rejection cylinder wall is chosen.

Rakopoulos et al. [18] carried out a second-law analysis for a multi-cylinder turbocharged diesel engine and all its components by using a single-zone thermodynamic model. It was demonstrated that exergy analysis offers a more well-rounded and comprehensive insight into the processes occurring in a diesel engine than its traditional first-law counterpart.

Ghazikhani et al. [21] carried out an experimental study about the effect of exhaust gas recirculation (EGR) on various exergy terms of an IDI diesel engine cylinder. Their results indicated that the application of EGR to engine mainly increases the total in-cylinder irreversibility due to the extension of the flame region, thereby raising the combustion temperature. Also, the results revealed that the behaviors of the total in-cylinder irreversibility and engine brake specific fuel consumption (BSFC) are the same, especially at high load conditions.

Rakopoulos et al. [22] carried out an experimental investigation on a six-cylinder, IDI, turbocharged and aftercooled, medium-high-speed marine-type diesel engine coupled to a hydraulic brake. The irreversibility components of every transient cycle were calculated for the diesel engine and its subsystems. Also, for the sake of comparison, the rate and the cumulative value of all the important exergy components were given for the first and last cycles of the transient event. The importance of combustion irreversibility as well as exhaust manifold irreversibility was revealed.

Amjad et al. [23] used a single-zone model to perform a numerical availability analysis of the combustion of n-heptane and natural gas blends in homogenous charge combustion ignition (HCCI) engines. They showed that as the mass percentage of natural gas in the fuel blend increases, irreversibility decreases and the second-law

efficiency increases. Adding the EGR to the intake charge of the dual-fuel HCCI engine, up to an optimum value, enhances the exergy efficiency. EGR values above this point could deteriorate engine performance.

Hosseinzadeh et al. [24] carried out a numerical study by comparing the thermal, radical, and chemical effects of EGR gases using a single-zone model to analyze availability in dual-fuel engines operating at 50% loads. They showed that the chemical effect of EGR causes an increase in the unburned chemical availability and a decrease in the work availability, in comparison with the baseline engine (without EGR); while the thermal and radical effects are positive on the availability terms, especially on the unburned chemical and work availabilities. They also demonstrated that by using low values for the radical and thermal components of EGR, the exergy efficiency increases.

Jafarmadar [25] carried out a numerical exergy analysis in pre and main chambers of an IDI diesel engine using three-dimensional modeling.

Jafarmadar [26] carried out a numerical analysis about the effect of engine load on the exergy terms of an IDI diesel engine using three-dimensional modeling. The results show that when the load increases from 25% to full load in steps by 25%, the percentage of combustion irreversibility decreases from 33.7% to 25% of fuel burn exergy. Also, exergy efficiency reaches its peak of 36.7% at 75% load.

Jerald [27] carried out an overview about quantitative levels of exergy destruction during the combustion process in a spark-ignition engine. He presented the detailed results for various operative conditions and design parameters.

Sivadas et al. [28] studied the effects of EGR on exergy destruction due to isobaric combustion for a range of conditions and fuels. Both cooled and adiabatic cases of EGR were studied.

Sayin et al. [29] carried out a comparative energy and exergy analyses of a four-cylinder, four-stroke spark-ignition engine using gasoline fuels of three different research octane numbers (RONs), namely 91, 93, and 95.3.

Jafarmadar et al. [30] carried out a numerical analysis about the exergy analysis in a low heat rejection IDI Diesel Engine by three-dimensional modeling. The comparison of the results for baseline and low heat rejection (LHR) cases shows that when the load increases from 25% to 100% (in 25% increments), heat loss exergy decreases by 68.73%, 80.24%, 91.38%, and 74.97%, respectively, in LHR engine, in comparison to baseline engine. Also, exergy efficiency increases by 17.2%, 12.4%, 6.07%, and 11.81% in LHR engine.

As can be seen in relevant literature, so far, no attempt has been made to use a three-dimensional model to study the effect of dwell duration in two-step injection schemes on various exergy terms in IDI diesel engines from the second-law viewpoint. In this study, a computational fluid dynamic (CFD) code along with an in-house code have been used to predict various exergy terms at different dwell times of a two-step injection scheme for an IDI engine operating at full load.

Initial and Boundary Conditions

Numerical calculations are conducted at closed system from intake valve closure (IVC) at 165°CA (crank angle)

before top dead center (BTDC) to exhaust valve open (EVO) at 180°CA after top dead center (ATDC). The numerical grid for modeling of geometry engine includes a maximum of 42,200 cells at 165°CA BTDC and generated by automated mesh generation of AVL FIRE Software. The present resolution was found to give adequately grid-independent results. Figure 1A shows the numerical grid which is designed to model the geometry of combustion chamber. Also, Figure 1B shows that grid dependency is based on the in-cylinder pressure and present resolution is found to give adequately grid-independent results.

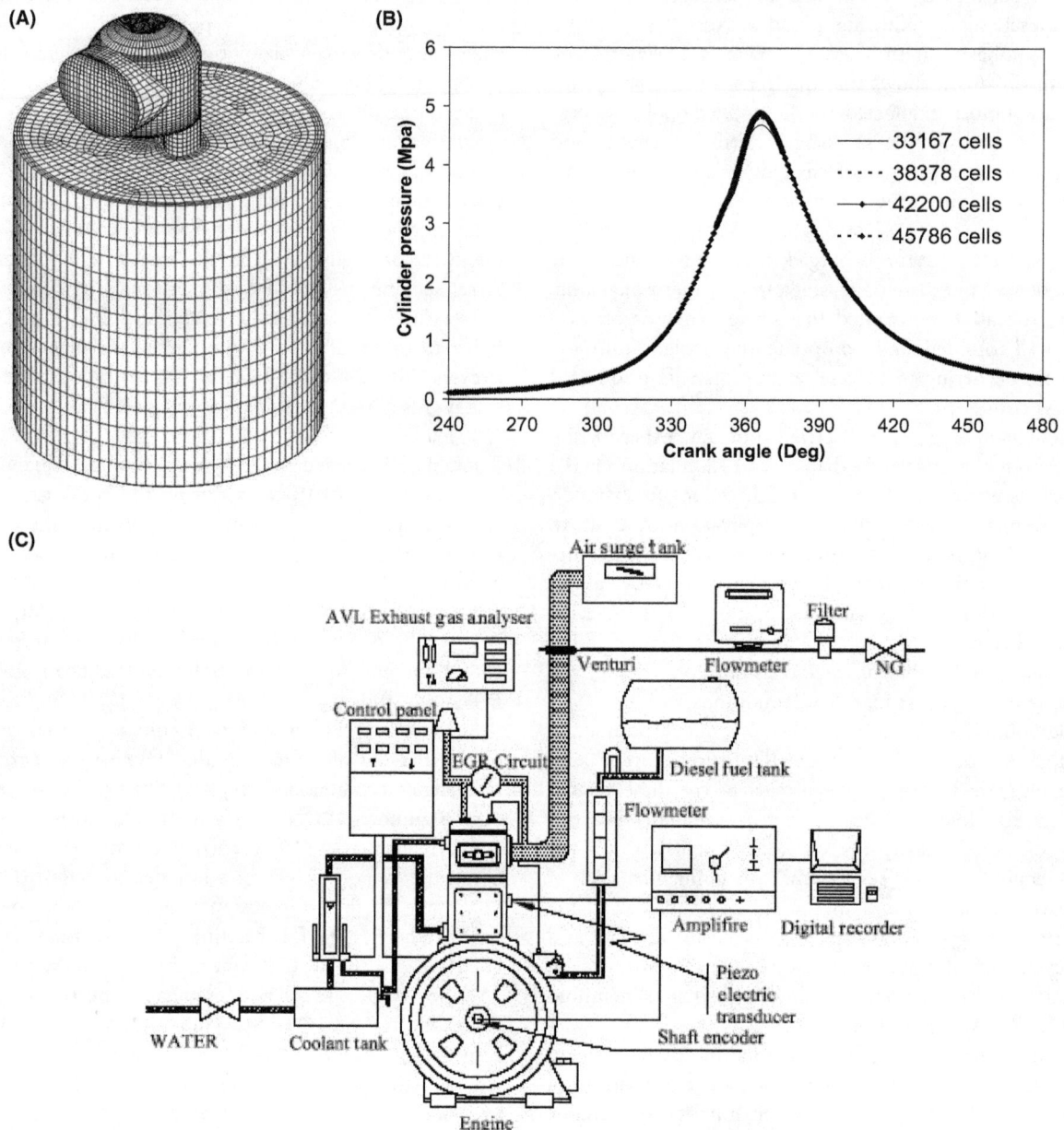

Figure 1. (A) Mesh of the Lister 8.1 indirect injection diesel engine. (B) Grid dependency based on the in-cylinder pressure. (C) Schematic diagram of experimental set-up.

Initial pressure in the cylinder is set to 86 kPa and initial temperature is designed to be 384 K. This value for pressure in-cylinder was chosen according to experimental data and for temperature in-cylinder is calculated by air-fuel ratio. All boundaries temperatures were assumed to be constant throughout the simulation, but allowed to vary at various combustion chamber surfaces. Boundary temperatures for two cases in the combustion chamber are as follows:

Head temperature: 550 K.
Piston temperature: 590 K.
Cylinder temperature: 450 K.

This work is studied at full load mode and the engine speed is 730 rpm. All boundary temperatures were assumed to be constant throughout the simulation, but allowed to vary with the combustion chamber surface regions.

First Law and Energy Analysis

The numerical model is carried out for Lister 8.1 IDI diesel engine with the specifications shown in Table 1.

The governing equations include species, momentum, and energy with appropriate turbulent model (RNG $k - \varepsilon$) [31] and solved from IVC to EVO. Standard WAVE breakup model [32] is used in order to simulate the primary and secondary atomization of the spray and resulting droplets. Also, drop parcels are injected in-cylinder with diameters equal to the diameter of nozzle exit. The Dukowicz model [33] is used for modeling of the heat up and evaporation of the droplets. A Stochastic dispersion model is employed to take the effect of interaction between the droplets and the turbulent eddies into account by adding a fluctuating velocity to the mean gas velocity [34]. This model assumes that the fluctuating velocity has a randomly Gaussian distribution. Modeling of the autoignition is based on Shell autoignition [35]. This generic mechanism includes six species as hydrocarbon fuel, oxidizer, total radical pool, branching agent, intermediate species, and products. Also, the important stages of autoignition such as initiation, propagation, branching, and termination were presented by generalized reactions [34, 36].

The Eddy Break-up model (EBU), which is based on turbulent mixing, is used for modeling of the combustion process [30] as follows:

$$\overline{\rho \dot{r}_{fu}} = \frac{C_{fu}}{\tau_R} \overline{\rho} \min\left(\overline{y}_{fu}, \frac{\overline{y}_{ox}}{S}, \frac{C_{pr} \cdot \overline{y}_{pr}}{1 + S} \right), \qquad (1)$$

where it is assumed in this model that in premixed turbulent flames, the reactants are contained in the same eddies and are separated from eddies containing hot combustion products. The rate of dissipation of these eddies determines the rate of combustion. Because chemical reaction occurs fast, it is assumed that the combustion is mixing controlled. The first two terms in the "minimum value of" operator determine whether fuel or oxygen is present in limiting quantity, and the third term is a reaction probability which ensures that the flame is not spread in the absence of hot products. The above equation also includes two constant coefficients (C_{fu}, C_{pr}) and τ_R is the turbulent mixing time scale for chemical reaction.

The governing equations for unsteady, compressible, turbulent reacting multi-component gas mixtures flow and thermal fields were solved from IVC to EVO using the commercial AVL FIRE CFD code [34].

Second Law and Exergy Analysis

The exergy of a system is defined as the maximum value of work obtainable when a system reaches thermal, mechanical, and chemical equilibrium with its environment by means of reversible processes while exchanging heat with only the environment. This state of equilibrium is defined as the dead state of the system and it depends on pressure, temperature, and composition of environment. In this study, the pressure and temperature are assumed equal $P_0 = 1$ bar and $T_0 = 298.2$ K, respectively. When there is no heat exchange between the system and the environment, that is, the system is at temperature T_0, thermal equilibrium is achieved. Similarly, mechanical equilibrium is achieved when there is no work exchange between the system and its environment. Chemical equilibrium is achieved when no system component can react with the environment. For the present case, this means that in the dead state all species of the working medium

Table 1. Specifications of Lister 8.1 IDI diesel engine.

Cycle type	Four stroke
Number of cylinders	1
Injection type	IDI
Cylinder bore × stroke	0.1141 × 0.1397 (m^2)
L/R	4
Displacement volume	1.43×10^{-3} (m^3)
Compression ratio	17.5 : 1
$V_{pre-chamber}/V_{TDC}$	0.7
Full load injected mass	6.4336×10^{-5} kg per cycle
Injection pressure	88.8 (bar)
Start injection timing	20° BTDC
Nuzzle diameter at hole center	0.003 m
Number of nuzzle holes	1
Nuzzle outer diameter	0.0003 m
Spray cone angle	10°
Valve timing	IVO = 5° BTDC
	IVC = 15° ABDC
	EVO = 55° BBDC
	EVC = 15° ATDC

have been either oxidized or reduced to N_2, O_2, CO_2, and H_2O. It is shown in Refs. [36, 37] that the total exergy, that is, thermo-mechanical plus chemical exergies of a closed system is equal to:

$$Ex = Ex_{ch} + Ex_{tm} = E - P_0 V - T_0 S - \sum_{i=1}^{kk} \mu_i^0 m_i \quad (2)$$

where μ_i^0 is the chemical potential of species i at the true dead state, and m_i is the mass of species i.

The exergy balance equation applied for the closed system inside the cylinder, on CA basis, is expressed as follows [36, 37]:

$$\frac{dEx}{d\theta} = \frac{dEx_w}{d\theta} - \frac{dEx_q}{d\theta} - \frac{dI}{d\theta} + \frac{dEx_f}{d\theta} \quad (3)$$

where Ex_w is exergy associated with work done by the system and it can be defined as follows:

$$\frac{dEx_w}{d\theta} = (P - P_0)\frac{dV}{d\theta} \quad (4)$$

This equation illustrates that the rate of work exergy is equal to the rate of useful work done by the system.

Also, Ex_q is exergy pertained to heat losses from across the system boundary. Its variation with CA is given as:

$$\frac{dEx_q}{d\theta} = \left(1 - \frac{T_0}{T}\right)\frac{dQ}{d\theta} \quad (5)$$

I is destruction exergy associated with combustion process and it can be defined as follows:

$$\frac{dI}{d\theta} = -\frac{T_0}{T}\sum_{i=1}^{kk} \mu_i \frac{dm_i}{d\theta} \quad (6)$$

where index i includes all reactants and products. For perfect gases $\mu_j = g_j$ and for the fuel $\mu_f = a_f$. The above equation reveals that both heat transfer and work production inside the cylinder only indirectly influence the irreversibility value.

Exergy of liquid fuels C_yH_z which are used in compression ignition engines is approximated as follows [38]:

$$a_f = LHV\left(1.04224 - 0.011925\frac{y}{z} + \frac{0.042}{z}\right) \quad (7)$$

and

$$\frac{dEx_f}{d\theta} = a_f \frac{dm_f}{d\theta} \quad (8)$$

is the burned fuel availability, with a_f being the fuel chemical exergy, and $\frac{dm_f}{d\theta}$ is the fuel burning rate.

The second-law efficiency can be defined as the ratio of indicated work over total input chemical exergy. For the closed part of the cycle in an engine, the second-law efficiency is defined as:

$$\eta_{II} = \frac{W_{net}}{M_{fi} a_f} \quad (9)$$

Results and Discussion

The calculations of exergy terms are carried out for the combustion chamber of a single-cylinder Lister 8.1 IDI diesel engine operating under different two-step injection schemes, at a constant speed of 730 rpm and full load conditions. For routine sampling of the cylinder pressure traces, a piezoelectric-type pressure transducer (Indi Modul 621) was flush-mounted with the combustion chamber. Cylinder pressure was measured at every 0.1 CAs. Figure 1C shows Schematic diagram of experimental setup. Figure 2 shows the computed and the measured [9] mean in-cylinder pressures for the baseline engine. The results presented in these figures are global (cylinder-averaged) quantities and are shown as a function of time (CA). The parameters of fuel injection timing and the amount of injected mass are adjusted according to experimental data. This figure indicates that the computational and experimental pressures during the compression, combustion, and expansion strokes are in excellent agreement. The peak cylinder pressure is 48.8 bars, which occurs at 366°CA (4°CA after TDC). The discrepancy between peak pressures obtained via experiment and computation is less than 0.2%.

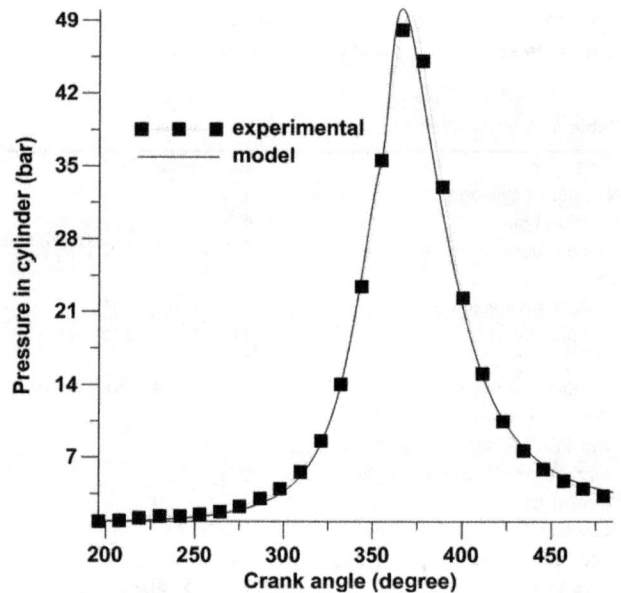

Figure 2. Comparison of measured [26] and calculated pressure for baseline at single-injection case for full load operation and engine speed 730 rpm.

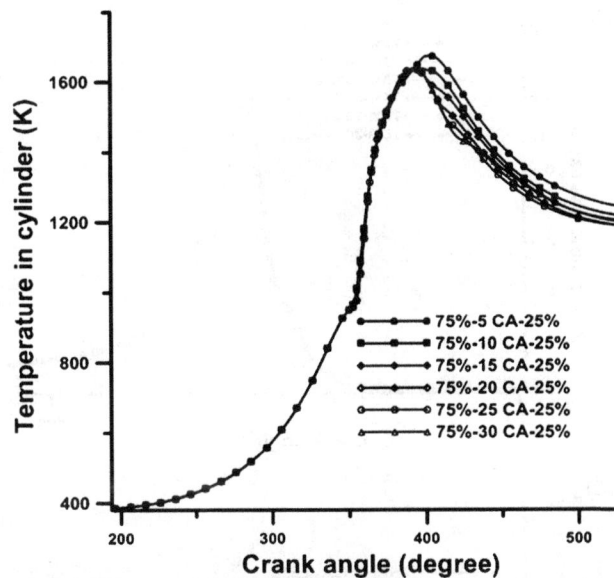

Figure 3. Variation in in-cylinder temperature with crank angle position at various schemes in full load operation and engine speed 730 rpm.

Temperature trend for different cases is illustrated in Figure 3. As can be seen in this figure, when the dwell time increases, the in-cylinder peak temperature diminishes as a result of late injection and the late combustion of fuel injected in the second pulse. The higher rise of peak temperature in the 75%-5°CA-25% case, compared to the other cases, is due to the injection of the second pulse during the ignition period and before the start of combustion. In other cases, similar peak temperatures are observed due to the same amount of fuel injection during the ignition delay period. In the 75%-30°CA-25% scheme, a second peak of temperature can be observed due to the late injection of the second pulse. It causes the in-cylinder temperature to go higher in the expansion stroke, relative to the 75%-20°CA-25% and 75%-25°CA-25% schemes. The peak temperature values for different cases, as the dwell time varies from 5°CA to 30°CA are 1674, 1636, 1934, 1637, 1637, and 1636 K, respectively.

The trend of work exergy rates under various schemes are illustrated in Figure 4A and B. It can be seen that in the compression stroke, the rates of work exergy are negative for all the cases as a result of the compression work. At the start of the combustion process, the rates of work exergy increase for all the cases due to the increase in the in-cylinder pressure and temperature. These figures clearly show that the peak values of work exergy rate remain almost constant for various schemes because of the almost same amount of fuel burned and the same in-cylinder peak pressure created. These figures also indicate that when the dwell duration increases from 5°CA to 30°CA in

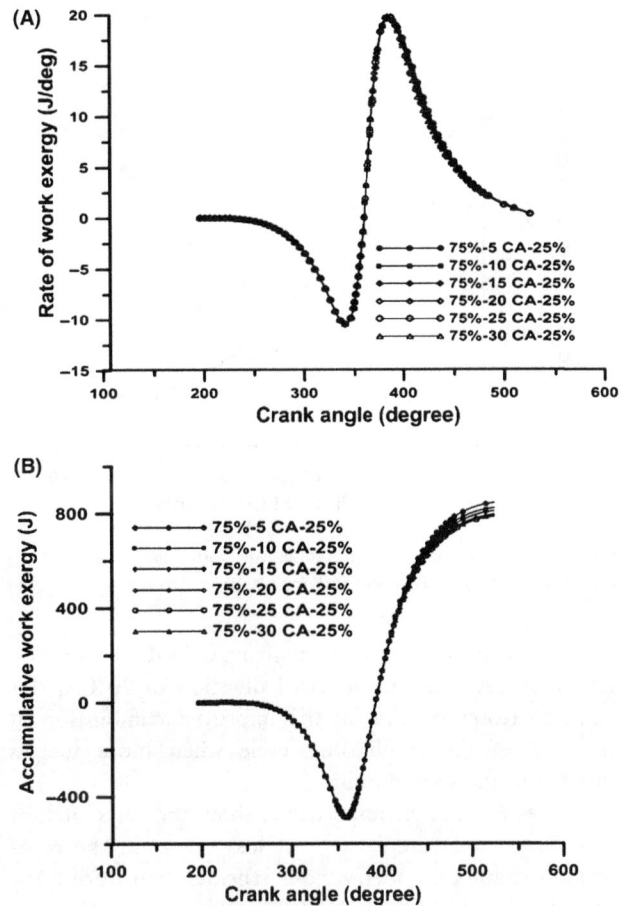

Figure 4. The variation in (A) rate of work exergy (B) accumulated work exergy with crank angle position at various schemes in full load operation and engine speed 730 rpm.

increments of 5°CA, the rates of work exergy following the start of combustion decreases slightly, except for the dwell time of 30°CA. During this dwell time, the combustion of the fuel injected by the second pulse at the end of the expansion stroke improves the combustion and increases the rate of work exergy. As is shown in Figure 3, this increase in work exergy rate is caused by the higher in-cylinder temperature gradients in the expansion stroke of the 75%-30°CA-25% scheme. According to Figure 4B, the amounts of cumulative work exergy for dwell times from 5°CA to 30°CA are 31.7%, 31.7%, 31.8%, 31.7%, 31.5%, and 31.3% of the exergy of fuel burned, respectively. These values are in good agreement with the result of Rakopoulos and Giakoumis [18]. They showed that the value of work exergy for IDI turbocharged diesel engine is 35%. Slight difference between the two values is due to turbocharging.

Figure 5 shows the variation in cumulative work exergy with dwell duration under various injection schemes. According to this figure, the cumulative work exergy

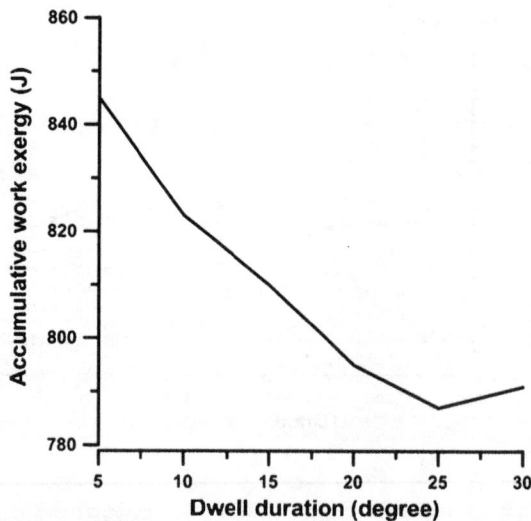

Figure 5. The variation in accumulative work exergy with dwell duration in full load operation and engine speed 730 rpm.

decreases continuously with the increase in dwell duration up to 25°CA. During the dwell duration of 30°CA, this value increases because of the improved combustion at the end of the combustion cycle when more fuel is injected in the second pulse.

Figures 6A and B, respectively, show the rates of heat loss exergy and cumulative heat loss exergy in the combustion chamber, under various schemes, and during the engine's closed cycle. At CAs larger than those at the start of combustion, because of the chamber's lower pressure and temperature, the rate of heat loss exergy decreases slightly as the dwell duration increases from 5°CA to 25°CA. During the dwell duration of 30°CA, the rate of heat loss exergy increases due to the improvement in the combustion process in the latter part of the expansion stroke. The peak rates of heat loss exergy are almost the same for various schemes, while the rates of heat loss exergy at the point of EVO are 0.822, 0.762, 0.741, 0.722, 0.710, and 0.756 J/deg, respectively, for dwell durations ranging from 5°CA to 30°CA. The increased rate of heat loss exergy in the 30°CA dwell duration is the result of higher temperature in the latter part of the expansion stroke. As is shown in Figure 6B, the amounts of cumulative heat loss exergy for dwell durations ranging from 5°CA to 30°CA are 525.42, 511.09, 503.56, 497.44, 491.05, and 496.1 J, which constitute 19.1%, 19.7%, 19.7%, 19.8%, 19.7%, and 19.7% of the burned fuel exergies, respectively. These values are in good agreement with the result of Rakopoulos and Giakoumis [18]. They showed that the value of heat loss exergy for IDI turbocharged diesel engine is 18.95%.

Figure 7 illustrates the changes of cumulative heat loss exergy with dwell duration under various injection

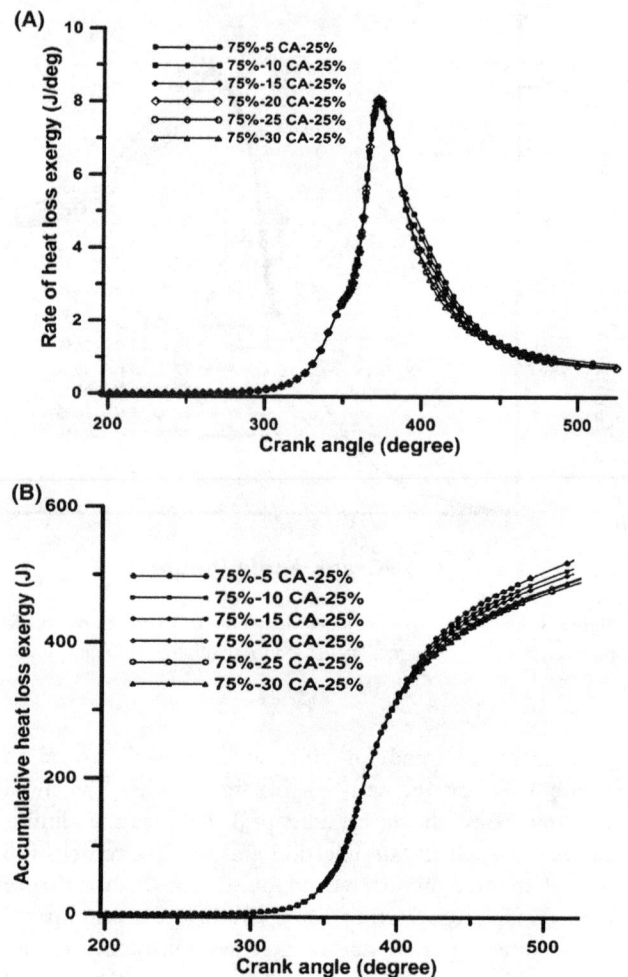

Figure 6. The variation in (A) rate of heat loss exergy (B) accumulated heat loss exergy with crank angle position at various schemes in full load operation and engine speed 730 rpm.

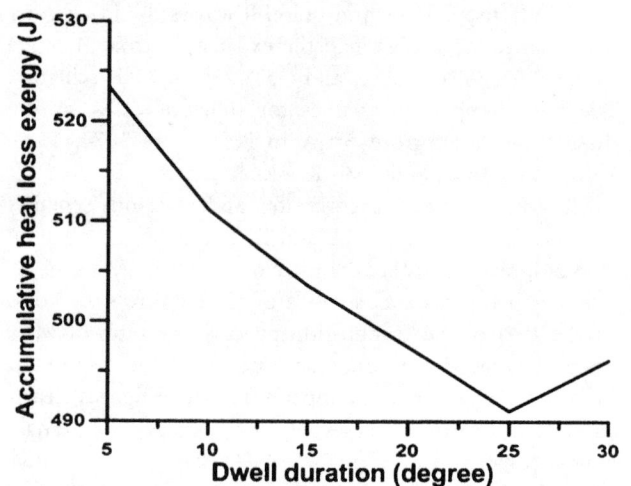

Figure 7. The variation in accumulative heat loss exergy with dwell duration in full load operation and engine speed 730 rpm.

schemes. According to this figure, the value of cumulative heat loss exergy decreases continuously with the increase in dwell duration up to 25°CA. During the dwell duration of 30°CA, this value increases because of improved combustion in the latter part of the combustion cycle due to the combustion of fuel injected in the second pulse.

Figures 8A and B, respectively, show the exergy rates of burned fuel and cumulative exergy rates of burned fuel in the chamber under various schemes during the engine's closed cycle. As is shown in Figure 8A, the first peak values in the premixed combustion phase in the dwell durations of 15, 20, 25, and 30°CA are higher than those in the 5 and 10°CA durations, as a result of longer ignition delays. In the diffusion combustion phase, similar peak values can be observed for all the schemes. The third peak

in the exergy rate of burned fuel indicates the combustion of the injected fuel in the second pulse. For the split injection case, the second peak in the exergy rate of burned fuel occurs at 410°CA and it indicates a rapid diffusion burn in the latter part of the combustion and affects the in-cylinder pressure, temperature, and soot oxidation. As is shown in Figure 8B, the amounts of cumulative burned fuel exergy for dwell durations ranging from 5°CA to 30°CA are 2667, 2595, 2550, 2512, 2495, and 2524 J, respectively. Figure 9 shows the changes of cumulative burned fuel exergy with dwell duration under different injection schemes. According to this figure, the value of cumulative burned fuel exergy decreases continuously with the increase in dwell duration up to 25°CA. During the dwell duration of 30°CA, this value increases because of improved combustion in the latter part of the combustion phase as a result of the combustion of fuel injected in the second pulse. Figure 10 shows the changes of thermo-mechanical exergy in the cylinder with CA positions under various injection schemes. In the compression stroke, before the injection of fuel, the thermo-mechanical exergy increases in the chamber, because of the work done by the piston. The value of thermo-mechanical exergy in the chamber increases at the beginning of the combustion process due to the rise in temperature and pressure, and this increase is more pronounced in lower dwell durations. At the end of the combustion phase and in the expansion stroke, the rate of thermo-mechanical exergy diminishes due to the gas temperature reduction during the expansion stroke. The increase in thermo-mechanical exergy during the dwell duration of 30°CA, as compared to the dwell durations of

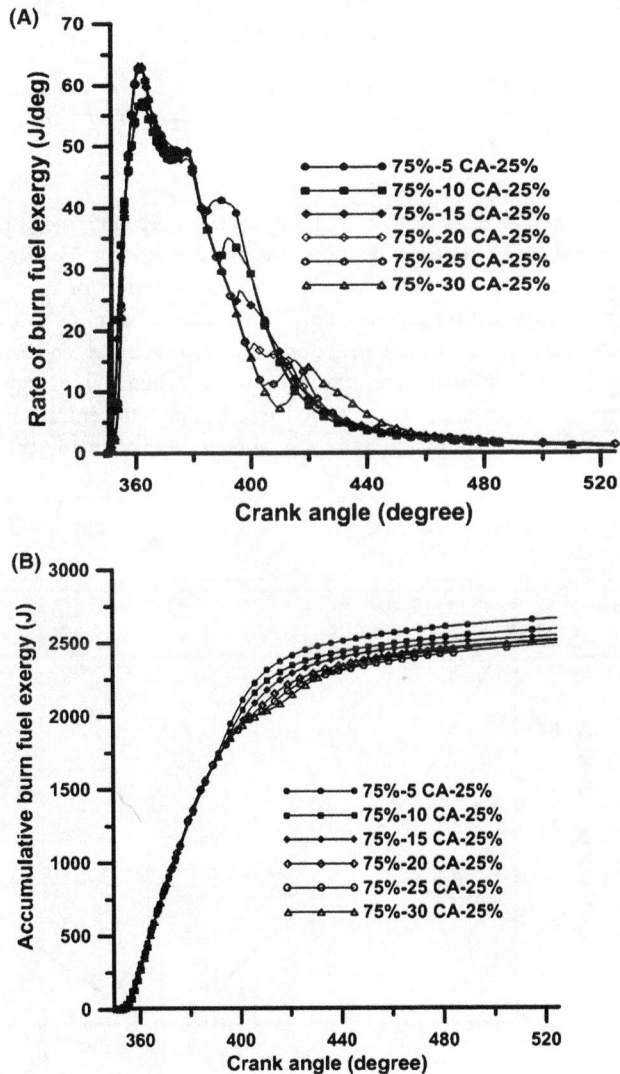

Figure 8. The variation in (A) rate of fuel burn exergy (B) accumulated fuel burn exergy with crank angle position at various schemes in full load operation and engine speed 730 rpm.

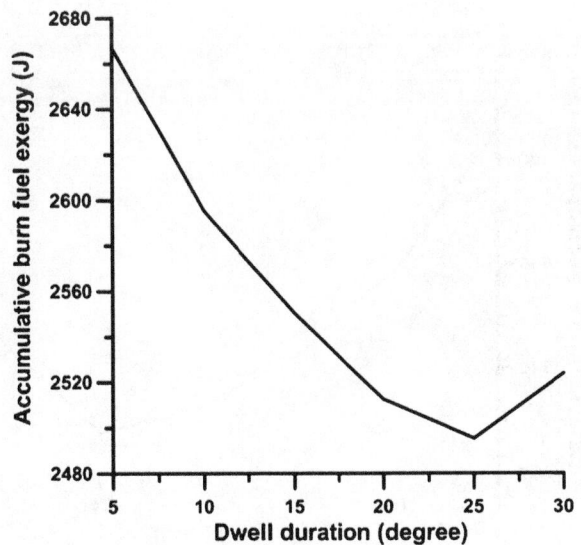

Figure 9. The variation in accumulative fuel burn exergy with dwell duration in full load operation and engine speed 730 rpm.

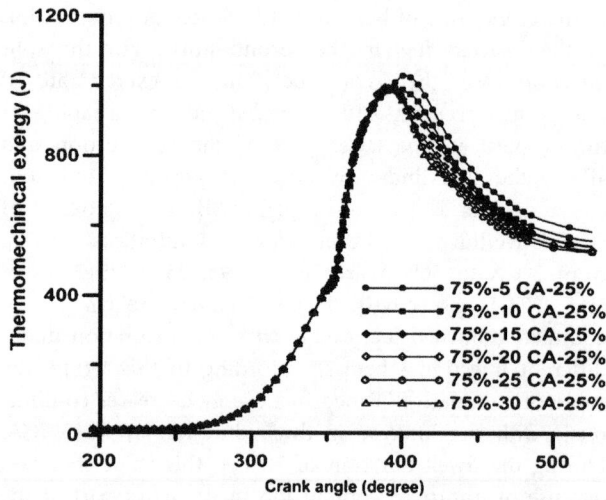

Figure 10. The variation in accumulated thermo-mechanical exergy with crank angle position at various schemes in full load operation and engine speed 730 rpm.

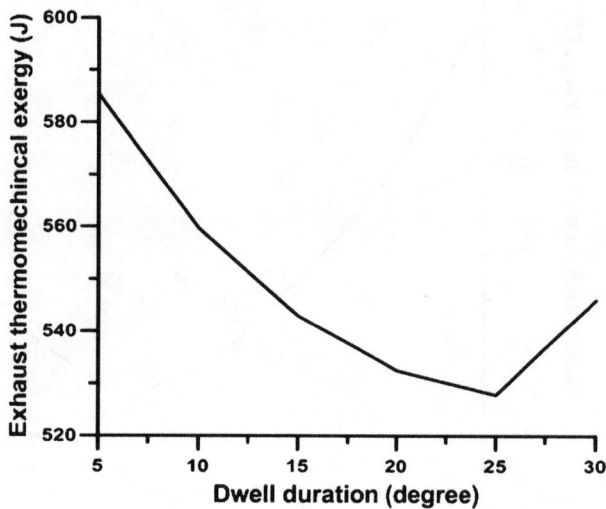

Figure 12. The variation in accumulative chemical exergy with crank angle position at various schemes in full load operation and engine speed 730 rpm.

25 and 20°CA, is due to the improved combustion at the end of the expansion stroke as a result of fuel injection in the second pulse. As Figure 11 illustrates, the exhaust thermo-mechanical exergy diminishes with the increase in dwell duration from 5°CA to 25°CA and then it increases again during the dwell duration of 30°CA. This increase is due to the higher pressure and temperature resulting from EVO during the dwell duration of 30°CA (as shown in Fig. 3). Figure 12 shows the changes of cumulative chemical exergy in the cylinder with CA positions in various injection schemes. As is shown in this figure, the chemical exergy does not change during the compression

stroke. The chemical exergy in the chamber increases at the start of the combustion process due to the rise in temperature and pressure and the concentration of complete combustion products; and this increase is more pronounced for shorter dwell durations. Toward the end of the combustion phase, the amount of chemical exergy remains constant, because of the constant concentration of combustion products. As is shown in Figure 13, the

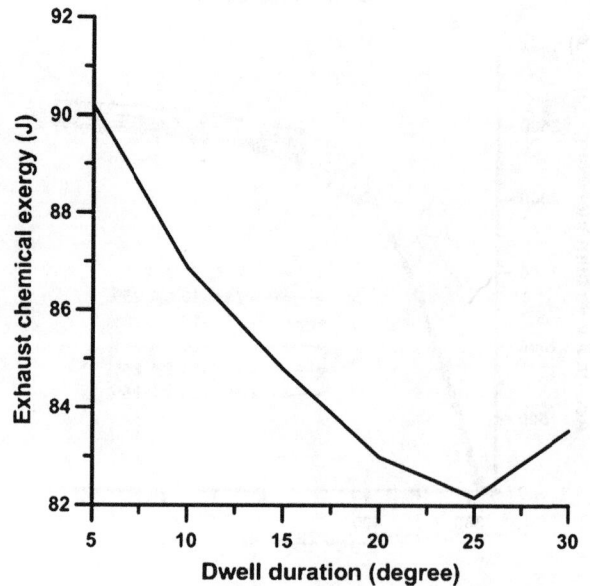

Figure 11. The variation in accumulative thermo-mechanical exergy with dwell duration in full load operation and engine speed 730 rpm.

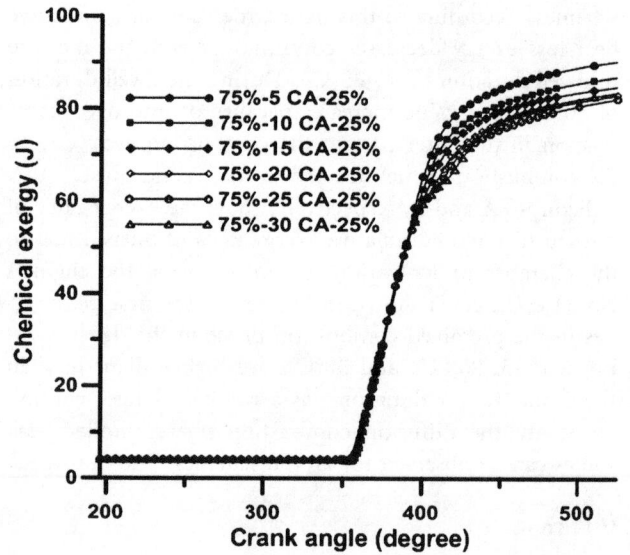

Figure 13. The variation in chemical exergy with dwell duration in full load operation and engine speed 730 rpm.

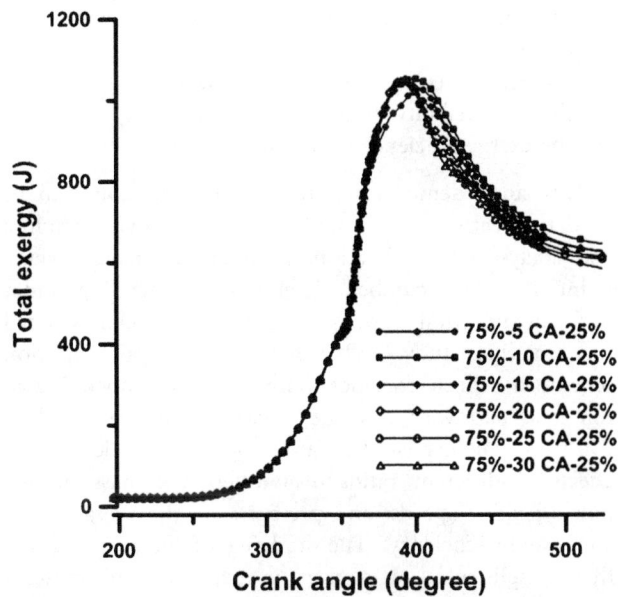

Figure 14. The variation in total exergy with crank angle position at various schemes in full load operation and engine speed 730 rpm.

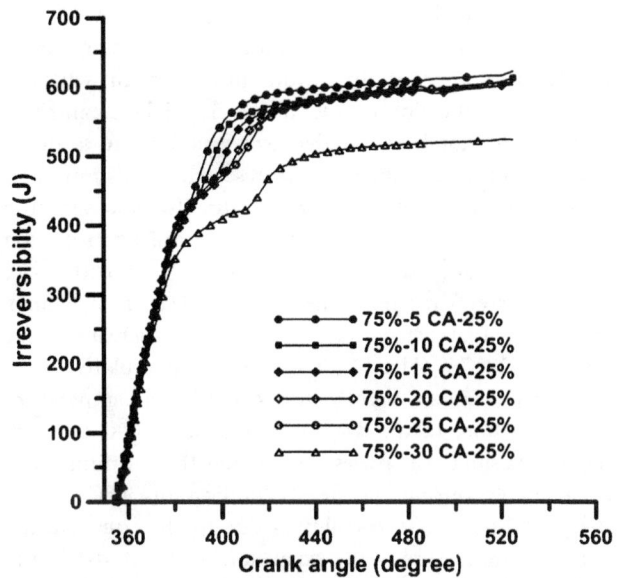

Figure 16. The variation in irreversibility with crank angle position at various schemes in full load operation and engine speed 730 rpm.

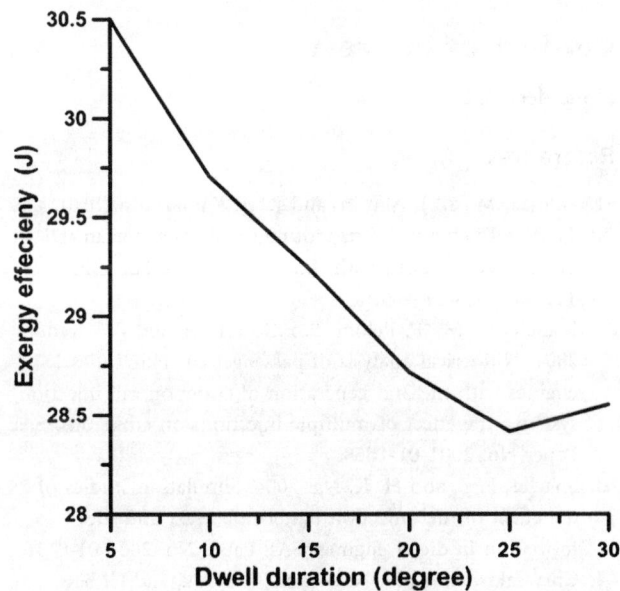

Figure 15. The variation in second-law efficiency with dwell duration in full load operation and engine speed 730 rpm.

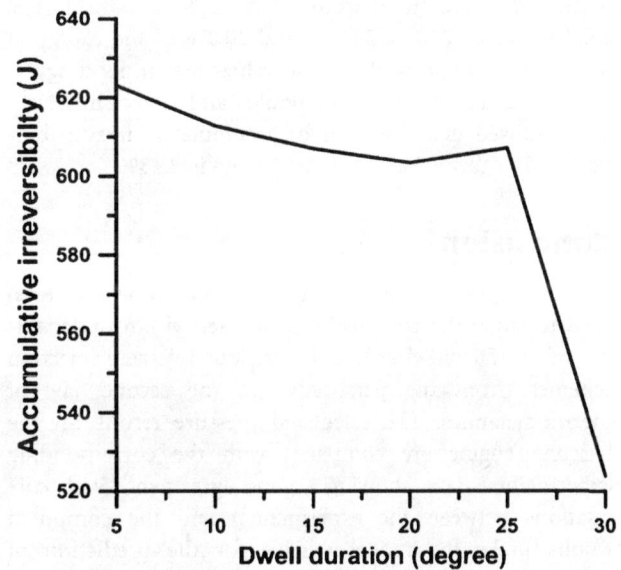

Figure 17. The variation in accumulative irreversibility with dwell duration in full load operation and engine speed 730 rpm.

amounts of cumulative chemical exergy for various schemes using dwell durations from 5°CA to 30°CA are 90.21, 86.86, 84.79, 82.99, 83.16, and 83.53 J, respectively.

Figure 14 shows the total exergy, which is defined as the sum of chemical and thermo-mechanical exergies. The changes of this value are similar to those of the chemical and thermo-mechanical exergies. The amounts of exhaust gas exergies for dwell times from 5°CA to

30°CA are 25.3%, 24.87%, 24.62%, 24.52%, 23.46%, and 24.98% of the exergy of fuel burned, respectively. These values are in good agreement with the result of Rakopoulos and Giakoumis [18]. They showed that the value of accumulative irreversibilities for IDI turbocharged diesel engine is 21.38%.

The variation in the second-law efficiency with dwell duration is shown in Figure 15. According to this figure, the value of the second-law efficiency diminishes with the increase in dwell duration from 5°CA to 25°CA, due to the reduction in work exergy, and then it increases during the dwell duration of 30°CA, due to the rise in work exergy. The combustion process makes the highest contribution to the total in-cylinder irreversibility in a diesel engine, which according to the research by Primus and Flynn [39], is more than 90%. Figure 16 illustrates the trend for the cumulative irreversibility arising from the in-cylinder combustion during an engine's closed cycle, under different injection schemes. It is clear from Figures 16 and 17 that when the dwell duration increases from 5°CA to 20°CA (Fig. 17), the value of cumulative irreversibility decreases slightly due to the lower temperature of combustion. In the dwell duration of 25°CA, cumulative irreversibility increases slightly and then it diminishes considerably during the dwell duration of 30°CA. This sharp decline of irreversibility is due to the rise of combustion temperature in the expansion stroke. A rise in the cylinder charge gas temperature reduces the relative heat transfer from the reacting gas to the yet unburned mixture, which causes the irreversibility of the combustion process to diminish. The amounts of cumulative irreversibilities for dwell times from 5°CA to 30°CA are 23.3%, 23.6%, 23.8%, 24.1%, 24.3%, and 20.8% of the exergy of fuel burned, respectively. These values are in good agreement with the result of Rakopoulos and Giakoumis [18]. They showed that the value of accumulative irreversibilities for IDI turbocharged diesel engine is 23.3%.

Conclusion

In this study, a three-dimensional CFD code has been used to study the combustion processes within the chamber of an IDI diesel engine, for various two-step injection schemes, from the perspective of the second law of thermodynamics. The calculated pressure results for the baseline engine are compared with the corresponding experimental data, showing a good agreement. Such correlations between the experimental and the computed results make the model reliable for the prediction of exergy terms under various two-step injection schemes. Various exergy terms including the fuel, heat loss, irreversibility, work, exhaust loss, and the chemical and thermo-mechanical exergies are presented for different two-step injection schemes. The findings of this study, when the dwell duration increases from 5°CA to 30°CA are as follows:

1. All exergy terms diminish with the increase in dwell duration from 5°CA to 25°CA and then increase during the dwell duration of 30°CA.
2. The cumulative heat loss exergy decreases by 5.2%.
3. The cumulative work exergy decreases by 6.32%.

4. The cumulative burned fuel exergy diminishes by 5.67%.
5. The exhaust exergy diminish by 6.82%.
6. The irreversibility diminishes by 1.45%.
7. The exergy efficiency diminishes by 6.32%.

This study demonstrates that CFD simulations can be used to greatly improve and facilitate the understanding and analysis of combustion, from the second-law viewpoint, in multi-chamber diesel engines operating under different injection modes. It should be acknowledged that in this study, only an optimum split injection scheme has been considered for an engine operating at full load and with a single maximum torque speed of 730 rpm. Further studies are required to address the effects of different ratios of injected fuel mass in two pulses, engine speeds, and pressures on the exergy terms and exergy efficiency. The studying of literatures shows that by split injection concept, a reduction can be established in NOx and particulate matter emissions [9, 10]. Therefore, this simulation is carried out as exergoenvironmental optimization in IDI diesel engine according to Ahmadi et al. [40].

Conflict of Interest

None declared.

References

1. Gomaa, M., A. J. Alimin, and K. A. Kamarudin. 2010. Trade-off between NOx, soot and EGR rates for an IDI diesel engine fuelled with JB5. World Acad. Sci. Eng. Technol. 62:449–450.
2. Bianchi, G. M., P. Peloni, F. E. Corcione, and F. Lupino. 2001. Numerical analysis of passenger car HSDI diesel engines with the 2nd generation of common rail injection systems: the effect of multiple injections on emissions. SAE Paper No. 2001-01-1068.
3. Shayler, P. j., and H. K. Ng. 2004. Simulation studies of the effect of fuel injection pattern on NOx and soot formation in diesel engines. SAE Paper No. 2004-01-0116.
4. Chryssakis, C. A., D. N. Assanis, S. Kook, and C. Bae. 2005. Effect of multiple injections on fuel-air mixing and soot formation in diesel combustion using direct flame visualization and CFD techniques. Spring Technical Conference, ASME No. ICES2005-1016.
5. Ehleskog, R. 2006. Experimental and numerical investigation of split injections at low load in an HDDI diesel engine equipped with a piezo injector. SAE Paper No. 2006-01-3433.
6. Sun, Y. D., and R. Reitz. 2006. Modeling diesel engine NOx and soot reduction with optimized two-stage combustion. SAE Paper No. 2006-01-0027.

7. Abdullah, N., A. Tsolakis, P. Rounce, M. Wyszinsky, H. Xu, and R. Mamat. 2009. Effect of injection pressure with split injection in a V6 diesel engine. SAE Paper No. 2009-24-0049.

8. Jafarmadar, S., and A. Zehni. 2009. Multi-dimensional modeling of the effects of split injection scheme on combustion and emissions of direct-injection diesel engines at full load state. IJE 22:369–378.

9. Jafarmadar, S., and A. Zehni. 2012. Multi-dimensional modeling of the effects of spilt injection scheme on performance and emissions of IDI diesel engines. IJE 25:135–146.

10. Showry, K., and A. Raju. 2010. Multi-dimensional modeling and simulation of diesel engine combustion using multi-pulse injections by CFD. Int. J. Dyn. Fluids ISSN 0973-1784, 6:237–248.

11. Iwazaki, K., K. Amagai, and M. Arai. 2005. Improvement of fuel economy of an indirect (IDI) diesel engine with two-stage injection. Energy Convers. Manage. 30:447–459.

12. Gibbs, J. W. 1961. The scientific papers. Vol. 1. Dover, New York, NY.

13. Dunbar, W. R., N. Lior, and R. A. Gaggioli. 1992. The component equations of energy and exergy. Trans. ASME J. Energy Resour. Technol. 114:75–83.

14. Dunbar, W. R., and N. Lior. 1994. Sources of combustion irreversibility. Combust. Sci. Technol. 103:41–61.

15. Obert, E. F. 1993. Internal combustion engines and air pollution. Intext Educ. Publ, New York, NY.

16. Rakopoulos, C. D., and E. G. Giakoumis. 1997. Simulation and exergy analysis of transient diesel engine operation. Energy 22:875–885.

17. Caton, J. A. 2000. On the destruction of availability (exergy) due to combustion processes – with specific application to internal combustion engines. Energy 25:1097–1117.

18. Rakopoulos, C. D., and E. G. Giakoumis. 1997. Development of cumulative and availability rate balances in a multi-cylinder turbocharged IDI diesel engine. Energy Convers. Manage. 38:347–369.

19. Kyritsis, D. C., and C. D. Rakopoulos. 2001. Parametric study of the availability balance in an internal combustion engine cylinder. SAE paper 2001-01-1263.

20. Rakopoulos, C. D., and D. C. Kyritsis. 2005. The influence of cylinder wall temperature profile on the second-law diesel engine transient response. Appl. Therm. Eng. 25:1779–1795.

21. Ghazikhani, M., M. E. Feyz, A. Joharchi. 2010. Experimental investigation of the exhaust gas recirculation effects on irreversibility and brake specific fuel consumption of indirect injection diesel engines. Appl. Therm. Eng. 30:1711–1718.

22. Rakopoulos, C. D., and E. G. Giakoumis. 2004. Availability analysis of a turbocharged diesel engine operating under transient load conditions. Energy 29:1085–1104.

23. Amjad, A. K., R. Khoshbakhi Saray, S. M. S. Mahmoudi, and A. Rahimi. 2011. Availability analysis of n-heptane and natural gas blends combustion in HCCI engines. Energy 36:6900–6909.

24. Hosseinzadeh, A., R. Khoshbakhti Saray, and S. M. Seyed Mahmoudi. 2010. Comparison of thermal, radical and chemical effects of EGR gases using availability analysis in dual-fuel engines at 50% loads. Energy Convers. Manage. 51:2321–2329.

25. Jafarmadar, S. 2013. Three-dimensional modeling and exergy analysis in combustion chamber of an indirect injection diesel engine. Fuel 107:439–447.

26. Jafarmadar, S. 2013. Multidimensional modeling of the effect of engine load on various exergy terms in an indirect injection diesel engine. Int. J. Exergy, in press.

27. Caton, J. A. 2012. Exergy destruction during the combustion process as functions of operating and design parameters for a spark-ignition engine. Int. J. Energy Res. 36:368–384.

28. Sivadas, H. S., and J. A. Caton. 2008. Effects of exhaust gas recirculation on exergy destruction due to isobaric combustion for a range of conditions and fuels. Int. J. Energy Res. 32:896–910.

29. Sayin, C., M. Hosoz, M. Canakci, and I. Kilicaslan. 2007. Energy and exergy analyses of a gasoline engine. Int. J. Energy Res. 31:259–273.

30. Jafarmadar, S., R. Tasoujiazar, and B. Jalilpour. 2013. Exergy analysis in a low heat rejection IDI diesel engine by three dimensional modeling. Int. J. Energy Res. 3 SEP 2013, DOI: 10.1002/er.3100

31. Han, Z., and D. Reitz. 1995. Turbulence modeling of internal combustion engines using RNG $k-\varepsilon$ models. Combust. Sci. Technol. 106:267–295.

32. Liu, A. B., and R. D. Reitz. 1993. Modeling the effects of drop drag and break-up on fuel sprays. SAE Paper No. 930072.

33. Dukowicz, J. K. 1979. Quasi-steady droplet change in the presence of convection. Informal report Los Alamos Scientific Laboratory, Los Alamos, NM. LA7997-MS.

34. AVL FIRE user manual V. 8.5. 2006.

35. Halstead, M., L. Kirsch, and C. Quinn. 1977. The auto ignition of hydrocarbon fueled at high temperatures and pressures – fitting of a mathematical model. Combust. Flame 30:45–60.

36. Van Gerpen, J. H., and H. N. Shapiro. 1990. Second law analysis of diesel engine combustion. Trans. ASME J. Eng. Gas Turb. Power 112:129–137.

37. Rakopoulos, C. D., and E. G. Giakoumis. 2006. Second-law analyses applied to internal combustion engines operation. Prog. Energy Combust. Sci. 32:2–47.

38. Rodriguez, L. 1980. Calculation of available-energy quantities. Pp. 39–59 *in* R. A. Gaggioli, ed.

Thermodynamics: second law analysis. American Chemical Society, Symposium Series No. 122, Washington, DC.

39. Primus, R. J., and P. F. Flynn. 1986. Pp. 61–68 *in* The assessment of losses in diesel engines using second law analysis. ASME WA-Meeting, Proc. AES, Anaheim, CA.

40. Ahmadi, P., A. Almasi, M. Shahriyari, and I. Dincer. 2012. Multi-objective optimization of a combined heat and power (CHP) system for heating purpose in a paper mill using evolutionary algorithm. Int. J. Energy Res. 36:46–63.

3

Esterification of hydrolyzed sea mango (*Cerbera odollam*) oil using various cationic ion exchange resins

Jibrail Kansedo & Keat Teong Lee

School of Chemical Engineering, Universiti Sains Malaysia Engineering Campus, Seri Ampangan, 14300, Nibong Tebal, Pulau Pinang, Malaysia

Keywords

Biodiesel, *Cerbera odollam*, esterification, free fatty acids, hydrolysis

Abstract

This study investigates the esterification of hydrolyzed sea mango (*Cerbera odollam*) oil using several cationic ion exchange resins. The best resins were selected based on their performance in a preliminary esterification process. The best resins were then subsequently used in the optimization of the process parameters. The esterification parameters studied were reaction temperature (40–160°C), reaction time (0–5 h), molar ratio of oil to methanol (0.5:1 to 1:14), and catalyst loading (1–14 wt%). Among the resins studied, Amberlyst 15 was found to be the most promising catalyst in the esterification of the hydrolyzed sea mango oil. Moderate reaction temperatures, 60–100°C, were found to be adequate in converting the hydrolyzed sea mango oil into esters. Further investigation revealed that the esterification reaction using the cationic ion exchange resins proceeds at a fast rate, whereby fatty acid methyl esters (FAME) yield of over 80% at moderate reaction temperature was achievable in less than 1 h of reaction time. Small amount of catalyst, which is less than 5 wt%, was also found to be sufficient in catalyzing the esterification process to an acceptable yield.

Introduction

The presence of free fatty acids (FFA) in conventional production of biodiesel – particularly in those that use sodium hydroxide (NaOH) or potassium hydroxide (KOH) as catalyst – is always deemed as nuisance due to its tendency to react with the catalyst itself to produce soap instead of esters [1]. Formation of soap in biodiesel production is unfavorable as it lowers the yield of esters (components of biodiesel) and creates emulsion especially when water is present. Emulsion causes the downstream separation and purification of the product (esters) to become more complicated. This adds unnecessary cost and is uneconomical for commercial production of biodiesel. Consequently, in conventional biodiesel production process, the oil feedstock used must not contain more than 0.5% of FFA [2]. In cases where the oil feedstock inevitably contains more FFA, other methods apart from homogenous basic catalysts must be used.

However, recent development and advancement in biodiesel production process have shown that the presence of FFA in oil feedstock could be actually beneficial. Hence, instead of reducing the FFA content in the oil feedstock, many new studies have attempted to utilize FFA to produce esters via esterification process. New methods such as noncatalytic process which utilizes superheated alcohol (sub- or supercritical alcohol) [3] and heterogeneous process which utilizes solid acid catalysts [4] were reported to take advantage of the high amount of FFA in oil feedstock for biodiesel production.

FFA are beneficial in biodiesel production due to its tendency to react easily with alcohol to form esters. Its relative easiness to react is due to its lower molecular stability compared to triglycerides (in which fatty acids bond strongly with glycerol). This lower molecular stability makes FFA to dissociate better in polar solvent to produce carboxylate (R-COO$^-$) and hydronium (H$^+$) ions compared to triglycerides molecules.

FFA are also advantageous in biodiesel production because of its similar polarity with alcohol by having a polar functional group (-COOH) on one of its molecular ends. Being partially polar makes FFA more soluble in alcohol compared to triglycerides. A high solubility of oil (FFA) in alcohol is important as it overcomes the mass transfer problem that has long been hindering the commercial production of biodiesel. A high solubility of oil (FFA) in alcohol is also necessary for a system where physical mixing to overcome the mass transfer problem is limited or cannot be implemented, for example, in a packed bed reactor.

Similar to transesterification process, the esterification process to convert FFA into esters can be carried out with or without catalyst [1, 2, 5]. Due to FFA' lower molecular stability, the activation energy of esterification process is generally much lower compared to transesterification process, making the former process more advantageous. This enables catalysts which are active at much lower temperature to be used in biodiesel production that can reduce the requirement of energy.

Ion exchange resin is an example of heterogeneous catalyst that can be used for esterification process. Ion exchange resins, particularly cationic ion exchange resins, are desirable due to their ability to catalyze the esterification process faster even at a lower temperature. The operating temperature by ion exchange resins normally range from 30°C to 150°C [6, 7], which is significantly lower than the requirement of 100–200°C [8, 9] by most other solid catalysts. By comparison, the operating temperature for conventional homogenous catalysts is around 60–70°C; however, homogenous catalysts in this case would not be suitable as they will react with the FFA to produce soap instead of fatty acid methyl esters (FAME).

Ion exchange resins, which are typically made of organic polymer substrate (e.g., cross-linked polystyrene), also have good mechanical strength, allowing them to be used in packed bed reactor for continuous production of biodiesel. Ion exchange resins are also easy to be regenerated and reused, making them more economical to be used as catalyst in biodiesel production.

With the intention of reaping the benefits of FFA in oil feedstock and its subsequent esterification process for biodiesel production, the objective of this project is to study and perform the esterification process of hydrolyzed sea mango (Cerbera odollam) oil using several cationic ion exchange resins. Several parameters – reaction temperature,

reaction time, molar ratio of oil to methanol, and catalyst loading – of the esterification process were investigated to determine the best condition to convert the hydrolyzed sea mango oil into esters using cationic ion exchange resins.

Material and Methods

Materials

Sea mango (C. odollam) oil was extracted from the kernels of mature/ripe sea mango fruits following a method by Kansedo and Lee [4, 10]. The fruits were collected around Penang Island, Malaysia (5°250′N 100°240′E). The composition and physical properties of the crude sea mango oil are shown in Table 1 [4]. Prior to use in the esterification process, the crude sea mango oil was hydrolyzed (without catalyst) using distilled water, at 200°C, volume ratio of oil to water of 20:80, and hydrolysis period of 10 h to increase its FFA content. The hydrolyzed sea mango oil was titrated using standard titration method and its acid value was measured to be 199.56 mg

Table 1. Composition and physical properties of crude sea mango oil [7].

Description	Method	Sea mango oil
Density (kg m^{-3})	ASTM	919.80
Viscosity at 40°C (mm^2 sec^{-1})	ASTM	29.57
Free fatty acids (%)	MPOB P2.5	6.40
Acid value (mg KOH g^{-1})	Standard titration	12.42
Water content (%)	Karl–Fisher	0.90
Fatty acid composition	GC	
C12: 0 (Lauric acid)		–
C14: 0 (Myristic acid)		–
C16: 0 (Palmitic acid)		24.86
C16: 1 (Palmitoleic acid)		0.75
C18: 0 (Stearic acid)		5.79
C18: 1 trans (Oleic acid)		0.24
C18: 1 cis (Oleic acid)		52.82
C18: 2 cis (Linoleic acid)		13.65
C18: 3n3 (Linolenic acid)		0.08
C20: 0 (Arachidic acid)		1.09
C20: 1 (Gadoleic acid)		0.19
C22: 0 (Behenic acid)		0.37
C24: 0 (Lignoceric acid)		0.16
Free fatty acids (FFA) (g L^{-1})		2.27
Monoglycerides (MG) (g L^{-1})		5.63
Diglycerides (DG) (g L^{-1})		2.99
Triglycerides (TG) (g L^{-1})		84.69
Average molecular weight (free fatty acids) (g mol^{-1})		276.35
Average molecular weight (monoglycerides) (g mol^{-1})		350.43
Average molecular weight (diglycerides) (g mol^{-1})		608.77
Average molecular weight (triglycerides) (g mol^{-1})		867.10

KOH g^{-1}, indicating a FFA composition of more than 90%.

The cationic ion exchange resins for the esterification process – Amberlyst 15, Dowex 50WX2, Dowex Marathon C, Dowex DR-G8, and Dowex DR-2030 – were all purchased from Sigma-Aldrich, Petaling Jaya Selangor, Malaysia. The cationic ion exchange resins were used in the esterification process without any pretreatment. The properties of the cationic ion exchange resins, obtained from their technical datasheets, are shown in Table 2. Sulfated zirconia catalyst, used for comparison purpose, was synthesized in the laboratory according to an optimized method by Kansedo and Lee [4]. Methanol (anhydrous grade), n-hexane (>99%), and FAME standard for gas chromatography analysis – methyl palmitate, methyl stearate, methyl oleate, methyl linoleate, and methyl heptadecanoate (internal standard) with purity above 99% – were all purchased from Fluka Chemie GmbH, Steinheim, Germany.

Esterification process

The esterification process (reaction) was carried out in a stainless steel tank reactor, equipped with magnetically driven stirrer with pressure and liquid-phase temperature indicators (Fig. 1). The reactor vessel has an internal volume of 220 mL and pressure rating of 5.5 MPa. The reactor temperature was controlled by an electrical heating device connected to a Proportional-integral-derivative (PID) controller.

In a typical reaction, 30 mL of preheated oil was premixed with catalyst (1–14 wt%) and methanol (1:10 by molar ratio) before being discharged into the reactor vessel. The initial pressure of the reactor vessel was raised to 2.0 MPa by an inlet nitrogen gas and sealed tightly before raising the reactor temperature to a predetermined setting (40–160°C). The reaction time was counted from the moment it reached the predetermined setting. After the specified reaction time, the heating device was switched off and the reactor vessel was immediately cooled with blowing air from a fan until it was cool enough to handle.

The reaction products consisted of two distinctive layers, oil layer (composed of methyl esters and unreacted oil) and aqueous layer (composed of byproduct water and excess methanol), which can be separated easily using a separating funnel. The separated oil layer was purified and dried using a rotary evaporator equipped with vacuum to remove excess methanol and moisture before being subjected to gas chromatography analysis for its methyl esters content.

Analysis of FAME

The analysis of sea mango esters was conducted using a Perkin Elmer Clarus 500 gas chromatography (PerkinElmer Inc., MA) using Nukol™ column (15 m × 0.53 mm, 0.5 μm film) with methyl heptadecanoate as internal standard. Helium gas was used as the carrier gas. Oven temperature at 110°C was initially held for 0.5 min and then increased to 220°C (held for 8 min) at a rate of 10°C per min. The temperature of the injector and detector was set at 220 and 250°C, respectively. The yield of sea mango oil into esters was calculated as the total mass (g) of methyl esters produced divided by the total mass (gram) of the fatty acids (hydrolyzed sea mango oil) in the feed [4, 10].

Table 2. Physicochemical properties of cationic ion exchange resins.

Properties	Amberlyst 15	Dowex DR-G8	Dowex DR-2030	Dowex 50WX2	Dowex Marathon C
Matrix	Styrene-divinylbenzene (macroporous)	Styrene-divinylbenzene (macroporous)	Styrene-divinylbenzene (macroporous)	Styrene-divinylbenzene (gel)	Styrene-divinylbenzene (gel)
Functional group	Sulfonics	Sulfonics	Sulfonics	Sulfonics	Sulfonics
Moisture content (%)	<3	<3	~3	74–82	53
Operating pH	0–14	0–14	0–14	0–14	0–14
Type	Strong cation (hydrogen form)	Strong cation (hydrogen form)	Strong cation (hydrogen form)	Strong cation (hydrogen form)	Strong cation (hydrogen form)
Particle size range/	18–23 mesh	16–50 mesh	16–40 mesh	100–200 mesh	23–27 mesh
Particle average diameter	600–850 μm		425–525 μm		550–650 μm
Total exchange capacity					
By wetted volume (meq mL^{-1})	1.8	4.5	1.7	0.6	1.8
By dry weight (meq g^{-1})	4.7	5.1	4.7	–	2.6
Maximum operating temperature (°C)	120	150	150	150	149
Relative density (g cm^{-3})	0.770	1.230	0.590	0.690	0.800

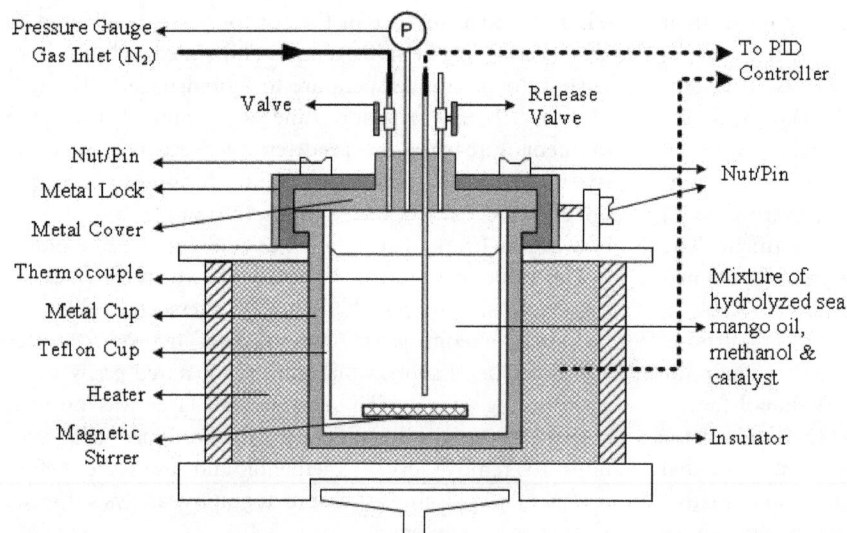

Figure 1. Experimental setup for the esterification process.

Results and Discussions

Selection of catalyst

Figure 2 shows the esterification of hydrolyzed sea mango oil using various cationic ion exchange resins and also sulfated zirconia catalyst. The catalytic performance of the cationic ion exchange resins and sulfated zirconia was measured by its yield of FAME. The catalytic performances of three cationic ion exchange resins, Amberlyst 15, Dowex DR-G8, and Dowex DR-2030, seem to be at

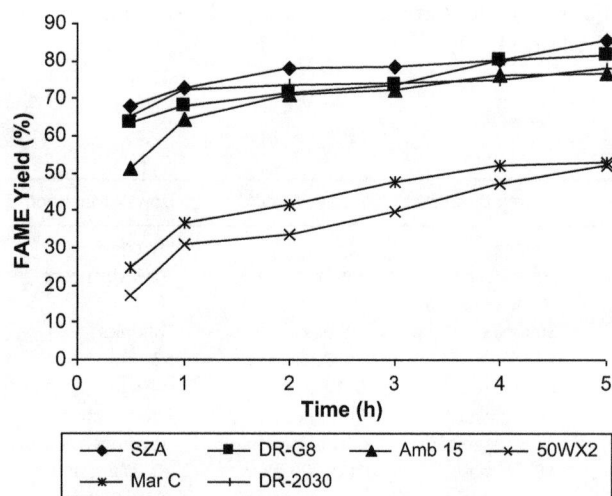

Figure 2. Esterification of hydrolyzed sea mango oil using various cationic ion exchange resins and sulfated zirconia catalyst (SZA). DR-G8, Dowex DR-G8; Amb 15, Amberlyst 15; 50WX2, Dowex 50WX2; Mar C, Dowex Marathon C; DR-2030, Dowex DR-2030 (reaction temperature = 100°C, molar ratio of oil to methanol = 1:3, catalyst loading = 5 wt%.

par with those by sulfated zirconia catalyst, whereas the catalytic performances of the rest of the ion exchange resins, Dowex 50WX2 and Dowex Marathon C, seem to be inferior.

The inferiority of Dowex 50WX2 and Dowex Marathon C compared to Amberlyst 15, Dowex DR-G8, and Dowex DR-2030 on the esterification process of the hydrolyzed sea mango oil could be due to their different physical and chemical properties. As shown in Table 2, Dowex 50WX2 and Dowex Marathon C were all in wet form (moisture content of 74–82% and 53% respectively), whereas the other three cationic ion exchange resins, Amberlyst 15, Dowex DR-G8, and Dowex DR-2030, were all in dry form (moisture content of about or less than 3%). The moisture in the cationic ion exchange resins could have inhibited the esterification process by creating a physical barrier between the reactants and the active sites (sulfonic group). The moisture (water) could also have attached to the active sites of the cationic ion exchange resins which prevented the reaction between fatty acids and methanol.

The good catalytic performance by Amberlyst 15, Dowex DR-G8, and Dowex DR-2030 compared to those by sulfated zirconia catalyst, a solid superacid catalyst, thereby confirmed their capability in converting fatty acids in the hydrolyzed sea mango oil into esters.

Optimization of parameters

Effect of temperature

Figure 3 shows the esterification of hydrolyzed sea mango oil by three of the best cationic ion exchange resins, Amberlyst 15, Dowex DR-G8, and Dowex DR-2030, at various reaction temperatures, ranging from 40°C to 160°C. Generally, the catalytic performances of the cationic ion

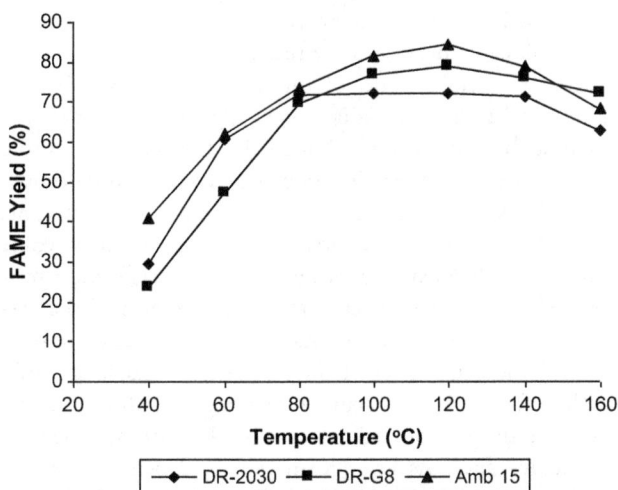

Figure 3. Effect of temperature on the esterification of hydrolyzed sea mango oil (molar ratio of oil to methanol = 1:3, catalyst loading = 5 wt%, reaction time = 1 h).

exchange resins were improved with increasing reaction temperature. This is expected as esterification process is an endothermic reaction, whereby energy is required in the process. The equilibrium of the reaction may have been shifted forward at higher temperature, resulting in more forward esterification reactions to produce more FAME.

According to Tesser et al. [11], esterification process using ion exchange resin as catalyst is assumed to proceed through a series of adsorption and desorption of reactants and products on the active sites of the catalyst. FFA are assumed to attach on the active sites (protonation) and react with methanol coming from the bulk liquid surrounding the catalyst particle to produce esters (surface reaction following Eley–Rideal mechanism). The whole adsorption and desorption process requires energy; therefore, the increase in reaction temperature could accelerate the adsorption and desorption process and also the surface reactions, and thereby increase the yield of FAME.

From Figure 3, among the three cationic ion exchange resins, Amberlyst 15 seems to have the upper hand in the esterification process compared to Dowex DR-G8 and Dowex DR-2030. Between Dowex DR-G8 and Dowex DR-2030, they seem to have a contrasting trend of FAME yield at different temperature. At lower temperature which is between 40°C and 80°C, Dowex DR-2030 produces higher FAME yield compared to Dowex DR-G8, but at higher temperature, which is more than 90°C, it was Dowex DR-G8 that produces the higher yield of FAME. This may suggest a distinctive optimum or suitable operating temperature for each type of cationic ion exchange resins for the esterification process. As indicated in Figure 3, the best operating temperature for Dowex

DR-G8 and Amberlyst 15 could be between 100°C and 120°C, whereas for Dowex DR-2030, the best operating temperature could be anywhere between 80°C and 140°C judged from their highest yield of FAME produced.

The performances (yield of FAME) of all three cationic ion exchange resins were found to significantly decrease at reaction temperature beyond 120°C. The decrease was particularly apparent for Amberlyst 15, whereby a steep decline in FAME yield was observed when the reaction temperature is raised above 120°C. On the other hand, the decrease in the catalytic activity of Dowex DR-G8 and Dowex DR-2030 becomes only apparent when the temperature is raised above 140°C, which is slightly higher than those for Amberlyst 15. This could be easily explained as each cationic ion exchange resin has its own maximum operating temperature. According to Table 2, the maximum operating temperature for Amberlyst 15 is 120°C, whereas for Dowex DR-G8 and Dowex DR-2030, the maximum operating temperature is 150°C. Beyond these temperatures, the cationic ion exchange resins (the matrix itself or its active sulfonic group) could be thermally destroyed or decomposed. The decomposition of the cationic ion exchange resin certainly reduces its ability to convert the fatty acids into esters, thereby lowers the yield of FAME.

Other possible explanation of the decrease in FAME yield at higher temperature as suggested by a reviewer is the excessive evaporation of methanol to the gas phase at higher temperature, especially above the boiling point of methanol (<65°C). This excessive evaporation of methanol at higher temperature caused decreased amount of methanol in the liquid phase that effectively reduces the reactions between oil and methanol.

The capability of Amberlyst 15 and Dowex DR-2030 to produce esters (although only achieving moderate yield) in the esterification process at lower temperature between 60°C and 80°C could be beneficial for operations which require reaction temperature below the boiling point of methanol. The operation below the boiling point of methanol, which is at 65°C under normal condition, is certainly advantageous as it could reduce the rapid loss of methanol by evaporation and therefore reduce the cost of biodiesel production.

Comparison with heterogeneous transesterification process that uses active metal oxides (which normally operates at much higher reaction temperature of 150–190°C), esterification process which operates at moderate low temperatures (as shown in this study) could be economical as it reduces energy consumption. In heterogeneous transesterification process, the triglycerides molecules are more structurally stable; therefore, more energy is required to break the triglycerides into fatty acids before it could react with alcohol in the transesterification process. FFA in esterification process, on the other hand, are less structurally stable; therefore, they could react more easily with alcohol to produce esters. Oils rich in FFA, for example, palm fatty acid distillate, a byproduct from palm oil refining which comprise of more than 80% of FFA [12], could easily be used in the esterification process for biodiesel production at much lower energy requirement (lower reaction temperature).

Effect of molar ratio of oil to methanol

Figure 4 shows the esterification of hydrolyzed sea mango oil by Amberlyst 15, Dowex DR-G8, and Dowex DR-2030 at 100°C using various molar ratio of oil to methanol from 0.5:1 to 1:10. As it can be observed from Figure 4, the catalytic performance by Amberlyst 15 seems to be significantly higher than those by Dowex DR-G8 and Dowex DR2030. It can also be observed that the catalytic performance of the three cationic ion exchange resins for molar ratio of oil to methanol of less than 1:1 is very low. This is expected, as according to the reaction stoichiometry, one mol of fatty acids requires one mol of alcohol to produce one mol of esters. Hence, when the amount of methanol used is less than the stoichiometric theoretical requirement, it will result to a very low yield of FAME.

Similar like any reversible process, the esterification process can be enhanced by shifting the equilibrium forward. This can be done by adding more methanol so that more esters could be formed. This is demonstrated in Figure 4, whereby significant increase in FAME yield is observed when the molar ratio of oil to methanol is increased from 1:1 to 1:3. However, beyond this point,

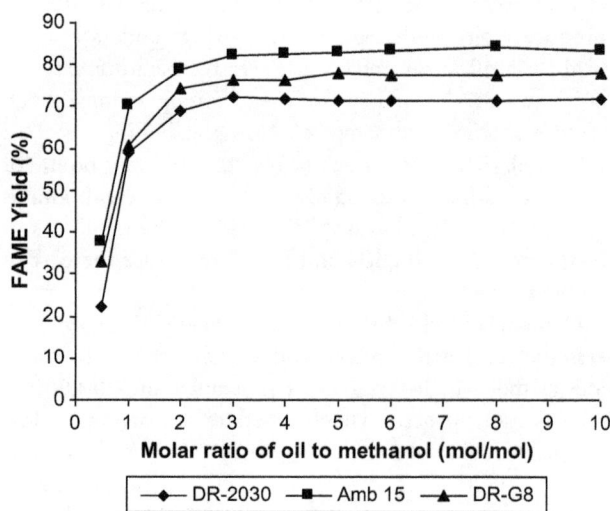

the FAME yield seems to reach a stagnant phase, whereby no significant increase in FAME could be observed albeit raising the molar ratio of oil to methanol beyond 1:3. This can be easily explained as at higher molar ratio, the catalyst becomes more diluted in the reaction media which reduces the reaction rate. This could also be caused by the disruption to the catalytic process that prevents further formation of esters. The disruption may come from water formed as byproduct in the esterification process whereby water attaches to the active sites of the cationic ion exchange resin. As water have better affinity toward the active sites of the cationic ion exchange resin with respect to fatty acids and methanol, thus it could displace the protonated fatty acids that are attached on the active sites. As ester formation is dependent on the surface reaction (exchange between protonated fatty acids attached on active sites and methanol coming from the bulk liquid), the displacement of protonated fatty acids by water therefore prevents formation of more esters.

Effect of catalyst loading

Figure 5 shows the esterification process by Amberlyst 15, Dowex DR-G8, and Dowex DR-2030 at 100°C with molar ratio of oil to methanol of 1:3, using different catalyst loading, from 1 wt% to 14 wt%. The best overall performance seems to be by Amberlyst 15, whereas the lowest seems to be by Dowex DR-2030. From Figure 5, it can be observed that addition of more catalyst (2–14 wt%) to the esterification process did not produce significant increase on the yield of FAME. This seems to be in contrast to what was expected as addition of more catalyst

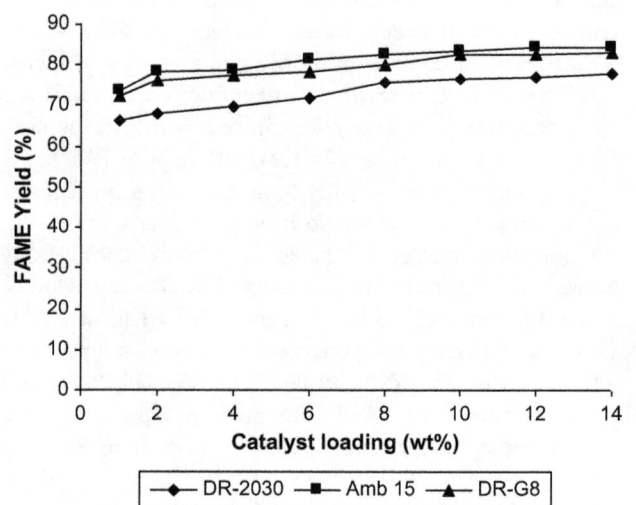

Figure 4. Effect of molar ratio of methanol to oil on the esterification of hydrolyzed sea mango oil (CATALYST loading = 5 wt %, reaction temperature = 100°C, reaction time = 1 h).

Figure 5. Effect of catalyst loading on the esterification of hydrolyzed sea mango oil (molar ratio of oil to methanol = 1:3, reaction temperature = 100°C, reaction time = 1 h).

should have enhanced the esterification process due to the availability of more active sites for the catalytic surface reactions.

The insignificant increase in FAME yield despite the addition of more catalyst could be explained by the constant amount of methanol supplied to the process. As mentioned in Figure 5, the molar ratio of methanol to oil for all catalyst loading is maintained at 1:3. With the same amount of methanol supplied to the process, it can be expected that the rate of reaction is also similar. This subsequently produces similar amount of FAME which is evident in Figure 5. On the other hand, the similar rate of reaction produces identical amount of water. This resulted in a nearly constant effect of catalyst poisoning by water on the active sites of the catalyst. However, with the addition of more catalyst, more active sites are still available for the reactions to occur. Significant increase of FAME, however, would not be expected due to the limitation of the methanol that is supplied to the process.

Figure 5 also indicates that small amount of catalyst could be adequate in converting the fatty acids into esters. As shown in Figure 5, Amberlyst 15 catalyst loading of 2 wt% is already sufficient to produce a modest 78% FAME yield. This is certainly promising as the requirement of less amount of catalyst can save the biodiesel production cost.

Effect of time

Figure 6 shows the effect of reaction time on the esterification of hydrolyzed sea mango oil using Amberlyst 15, Dowex DR-G8, and Dowex DR-2030, at 100°C, molar ratio of oil to methanol of 1:3, and catalyst loading of 5 wt%. As can be seen in Figure 6, increasing the reaction time from 1 to 5 h did not significantly improve the yield of FAME.

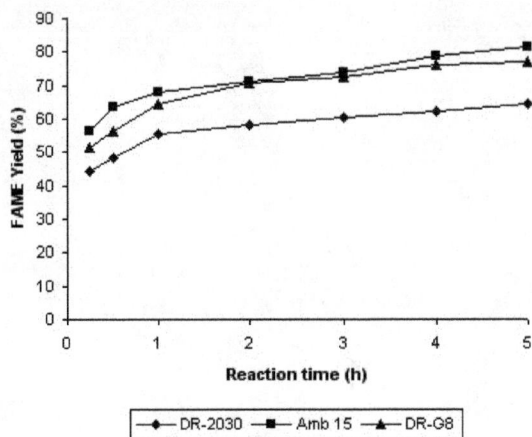

Figure 6. Effect of reaction time on the esterification of hydrolyzed sea mango oil (reaction temperature = 100°C, molar ratio of oil to methanol = 1:3, catalyst loading = 5 wt%).

Significant increase in FAME yield for all cationic ion exchange resins was only observed within the 1 h of reaction time. This may indicate that the esterification process is a fast process, whereby adsorption and desorption of reactants and products, to and from the active sites, could happen in a rapid manner. As the interaction between reactants is entirely driven by ions exchange, whereby positively charged molecules interact with negatively charged molecules, the esterification process could happen in rapidly manner, unlike those in transesterification process where reacting molecules such as triglycerides are in neutrally charged form. In heterogeneous transesterification process, the reaction time required for the conversion of oil into esters normally ranged from 1 to 5 h [13–15]. In some other cases, the reaction time required was reported to be as long as 8 h [16, 17], and even up to 18 h [18]. The long reaction time is definitely uneconomical as it is energy intensive. Therefore, the fast reaction rate by esterification process using cationic ion exchange resins, as shown in this study, could be economically more feasible for more economic biodiesel production.

Optimization

Based on the experimental results in this project, the esterification process is best carried out using Amberlyst 15. The best reaction condition is as follow: reaction temperature = 100°C, reaction time = 1 h, molar ratio of oil to methanol = 1:3, and catalyst loading = 2 wt%.

Conclusions

This study demonstrated that it is possible to produce esters (FAME) via esterification process at moderate temperature (60–100°C) using cationic ion exchange resins for biodiesel production. Amberlyst 15 was shown to have the best catalytic activity in the esterification process compared to other cationic ion exchange resins, namely, Dowex DR-G8, Dowex DR-2030, Dowex Marathon C, Dowex 50WX2, and Diaion PK228. The requirement of moderate temperature (60–100°C), much lesser amount of methanol (1:3), and also small amount of catalyst (less than 2 wt%) is advantageous to the esterification process over conventional transesterification process. Esterification process which requires a rather short reaction time (less than 1 h) compared to 3–5 h for conventional heterogeneous transesterification process could be a viable way to produce esters for utilization as biodiesel.

Acknowledgments

The authors would like to acknowledge Universiti Sains Malaysia (PRGS No. 8044027, Research University Grant

No. 814062) and Malaysia's Ministry of Higher Education (MyPhD) for providing funds in this project.

Conflict of Interest

None declared.

References

1. Van Gerpen, J. 2005. Biodiesel processing and production. Fuel Process. Technol. 86:1097–1107.
2. Ma, F., and Milford. A. Hanna. 1999. Biodiesel production; a review. Bioresour. Technol. 70:1–15.
3. Tan, K. T., K. T. Lee, and A. R. Mohamed. 2010. Effects of free fatty acids, water content and co-solvent on biodiesel production by supercritical methanol reaction. J. Supercrit. Fluid 53:88–91.
4. Kansedo, J., and K. T. Lee. 2012. Transesterification of palm oil and crude sea mango (Cerbera odollam) oil: the active role of simplified sulfated zirconia catalyst. Biomass Bioenergy 40:96–104.
5. Borges, Markus. E., and L. Díaz. 2012. Recent developments on heterogeneous catalysts for biodiesel production by oil esterification and transesterification reactions: a review. Renew. Sustain. Energy Rev. 16:2839–2849.
6. Feng, Y., B. He, Y. Cao, J. Li, M. Liu, F. Yan, et al. 2010. Biodiesel production using cation-exchange resin as heterogeneous catalyst. Bioresour. Technol. 101:1518–1521.
7. Park, J.-Y., D. K. Kim, and J. S. Lee. 2010. Esterification of free fatty acids using water-tolerable Amberlyst as heterogeneous catalyst. Bioresour. Technol. 101(Suppl): S62–S65.
8. Chouhan, A. P. S., and A. K. Sarma. 2011. Modern heterogeneous catalysts for biodiesel production: a comprehensive review. Renew. Sustain. Energy Rev. 15:4378–4399.
9. Semwal, S., A. K. Arora, Rajendra. P. Badoni, and D. K. Tuli. 2011. Biodiesel production using heterogeneous catalysts. Bioresour. Technol. 102:2151–2161.
10. Kansedo, J., and K. T. Lee. 2013. Process optimization and kinetic study for biodiesel production from non-edible sea mango (Cerbera odollam) oil using response surface methodology. Chem. Eng. J. 214:157–164.
11. Tesser, R., D. Luca Casale, M. D. Verde, and E. S. Serio. 2010. Kinetics and modeling of fatty acids esterification on acid exchange resins. Chem. Eng. J. 157:539–550.
12. Top, A. G. M. 2010. Production and utilization of palm fatty acid distillate (PFAD). Lipid Technol. 22:11–13.
13. Miao, X., R. Li, and H. Yao. 2009. Effective acid-catalyzed transesterification for biodiesel production. Energy Convers. Manage. 50:2680–2684.
14. Sinha, S., A. K. Agarwal, and S. Garg. 2008. Biodiesel development from rice bran oil: transesterification process optimization and fuel characterization. Energy Convers. Manage. 49:1248–1257.
15. Abd Rabu, R., I. Janajreh, and D. Honnery. 2013. Transesterification of waste cooking oil: process optimization and conversion rate evaluation. Energy Convers. Manage. 65:764–769.
16. Zhang, L., B. Sheng, Z. Xin, Q. Liu, and S. Sun. 2010. Kinetics of transesterification of palm oil and dimethyl carbonate for biodiesel production at the catalysis of heterogeneous base catalyst. Bioresour. Technol. 101:8144–8150.
17. Dias, J. M., M. C. M. Alvim-Ferraz, M. F. Almeida, J. D. M. Díaz, M. S. Polo, and J. R. Utrilla. 2013. Biodiesel production using calcium manganese oxide as catalyst and different raw materials. Energy Convers. Manage. 65:647–653.
18. Soriano, N. U., Jr., R. Venditti, and D. S. Argyropoulos. 2009. Biodiesel synthesis via homogeneous Lewis acid-catalyzed transesterification. Fuel 88:560–565.

4

Evaluation of the economics of conversion to compressed natural gas for a municipal bus fleet

Lin Yang, Wallace E Tyner & Kemal Sarica

Agricultural Economics, Purdue University, 403 West State St., West Lafayette, Indiana 47907

Keywords

Bus fleet economic analysis, bus fleet environmental analysis, compressed natural gas (CNG), hybrid diesel-electric.

Abstract

Domestic natural gas production has increased markedly in the United States, and now compressed natural gas (CNG) has the potential to become a cleaner and less expensive energy source than diesel fuel for use in the public transportation sector, especially for city bus fleets. This paper provides an economic analysis of possible CNG conversion for Lafayette, IN CityBus Corporation. It uses benefit–cost analysis to compare the total cost of three potential options for bus replacement: standard diesel, hybrid diesel-electric, and CNG. A spreadsheet model was used to estimate the total cost of these three fleet options over a 15-year project horizon. Results suggest that the CNG option has the lowest net present value (NPV) cost, and that cost savings would be larger if the corporation could obtain a grant for the CNG fueling station or if the project life span could expand to 20 years. From the environmental perspective, the CNG option would reduce greenhouse gas and particulate emissions particularly in comparison with the diesel option. Monte Carlo simulation was used to examine the inherent riskiness of the three fleet options. CNG is always lower cost than hybrid diesel-electric. Depending on assumptions regarding the underlying price distributions, the CNG option has a 51–100% chance of being lower cost than the standard diesel option.

Introduction

Natural gas is a mixture of hydrocarbon compounds, primarily composed of methane (CH_4) [1]. In recent years, US natural gas production has increased dramatically. The extraction of many unconventional natural gas sources consisting mainly of tight gas, coal-bed methane, and shale gas now has significantly increased total domestic production [2]. According to Annual Energy Outlook, the total natural gas production in the United States was 21.6 trillion cubic feet in 2010 and is expected to increase significantly until 2035. Because of the recent increases in natural gas production, especially the rapid growth of shale gas production, the domestic natural gas price in the United States in 2013 is notably lower than crude oil price on an energy equivalent basis [3]. The price difference between crude oil and natural gas has become larger since 2009. Previously, the prices were linked, but today they are largely decoupled.

The transportation sector uses about 28% of the total energy in the United States [4]. Natural gas has the potential to be an alternative fuel to replace some uses of crude oil. For public fleet companies, only when all the additional capital costs of vehicles and natural gas fueling stations for compressed natural gas (CNG) are fully compensated by the savings of fuel costs over vehicle lifetimes will companies make the decision to switch to natural gas vehicles.

The purpose of this study is to perform an economic and environmental analysis of a bus system using CNG, diesel, and diesel-hybrid buses. We use a case study of a real municipal bus company, CityBus Corporation of Lafayette/West Lafayette, IN, to do the analysis. While many of the assumptions are somewhat specific to CityBus, the general approach should apply to similar municipal bus systems around the country. The main goal of CityBus Corporation is to reduce the total cost of maintaining and operating its fleet. One of the biggest parts of the total cost is fuel cost. Most of the transit buses in the United States use diesel for fuel [5]. CNG and diesel electric hybrids are being explored as means of decreasing fuel cost.

CityBus Corporation has already implemented on a limited basis of the hybrid diesel-electric bus, which has a higher fuel economy than a standard diesel bus, but considerably higher capital costs. Thus, for CityBus the options being considered are CNG, hybrid diesel-electric, and standard diesel. We will compare some of our results with those from recent literature in the results section.

Material and Methods

Most of the data analyzed in this study was provided by CityBus Corporation. CityBus is the operating name of the Greater Lafayette Public Transportation Corporation (GLPTC). GLPTC is a nonprofit corporation serving the adjacent cities of Lafayette and West Lafayette. It also provides bus service for students and faculty of Purdue University on a contract basis with the University. By facing the rise of fuel cost and the increasing concerns of emissions caused by fleet operations, CityBus Corporation expects to find a long-term solution to maintain the current level of services. This study aims to help them to find the most effective way to reduce operating cost as well as make the fleet "greener." The data provided by CityBus were analyzed using Microsoft Excel to estimate the total present value cost of three fleet options. Cost estimates section explains parameters used and the resulting estimates for operating cost, capital cost in the base case study, and environmental cost in the sensitivity analysis.

In order to compare the total cost over the full lifetime of the project, the net present value (NPV) is used. NPV is defined as the sum of the present values of the project cash flow [6]. Due to the fact that the benefits and costs in the future have a lower value than the flows today, cash flows need to be discounted at a certain "discount rate" to calculate the NPV. In our study, the nominal discount rate and the inflation rate are assumed as 0.05 and 0.025, respectively, and sensitivity analysis was done on the discount rate. All of the analysis was done in nominal terms.

The cost estimates explained in the Cost estimates section were based on a number of assumptions and parameters. Some of these assumptions and parameters could change over time. In order to assess the uncertainties that might affect our benefit–cost analysis, this study applies Monte Carlo simulation to produce distributions of key inputs that reflect the inherent uncertainty in input values. Monte Carlo simulation is a stochastic method that takes input distributions for key inputs instead of fixed values. It then calculates the spreadsheet many times, each time drawing from the input distributions. For each iteration, it stores the output values (NPV, etc.) and creates a

distribution of values for the outputs based upon the spreadsheet calculations and the uncertain inputs [6]. Monte Carlo simulation section explains the process of using Monte Carlo simulation in price projections for diesel and CNG fuel and the corresponding results of the total cost of these three fleet options. Results section explains the results of this study and the conclusions are provided in Conclusions section.

Calculations

Cost estimates

There are mainly two kinds of costs for this analysis: operating cost and capital cost. Environmental cost is also included in the study to evaluate the environmental effect of these three fleet options.

Capital costs

There are three main components of capital cost: the costs of purchasing new buses, rebuild cost, and the cost of building a new CNG fueling station for the fleet option with CNG buses.

The base case project period (vehicle lifetime) of three kinds of buses is the same at 15 years. We used the actual CityBus vehicle replacement schedule, so the analysis is not a head to head comparison of pure systems, but a comparison of how CityBus would be impacted by the decision to replace buses with each alternative bus type. Each year, some of the existing buses need to be retired and replaced by the same number of new buses to keep the total number of buses in the fleet the same. CityBus provided us the price which CityBus Corporation would pay for each type of new bus, which is $600,000 for hybrid diesel-electric bus, $450,000 for CNG bus, and $400,000 for standard diesel bus, respectively, in 2012 dollars (Personal communications with the CityBus Chief Executive Officer, Marty Sennett, and other CityBus officials, 2012). Meanwhile, based on the recent data from CityBus Corporation, they predicted the annual price increase rate of bus price is 5% in nominal terms (Personal communications). In the life span of the project, 65 of the existing buses in the fleet (73 buses) need to be replaced. However, because of the limited budget of the CityBus Corporation and the higher price of hybrid buses, if the corporation wants to change to hybrid buses, fewer hybrid buses can be added to the fleet, and the rest need to be replaced by standard diesel buses. So in the hybrid bus option, or option 1, only 32 hybrid buses will be purchased, and the remaining 33 will be replaced by standard diesel buses. The details of the replacement for the three fleet options are shown in Table 1.

Table 1. Number of buses purchased each year for the three fleet options.

Year	Option 1 hybrid project		Option 2 CNG project	Option 3 diesel only
	Hybrid bus	Diesel bus	CNG bus	Diesel bus
1	4	3	7	7
2	2	1	3	3
3	2	1	3	3
4	2	2	4	4
5	2	2	4	4
6	2	2	4	4
7	2	2	4	4
8	2	2	4	4
9	2	2	4	4
10	2	2	4	4
11	2	2	4	4
12	2	3	5	5
13	2	3	5	5
14	2	3	5	5
15	2	3	5	5
Total	32	33	65	65

CNG, compressed natural gas.

For every transit bus, in the middle of its life span, at the eighth year after it is purchased, the engine and transmission need to be rebuilt. Although the rebuild cost of engine and transmission system in the hybrid bus is the same as the other two types, which is $19,000 in 2012, the battery in the hybrid diesel-electric bus also needs to be replaced at this time. So the fleet needs to pay the additional battery replacement cost for hybrid diesel-electric buses, which is $40,000 in 2012. For all these three types of buses, the annual increase rate of rebuild cost provided by CityBus Corporation is 3% in nominal terms (Personal communications). We did not reduce the battery replacement cost through time, as we do not have confidence in the available projected rates of technical change in battery costs. However, as will be clear below, even if we had used battery replacement costs half the stipulated value, it would not have changed the basic results.

Another important capital cost is the cost to build a new CNG fueling station. Before the CNG transit bus is implemented in the fleet, a CNG fueling station must be built. This cost, which is estimated at $2 million (Personal communications), would happen at the beginning of the first year and needs to be amortized under a certain amortization rate to evaluate the annualized cost. Because the discount rate and the amortization rate used in this study are the same at 5%, the amortization of this cost has no impact on the NPV calculation of the cost in the CNG option. If more favorable financing terms were to become available to CityBus, it would make the CNG option more attractive.

Moreover, due to the environmental benefits of using CNG buses, CityBus Corporation may have the opportunity to get a grant from the federal government or Indiana State government to cover part of the cost of building the CNG fueling station. The study evaluates three different grant fractions, 10%, 50%, and 100% in the sensitivity analysis.

Fuel cost

In the fuel cost part, the total mileage of the fleet is fixed at 1.8 million miles every year. After communicating with CityBus, the distance of the route from the fuel station is usually not so long that each kind of bus can finish the route without coming back to the fuel station (Personal communications). So in this study we assumed that all these three kinds of transit buses in the fleet traveled the same distance, which means the percentage of each kind of bus in the fleet equals the percentage of mileage travelled by each kind of bus every year. Fuel cost of each type of bus is calculated by the price of the fuel multiplied by the fuel used for that type of bus. The annual fuel cost of each fleet option is calculated as the summation of each kind of bus's fuel cost in that year.

In this analysis, the initial diesel and CNG price in 2012 are $3.11 per gallon and $1.5 per DGE, respectively (Personal communications). After regressing the historical annual prices of diesel and crude oil from 2000 to 2011 [7], the correlation between them is 0.996, which means crude oil price and diesel price have the same price growth rate. We used U.S. Department of Energy (DOE) [4] projections of crude oil price from 2010 to 2035. DOE also forecasts retail diesel prices, but we wanted to use the wholesale value, without taxes and distribution costs, so it was better to calculate the rates of change directly from the crude oil forecast. The DOE reference case was used as the growth rate of diesel price in the base case study. The CNG price is mainly comprised of two parts: natural gas wellhead price and transmission/distribution cost. After regressing the U.S. historical annual transmission/distribution cost, which equals CNG retail price minus natural gas wellhead price, from 2000 to 2011[7] and the consumer price index (CPI) from 2000 to 2011[8], the result shows that the correlation between transmission/distribution cost and CPI is 0.939, which means the growth rate of transmission/distribution cost is highly correlated with the general inflation rate. The growth rate of the wellhead price is taken from the Henry Hub spot natural gas price projection from U.S. Energy Information Administration (EIA) Annual Energy Outlook 2012 [4]. According to Clean Cities Alternative Fuel Price Report, the conversion factor between CNG price and wellhead price is 7.236, which means 1 mmBtu natural gas converts to 7.236 diesel gallon equivalent (DGE) CNG fuel [9]. Using this factor we can

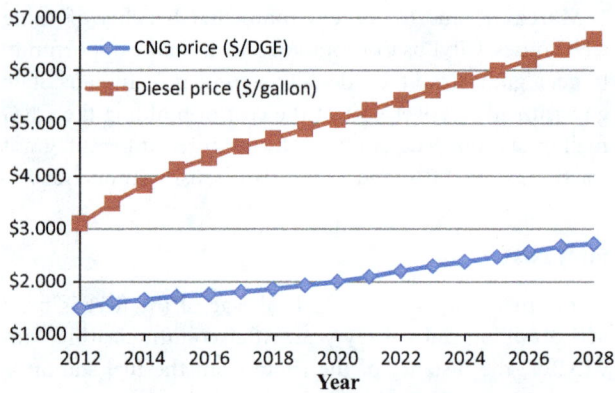

Figure 1. Price projections of diesel and CNG from 2013 to 2028.

estimate that, in 2012, the wellhead price of CNG is $0.52 per DGE and the transmission/distribution cost of CNG is $0.98 per DGE. Figure 1 presents the price projections of diesel (/gallon) *and CNG* (/DGE) in nominal terms. From 2013 to 2018, the average annual growth rate of CNG price is 3.9%, while the price growth rate of diesel is 4.9%.

Environmental costs

Two kinds of emissions included in this study are carbon dioxide (CO_2) equivalent and PM10 (particulate matter [PM] with a diameter of 10 μm or less). CityBus Corporation uses ultra-low sulfur diesel (ULSD) as the fuel to follow a mandate which took effect in 2010 in the United States requiring all the diesel fuel refined to be ULSD (Personal communications). An important emission caused by transit buses is PM. Passengers and bus drivers are among the most vulnerable groups of people with immediate and long-lasting exposure to these small particles [10]. In our analysis, we decide to use the PM10 emission estimates by Wayne [11], which is from the average PM10 emissions for the buses during 2007 to 2009. The PM10 emissions for hybrid, CNG, and standard diesel bus are 0.020, 0.013, and 0.022 g/mile, respectively.

With these emission factors, the amount of Greenhouse gas and PM10 emissions of each fleet option could be estimated. Greenhouse gas emissions are typically calculated in the units of carbon dioxide equivalent (CO_2e). In this study, three of the main Greenhouse gases are calculated, which are CO_2, CH_4, and N_2O. In order to convert them to CO_2e, the gas's global warming potential (GWP) need to be used. From the 2011 report of U.S. Environmental Protection Agency (EPA), if the GWP of CO_2 is set at 1, the GWP of CH_4 and N_2O are 21 and 310, respectively [12]. According to data from U.S. EPA, CNG transit bus releases 1.966 g CH_4 per mile and 0.175 g N_2O per mile. For the standard diesel bus, the emission factor is 0.0051 g per mile

for CH_4 and 0.0048 g per mile for N_2O [12]. For the diesel hybrid bus the factors are 0.0039 and 0.0037, respectively.

We use the shadow price of avoiding these two pollutants to calculate the environmental cost. For CO_2 equivalent emissions, a carbon tax could be applied as its shadow price. Currently, there is no carbon tax leveled nationwide in the United States. Previous studies [13, 14] suggested a carbon tax at a range from $55 to $110 per ton of carbon, which is equivalent to $15 to $30 per ton of CO_2 when divided by the factor 3.67. For PM10 emission, the report "Transit bus life cycle cost and emissions estimation" [15] estimated that the shadow price of PM10 emission is $6367 per ton in 2006 dollars. After adjustment by the historical inflation from 2006 to 2011, the social cost of PM10 is $7384 in 2012.

Monte Carlo simulation

The software @Risk is used to perform the Monte Carlo simulation in this study. It is a widely used software package operating as an add-in to Microsoft Excel. A key source of uncertainty in this analysis is the trajectory of diesel and natural gas prices. The technology development of extracting shale gas, its transportation, and storage cost as well as the speed of economic recovery of the United State may all influence on the growth rate of CNG prices. For the diesel price, the fluctuation of the crude oil is the key driver of diesel price. Uncertainties of diesel and CNG price may also appear when the regulations and policies on fuel emission standards change in the future.

Our analysis uses the projected price growth rate of diesel and CNG calculated from the reference case of price projection from U.S. DOE and U.S. EIA, respectively. The range of the projected growth gate in the Monte Carlo simulation comes from the low case and high case of these projections.

In this study, two methods are used in the Monte Carlo simulation to estimate the price growth rate, which are named Monte Carlo 1 and Monte Carlo 2. The method Monte Carlo 1 is that a certain year's price change is only related to the price in that year and the price in the base year, or the price at the beginning of 2012. In Monte Carlo 2, 1 year's price is related to the price of the previous year. In other words, Monte Carlo 2 is a lag price structure, which would be expected to have higher variance. The basic reason is that prices from year to year are highly correlated, so low draws early on will result in low prices throughout the sequence and vice versa for high draws. When prices are not correlated from year to year, variance is expected to be lower because there is a greater chance of offsetting random draws.

The normal distribution is used in both of these two methods [10, 16]. The price growth rate of the reference case is used as the mean in the distribution. Standard deviation is calculated by the range of price growth rate from high and low case of the projection, while using the range as the 99% confidence interval.

In Monte Carlo 1, the range of price growth rate between high case and low case is used as the truncation. But in Monte Carlo 2, each year's price is based on the price of the previous year, which would make the standard deviation quite large. In order to make the result more realistic, we decide to narrow down the truncation of the distribution and use 70% of the confidence interval to estimate the truncation each year. The results of Monte Carlo simulation will be shown in the Results section.

Results

The benefit–cost analysis in this study was done for a base case and five alternative cases. The base case study compared the operating cost and capital cost of three fleet options over a 15-year project life span, while both environmental cost and investment cost grant are excluded. The five cases for sensitivity analysis were the following:

- Environmental cost included
- Investment Grant included
- Monte Carlo simulation on fuel price projection
- Twenty-year project life span
- Breakeven analysis on CNG price growth rate and discount rate.

Base case

Figure 2 summarizes the fuel cost of these three fleet options each year in nominal terms. The fuel cost in option 2 is much lower than the fuel cost in the other two options. Fuel cost savings of CNG bus is one of the biggest advantages for choosing it as alternative vehicle of the fleet. In option 1, although the hybrid bus has a much higher fuel economy than a standard diesel bus, because only 32 of the hybrid buses can be purchased in the next

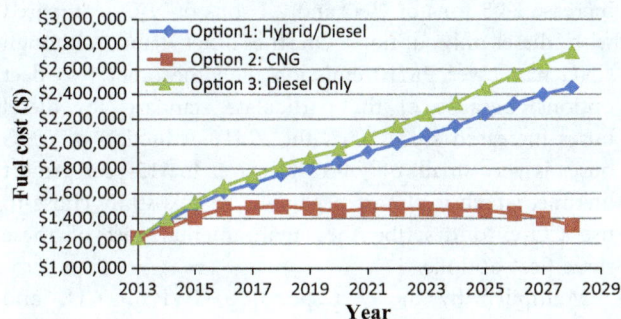

Figure 2. Fuel costs of three fleet options in the project life span in nominal term.

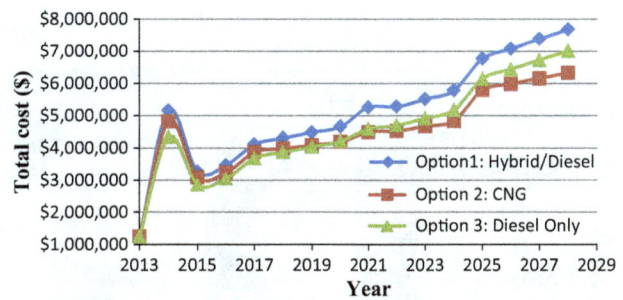

Figure 3. Total costs of three fleet options in the project life span in nominal term.

15 years, the fuel cost savings is not as much as CNG option. Toward the end of the planning horizon, the total fuel cost for the CNG option actually falls because the fraction of CNG buses in the total is steadily increasing. This fact indicates that the fuel cost differences between option 2 and the other two options will become even larger beyond the 15-year planning horizon. Meanwhile, in the last year of the project, the cost savings between option 1 and option 3 is only $283.651, while in option 2, the fuel cost in the last year is almost 1.4 million ($1,394,091) less than option 3, which saves almost half of the cost of option 3. This is a huge cost savings, and the tendency of Figure 2 also points out that in the years after the project life span (after 2028) fuel cost saving in using CNG buses would be even higher than $1.4 million per year.

The total cost is the sum of fuel cost and capital cost. Figure 3 provides the total cost of these three fleet options over the project life span. These costs assume the CNG capital cost is amortized. At the beginning of 2014, seven of the existing standard diesel buses will be retired, while in the remaining years, the number of retired buses will be only 3–5 each year. The higher capital cost for purchasing new buses in 2014 leads to the fact that there is a surge in the total cost at the beginning and then a drop. In all project years, the total cost of option 1 is higher than the cost of the other two options, which is mainly due to the higher price of the new hybrid diesel-electric buses. Compared with option 3, option 2 has a higher annual total cost in the first several years of the project. However, because of the lower price of CNG fuel, the difference between the annual total cost of these two options decreases. After 2020, the total cost of option 2 becomes lower than that of option 3. Moreover, the cost savings of using CNG buses increase in the years after 2020.

Figure 4 summarizes the components and NPV of the total cost of these three fleet options in the project. NPV of the total cost in option 1 is more than $5 million higher than the other two options. Because of the lower price and price growth rate of CNG fuel, option 2 has the

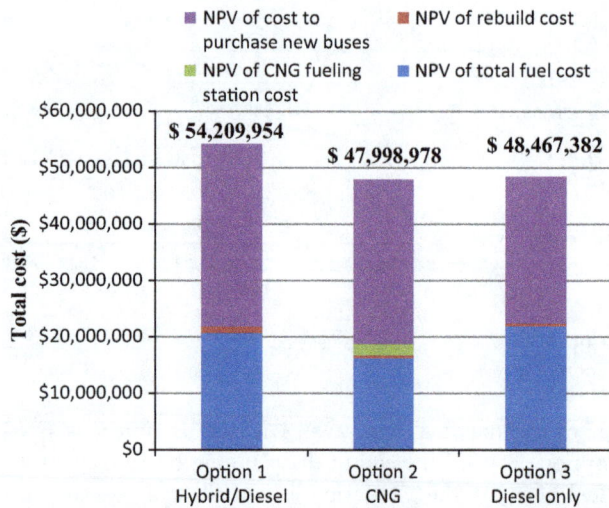

Figure 4. NPV of total cost of three fleet options.

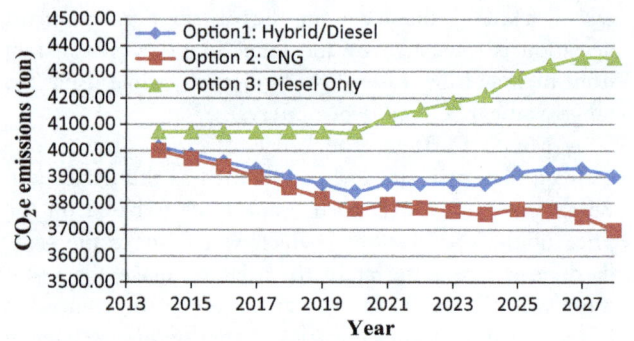

Figure 5. CO_2e emissions of three fleet options in the project life span.

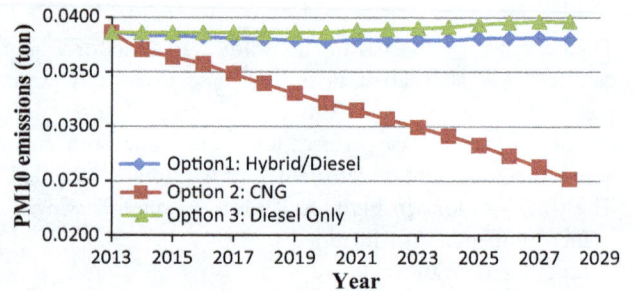

Figure 6. PM10 emissions of three fleet options in the project life span.

lowest total cost, for which NPV is $468,404 lower than that of option 3. Thus, without consideration of other factors, the CNG option is the least cost.

These results are consistent with existing literature in this area. We used the Transportation Research Board NRC [17] bus life-cycle model to do a comparison of the three systems. That model does a one to one comparison of the CNG, diesel, and diesel electric hybrid buses. Thus, one would expect to get different numbers as our comparison is tailored to the CityBus specifics. However, we used the same default values as in this study to come as close as possible. The NPV cost results from that model for 65 buses were $62,857,667 for diesel, $62,668,509 for CNG, and $74,724,408 for diesel-electric hybrid. Thus, CNG was the lowest cost system, and the diesel-electric hybrid the highest. We were not able to project the differential growth rates of diesel and CNG prices in that system, which is one reason the CNG-diesel difference is not as large. Also, we did not include base maintenance costs for any system as they are about the same, but this system included all the maintenance costs, leading to higher total costs. However, the important thing is that the general order of the conclusions is the same.

Also, another important study in the recent literature is Krutilla and Graham [18]. They concluded that the diesel electric hybrid technology is not competitive under most assumption with the standard diesel, the same as our conclusion. They also included a reduction in battery replacement cost over time in their analysis.

Environmental cost

The results in this section estimate how much Greenhouse gas and PM emissions are reduced when the fleet

changes to use alternative buses in the next 15 years. CO_2e and PM10 emissions of all three fleet options in the project are estimated and shown in Figures 5 and 6. These two figures have similar results on the emission amounts of CO_2e and PM10. Option 3 has the highest emissions, and the emission amount is even increasing after 2020, when the existing hybrid diesel-electric buses are beginning to be retired and replaced by standard diesel buses. In the other two options, CO_2e and PM10 emissions both decrease, while the CNG option has a higher rate of decline. For the Greenhouse gas emissions, the CNG option would reduce 6300 tons CO_2, but increase 24.5 tons of CH_4 and 2.1 tons of N_2O, compared with diesel-only option. On the other hand, although CNG has lower PM10 emission than the other two fleet options, because of the particulate standard for diesel buses increased since 2007, the PM10 reduction of CNG buses is very small, only 0.11 ton over the 15-year project lifetime, which is almost negligible. So this study chose to use CO_2e to describe the environmental cost of these three fleet options.

Multiplied by the corresponding GWP of CH_4 and N_2O, the emission of CH_4 and N_2O are changed into CO_2 equivalent (CO_2e) emissions. The results indicate that the total CO_2e emissions for these three fleet options over the 15-year project lifespan are 62,754, 61,432, and 66,560 tons, respectively.

Using the shadow price of CO_2e described in Cost estimates section, at the lower bound of the range, $15 per ton of CO_2e, the NPV of environmental cost of these three fleet options are $791,358, $776,051, and $836,325, respectively. At the upper bound, which is $30 per ton of CO_2e, the environmental cost NPV are $1,582,716, $1,552,102, and $1,672,650, respectively. In other words, the environmental cost NPV for CNG option ranges between $60,274 to $120,548 less than the diesel-only option and $15,307 to $30,614 less than the hybrid/diesel option. Although the environmental cost is not paid by CityBus Corporation under the current policy, they are a true cost to society. The results quantify the health benefits and air pollution reductions to society, as the fleet replaces standard diesel buses with alternative buses.

Monte Carlo simulation

The simulation results of total cost of these three fleet options in the base case study are shown in Tables 2 and 3 by using the method Monte Carlo 1 and Monte Carlo 2, respectively. Option 2, or CNG option, has the lowest mean value of total cost in both of these two simulation methods, which are consistent with the results in the base case study. Because of the relative stability of CNG future price compared with diesel price, CNG option has smaller variance than the other two options. This smaller variance also means a smaller risk of investing in CNG buses.

Moreover, in both Monte Carlo 1 and 2 simulation methods, the probabilities that the total cost of option 1

is lower than that of option 2 are zero, which means the cost difference between them is so large (the mean value is about $6.2 million) that option 1 has zero probability of being lower than option 2. This result indicates that from an economic perspective, the CNG option is always preferred over the hybrid/diesel option. For the cost difference between option 2 and option 3, the possibility of changing its sign is different between these two simulation methods. In the Monte Carlo 1 method, the probability that the total cost of diesel-only option is lower than CNG option is 21%. But in Monte Carlo 2, the number is 49%, which means diesel-only option has a higher probability of costing less than the CNG option while using Monte Carlo 2 simulation method.

Compared with Monte Carlo 1 method, Monte Carlo 2 seems more plausible, because it develops a price trajectory in which each year's price is dependent on the price of previous year. Monte Carlo 1 has a smaller variance and seems a little bit conservative. However, the range of diesel price in the last year of the project is from $1.13 to $24 in the Monte Carlo 2 method. The higher variance in Monte Carlo 2 is also reflected in the NPV output distributions as would be expected, which could explain why option 3 has a higher possibility to cost less than option 2 in the Monte Carlo 2 method. Although the probability of these extreme cases is very small, the price variability in Monte Carlo 2 may be high, and in the real world, the price variability would fall in between these two methods.

Total cost with grant included

The base case did not consider any grant or subsidy the company might get for investment in the CNG fueling sta-

Table 2. Monte Carlo 1 simulation results on total cost.

	Mean	Range	SD
Option 1 total cost	$54,271,400	($50,576,090, $57,605,380)	$1,004,900
Option 2 total cost	$48,061,970	($45,830,300, $50,331,600)	$638,914
Option 3 total cost	$48,529,690	($44,705,660, $52,246,450)	$1098,218
Difference between option 1 and 2	$6209,437	($4231,903, $7987,525)	$490,874
Difference between option 2 and 3	$467,722	(−$1318,911, $2621,786)	$575,880

Table 3. Monte Carlo 2 simulation results on total cost.

	Mean	Range	SD
Option 1 total cost	$54,184,620	($43,033,560, $85,348,700)	$4,995,222
Option 2 total cost	$47,999,920	($42,318,920, $57,732,980)	$2,417,925
Option 3 total cost	$48,468,620	($36,361,950, $74,633,020)	$5,458,817
Difference between option 1 and 2	$6,196,010	($603,036, $25,421,120)	$2,707,017
Difference between option 2 and 3	$468,699	(−$5,963,777, $17,138,620)	$3,125,033

Table 4. Probabilities of option 3 cost less than option 2 with grants.

	No grant	10% grant	50% grant	100% grant
Monte Carlo 1	20.7%	11.9%	0.5%	0.0%
Monte Carlo 2	49.2%	46.8%	36.7%	22.7%

Table 5. Probabilities of option 3 total cost less than option 2 (comparison between base case and 20-year lifespan case).

	Monte Carlo 1	Monte Carlo 2
15-year lifespan	20.7%	49.2%
20-year lifespan	0.0%	35.5%

tion. One of the most likely grants CityBus Corporation can get is for building a new CNG fueling station, which would reduce the total cost of option 2. This section analyzes the impact on total cost if the company can get a grant for building the CNG fueling station. Three scenarios of grant fraction for the capital cost of CNG fueling station are assumed in the analysis: 10%, 50%, and 100%.

In the base case study, without any grant, the NPV of the total cost in option 2 is lower than the other two options. If the corporation can get a grant, option 2 would become more attractive in the final decision. Essentially the present value of total cost gets reduced by the amount of the grant as the capital cost for CNG is at the beginning of the project. In other words, the 10% grant reduces CNG cost $200,000 and similarly for the other percentages.

Table 4 shows the probabilities that the total cost of option 3 is lower than that in option 2 if the grant is included into the analysis. As the grant fraction for the capital cost becomes larger, the total cost in option 2 becomes smaller, which, in turn, causes the probability in Table 4 to become smaller. The probability that diesel is cheaper than CNG in Monte Carlo 1 even becomes zero in the 100% grant scenario.

Twenty-year project life span

Compressed natural gas fueling stations have a lifetime of 20–25 years. In this section, we choose 20 years as the lifespan of the project, instead of the 15 years selected by the CityBus Company. In this sensitivity analysis, the capital cost of CNG fueling station is annualized over 20 years. The vehicles' lifetime remains 15 years. The bus replacement cycle and rebuild for the last 5 years is assumed to be continued as it was done for the first 15 for all three fleet options.

Total cost savings of the CNG option compared with diesel-only option is $2,923,355 in the 20-year project lifespan, which is a lot larger than the cost savings in the base case study at $468,404. The result reinforces the economic advantage of using the CNG buses in the fleet.

The simulation results of the mean values in the 20-year project lifespan case are consistent with the results shown previously. Table 5 presents the probabilities that the total cost of option 3 is lower than option 2. Due to the fact that the cost savings of the CNG option is larger over a 20-year lifespan, the probabilities in Table 5 become smaller in the 20-year lifespan. In Monte Carlo 1, the diesel-only option has zero probability of being lower than the CNG option in the 20-year lifespan case. Because the 20-year lifetime more closely matches that of the fueling station, we believe the results for the 20-year lifetime are more realistic.

Additional sensitivity analysis on price projection

Breakeven analysis on CNG price and discount rate

Breakeven analysis is a technique for conducting sensitivity analysis on key variables. In this section we do sensitivity analysis on the breakeven rate of CNG price increase and on the discount rate used for the analysis.

In the base case study, the average annual growth rate of diesel and CNG price are 3.9% and 4.9%, respectively. The total cost difference between CNG option and diesel-only option is $468,404. In this section, we assumed that the average annual price growth rate of diesel remains the same at 4.9%, and solved for the breakeven annual price growth rate of CNG which can make the NPV of the total cost of the CNG option and diesel-only the same. After using what-if analysis in our spreadsheet model, we get a breakeven trajectory of CNG price and we can find out that the last year CNG price increases from $2.669 to $2.999. It indicated that when CNG price has an average annual growth rate of 4.7%, the total cost of CNG fleet option and diesel-only option would be the same. In other words, CNG would have to go up almost as fast as diesel for the two options to have the same NPV total cost.

We assumed that the base case nominal discount rate in this study is 5%. In order to find out the breakeven discount rate which causes the total cost of CNG fleet option and diesel-only fleet option to be the same, what-if analysis is used. The breakeven discount rate is 6.8%. This result indicates that if the discount rate in this study increases to 6.8%, the total cost savings of CNG fleet option compared with diesel-only fleet option would become zero. At a higher discount rate, the diesel option would become preferred.

All the results of the breakeven analysis in this section exclude the environmental benefits of the CNG option.

Conclusions

Because of the lower fuel price and pollution reduction, the CNG bus is considered to have good potential as an alternative vehicle used in the public fleet in the United States. Among the three candidate fleet options for the CityBus Corporation to choose, hybrid/diesel option, CNG option, and diesel-only option, the total cost of CNG fleet option is the lowest over the project life span of 15 years, even in the case with no grant for building the CNG fueling station. Due to the higher cost of purchasing a new hybrid bus, which is 50% higher than a standard diesel bus, hybrid/diesel option is easily the highest total cost, about $6.2 million higher than CNG option and $5.7 million higher than diesel-only option in the base case study.

Moreover, from the environmental perspective, the implementation of CNG buses in the fleet would also produce less emission and provide benefit to the environment of the local society. In our analysis, CNG option would reduce about 5128 tons CO_2e emissions compared with diesel-only option over the whole 15-year project lifetime. For the hybrid/diesel option, however, due to the fact that only 32 of the hybrid diesel-electric buses are purchased in the project, the effect of emission reductions is not as good as that in the CNG option.

The wide range of NPV generated by the Monte Carlo simulation reflects the high uncertainty of the total costs of these three fleet options due to the uncertain future diesel and CNG prices. However, because the future price of CNG is likely to be more stable than the price of diesel, the total cost of CNG option has the smallest variance, which would help the fleet to decrease the chance of financial losses in the future.

Over the project lifespan of 15 years, the total cost of the CNG option is $468,404 lower than the diesel-only option in the base case, and has a range of 51–79% probability of a lower system cost. Moreover, if CityBus can get a grant for the CNG fueling station or the project life expands to 20 years, the probability would increase significantly. Meanwhile, a higher discount rate or a higher growth rate of CNG price would make the diesel-only option become preferred.

Based on all these results from our analysis, the CNG option is somewhat preferred from both a financial and environmental perspective. It is possible that other factors not included in this analysis also could play in the decision.

Conflict of Interest

None declared.

References

1. U.S. Energy Information Administration 2012b. Natural gas glossary. Available at http://www.eia.gov/tools/glossary/?id=natural%20gas.
2. Mellquist, N., M. Fulton, and B. M. Kahn 2011. Natural gas and renewables: the coal to gas and renewables switch is on. Deutsche Bank climate change investment research, October 2011.
3. U.S. Energy Information Administration 2011. Annual Energy Review 2010. Available at http://www.eia.gov/totalenergy/data/annual/pdf/aer.pdf.
4. U.S. Energy Information Administration 2012a. Annual energy outlook 2012. Available at http://www.eia.gov/forecasts/aeo/pdf/0383(2012).pdf (accessed 3 September 2012).
5. American Public Transit Association 2011. 2011 public transportation fact book. Available at http://www.apta.com/resources/statistics/Documents/FactBook/APTA_2011_Fact_Book.pdf.
6. Boardman, A. E., D. H. Greenberg, A. Vining, and D. Weimer. 2006. P.576in Cost–benefit analysis: concepts and practice. Prentice Hall Publishers, Upper Saddle River, NJ.
7. U.S. Energy Information Administration 2012d. Natural gas spot and futures prices and crude oil spot prices. Available at http://www.eia.gov/dnav/ng/ng_pri_fut_s1_d.htm and http://www.eia.gov/dnav/pet/hist/LeafHandler.ashx?n=pet&s=rwtc&f=m (accessed 12 September 2012).
8. U.S. Department of Labor-Bureau of Labor Statistics2012. Consumer price index database. Available at http://www.bls.gov/cpi/data.htm (accessed 13 September 2012).
9. U.S. Department of Energy 2012. Clean cities alternative fuel price report April 2012. Available at http://www.afdc.energy.gov/uploads/publication/afpr_apr_12.pdf (accessed 15 September 2012).
10. Cheng, E., L. Grigg, E. Jones, A. Smith, and M. Berger. 2011. Compressed natural gas vehicles for the city of Milwaukee's Department of Public Works: a cost-benefit analysis. In Workshop in public affairs. Available at http://minds.wisconsin.edu/handle/1793/52610 (accessed 25 July 2012).
11. Wayne, W. S., J. A. Sandoval, and N. N. Clark. 2009. Emissions benefits from alternative fuels and advanced technology in the U.S. transit bus fleet. Energy Environ. 20:497–515.
12. U.S. Environmental Protection Agency. 2011. Emission factors for greenhouse gas inventories. Available at http://www.epa.gov/climateleadership/documents/emission-factors.pdf (accessed 20 June 2013).
13. Metcalf, Gilbert E. 2009. Designing a carbon tax to reduce U.S. greenhouse gas emissions. Rev. Environ. Econ. Policy 3:63–68.

14. Aldy, J. E., E. Ley, and I. Parry. 2008. A tax-based approach to slowing global climate change. Nation. Tax J. LXI:493–517.

15. Clark, N. N., F. Zhen, W. S. Wayne, and D. W. Lyons. 2007. Transit bus life cycle cost and year 2007 emissions estimation. U.S, Department of Transportation, Federal Transit Administration 2007.

16. Clemen, R. T., and T. Reilly. 2000. Making hard decisions with DecisionTools. Duxbury Thomson Learning. Available at http://ebookee.org/Making-Hard-Decisions-with-Decision-Tools_1599646.html (accessed 20 August 2012).

17. Transportation Research Board of the National Academies. 2009. Assessment of hybrid-electric transit bus technology. in TCRP Report 132. TRB/NAS, Washington, DC.

18. Krutilla, K., and J. D. Graham. 2012. Are green vehicles worth the extra cost? The case of diesel-electric hybrid technology for urban delivery vehicles. J. Policy Anal. Manage. 31:501–532.

5

Enhanced biogas production from coffee pulp through deligninocellulosic photocatalytic pretreatment

Grisel Corro[1], Umapada Pal[2] & Surinam Cebada[1]

[1]Instituto de Ciencias, Benemérita Universidad Autónoma de Puebla, 4 sur 104, 72000 Puebla, Mexico
[2]Instituto de Física, Benemerita Universidad Autonoma de Puebla, Apdo. Postal J-48, 72570 Puebla, Mexico

Keywords
Anaerobic digestion, biogas, biomass, lignocelluloses degradation, photocatalysis, supported metal catalysts

Abstract

Production of biogas utilizing agricultural wastes is one of the most demanding technologies for generating energy in sustainable manner considering environmental concerns. However, though the agricultural wastes are available in abundance, the technologies used for the production of biogas such as biological and thermochemical processes are not very efficient. In the present article, we describe a process for the production of biogas from coffee pulp utilizing Cu/TiO_2 as an efficient photocatalyst in its solar photocatalytic pretreatment, producing biogas of high caloric value at enhanced rate. The pretreatment process enhances the coffee pulp–cattle manure codigestion, producing increased amounts of methane, propane, and other combustible components of biogas. The photocatalytic pretreatment was performed using $10\%Cu/TiO_2$ as photocatalyst, bubbling air as oxidizing agent, and solar radiation as light source. The process enhances the degradation rate of lignocellulosic components of coffee pulp, consequently increasing the biogas production through anaerobic codigestion. The mechanism of photocatalytic degradation of lignocelluloses has been discussed. The results presented in this work indicate the photocatalytic pretreatment is a useful process to increase biogas generation from lignocelluloses-rich natural wastes.

Introduction

Methane production from a variety of biological wastes through anaerobic digestion technology is growing worldwide and is considered ideal in many ways due to its economic and environmental benefits [1–10]. Methane fermentation is the most efficient technology for energy generation from biomass in terms of energy output/input ratio (28.8 MJ/MJ) among all the technologies used for energy production through biological and thermochemical routes [11].

Coffee is the second largest traded commodity in the world which generates large amounts of by-products and residues during processing. Industrial processing of coffee cherries is performed to separate coffee beans by removing shell and mucilaginous part. In wet industrial processes, a large amount (about 29% dry-weight of the whole coffee berry) of coffee pulp is produced as the first by-product. The organic components present in coffee pulp include cellulose (63%), lignin (17%), proteins (11.5%), hemicelluloses (2.3%), tannins (1.80–8.56%), pectic substances (6.5%), reducing sugars (12.4%), nonreducing sugars (2.0%), caffeine (1.3%), chlorogenic acid (2.6%), and caffeic acid (1.6%) [12–14]. Coffee wastes and by-products produced during coffee berry processing constitute a source of severe contamination and pose serious environmental problems in coffee-producing countries. Therefore, disposal of coffee pulp is becoming an emerging environmental problem worldwide due to its putrefaction. Due to anaerobic conditions of open pulp-storage or composting areas, an uncontrolled emission of methane (CH_4) and nitrous oxide (N_2O) from these places cannot be excluded [15–17]. Hence, the utilization and management of coffee wastes in large scale still remains a challenge worldwide not only due to the generation of earlier gases but also for their high contents of caffeine, free phenols, and tannins, which are known toxic agents for many biological processes [18].

Use of anaerobic digestion process for the treatment of organic fraction of municipal solid wastes has been popular worldwide during the past decades. Indeed, the biological degradation process of organic materials under anoxic conditions lead to the production of methane, which can be used as an efficient renewable energy source in comparison with aerobic stabilization processes that require higher energy consumption [19].

Codigestion can be defined as the simultaneous digestion of two or more biomass wastes which contribute to different organic matter. The benefits of codigestion include dilution of potential toxic compounds, improved balance of nutrients, synergistic effect of microorganisms, higher loading of biodegradable organic matter, and higher biogas yield [20].

Generation of methane from solid organic waste through anaerobic digestion is significantly affected by the mass transfer in each of the involved biological steps, as well as by the availability of biodegradable organic matter [21]. The most important step in this regard is the hydrolysis of complex organic molecules to soluble compounds. Several physical [22, 23], chemical [24, 25], and biological processes [26, 27], or their combinations have been utilized to improve the efficiency of anaerobic treatments [28, 29].

Lignocellulosic biomass is the major structural component of wood and plant fibers, which is also the most abundant polymer synthesized by nature. However, its use as a feedstock for fuels and chemicals has been limited due to its highly crystalline structure, inaccessible morphology, and limited solubility. Moreover, it is difficult to degrade biologically. Pretreatment of the agricultural residues by mechanical size reduction, heat treatment, and/or chemical treatment usually improves its digestibility. Explored chemical pretreatment methods include bicarbonate treatment [30], alkaline peroxide treatment [31], ammonia treatment [32]. Electron-beam radiation has also been used as pretreatment of biomass to improve its digestibility [33–36]. However, these pretreatment methods are complicate, need special equipment, and hence not cost-effective.

On the other hand, application of the photocatalytic method for the destructive removal of natural organic substances has been extensively studied [37, 38]. As example, several researchers have utilized TiO_2 under UV illumination to decompose refractory lignin [38, 39]. Tanaka et al. have applied TiO_2 and UV irradiation to degrade wastewater lignin [40]. The time for complete delignification could be shortened by increasing the amount of TiO_2. On the other hand, incorporation of noble metal into TiO_2 has also seen to enhance its photocatalytic efficiency for the degradation of refractory compounds [41–45].

In general, photocatalytic oxidation can be considered as an example of innovative technologies collectively known as advanced oxidation processes (AOP) that rely on the generation of highly reactive radicals. Those reactive species are subsequently used to degrade organic pollutants. The principles of photocatalytic oxidation and its environmental applications have been reviewed extensively [36–43, 46]. Basically, the heterogeneous photocatalytic processes depend on the utilization of UV or near-UV radiation to photo-excite the semiconductor catalysts in presence of oxygen. Under these circumstances, oxidizing species such as surface-bound hydroxyl radical ($\cdot OH$), superoxide ($\cdot O_2^-$), hydroperoxy radical ($HO_2\cdot$), and free holes are generated. The active species could initiate a series of redox reactions to degrade adsorbed molecules [47]. Of the several semiconducting oxides utilized as photocatalysts, TiO_2 in anatase phase has seen to be the most efficient due to its high stability, high photocatalytic efficiency, low toxicity, and low cost. High optical absorbance in the near-UV region remains the major advantage of TiO_2 for this purpose. However, due to the technical problem arising from the dissolution/dispersion of TiO_2, it is difficult to separate from the solvent or reaction mixture. While this is a practical problem for the reuse of the catalyst, the presence of fine TiO_2 particles in the supernatant complicates the estimation of catalytic products.

In this investigation, we have prepared a Cu/TiO_2 catalyst, which is of semiconducting nature, having strong absorbance in between 235 and 400 nm, and an intense absorption band in the visible region (400–800 nm), suggesting its utilization for photocatalytic degradation of the complex organic substances present in waste coffee pulp. The photocatalytic degradation of lignin may generate simple organic molecules, which may be suitable for further anaerobic digestion. The photocatalytic activity of the calcined Cu/TiO_2 catalyst under direct solar radiation for enhanced solubility of refractory compounds of lignocellulosic biomass present in coffee pulp has been studied. The use of solar energy as UV–vis irradiation source brought down the biogas production cost in comparison with the other biomass pretreatment methods developed to date [30–36]. To the best of our knowledge, there is no report in the literature so far on the photocatalytic degradation of lignocellulosic biomass contained in coffee pulp.

Materials and Methods

Collection and preparation of substrates

Our experiments were performed with coffee pulp collected during the coffee bean processing in a semitropical region of Puebla, Mexico. The cattle manure was collected from a dairy farm.

Catalyst preparation

The Cu/TiO_2 catalyst was prepared by the impregnation method using titanium dioxide (Baker 99.99%) and appropriate amounts of aqueous solution of $Cu(NO_3)_2$ (Aldrich 99.99%) to obtain a 10 wt% Cu on the TiO_2. The suspension was stirred at room temperature for 4 h. After drying at 120°C overnight, the sample was calcined in air at 800°C for 4 h, after which it is called as 10% Cu/TiO_2 catalyst. After calcination, the catalyst presented a very high indentation hardness and resistance to pulverization. A reference TiO_2 support was prepared in the same way using only distilled water (without Cu $(NO_3)_2$).

Catalyst characterization

Adsorption/desorption isotherm of the samples were measured using a Quantachrome Nova-1000 sorptometer (Quantachrome Instruments, Florida, USA). Specific surface area (Sg) of the samples was estimated from their N_2 physisorption at 77 K, using BET analysis methods.

The diffuse reflectance spectra (DRS) of the composite catalyst and the reference sample were obtained for their dry-pressed disks (~15 mm diameter) using a Varian Cary 500 UV–Vis spectrophotometer (Agilent Technologies, Santa Clara, CA, USA) with DRA-CA-30I diffuse reflectance accessory using $BaSO_4$ as standard reflectance sample. The crystallinity and structural phase of the samples were verified through powder X-ray diffraction (XRD) technique using the Cu Kα source ($\lambda = 1.5406$ Å) of a Bruker D8 Discover diffractometer (Bruker Corporation, Bruker, Mexico).

Photocatalytic pretreatment of coffee pulp

The photocatalytic pretreatment of coffee pulp was carried out in a quartz photoreactor, newly designed and built in our laboratory. The detailed schematic drawing of the experimental setup is shown in Figure 1.

The sunlight was utilized as UV radiation source to expose the catalytic reaction mixture. Ambient air from a compressor was fed into the reactor at the rate of 0.5 L min^{-1}.

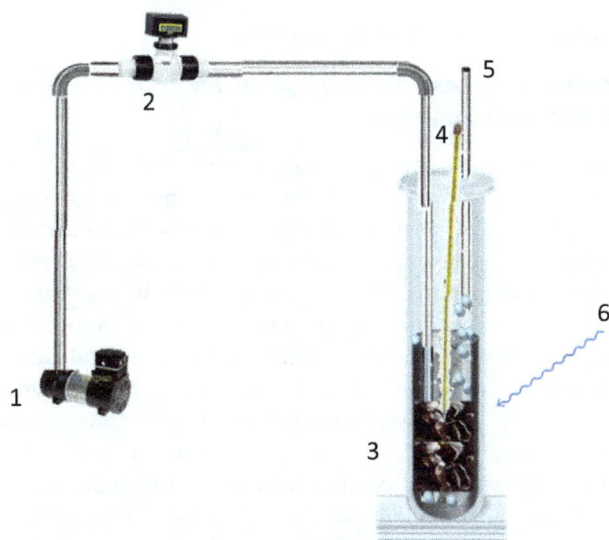

Figure 1. Schematic layout of the photoreactor used for the photocatalytic pretreatment of coffee pulp. 1: air compressor; 2: mass flow controller; 3: coffee pulp, water, and catalyst mixture; 4: thermometer; 5: air outlet; 6: solar radiation exposure.

The quartz reactor (called reactor A) of 10 cm inner diameter was filled with a mixture of 10%Cu/TiO_2 (100 g), coffee pulp (100 g) and water (500 mL). Small transparent fabric sachets were used as catalyst containers, which permitted a total and immediate separation of catalyst from the reaction mixture after the photocatalytic pretreatment process.

A second reactor (called reactor B) containing a similar reaction mixture without the catalyst was also studied under the same solar irradiation and air-flow conditions. The temperature of the reactor was maintained at 30°C. The reactors were exposed to solar radiation from 9:00 h to 16:00 h during 30 days (April). On average, the intensity of the exposed solar radiation was about 1000 W m^{-2}. After these 30 days of exposure, the catalyst was separated from the reaction mixture, which was transferred to the digester used for anaerobic codigestion tests.

The catalyst was easily separated from the mixture due to its high crystallinity and indentation hardness. However, the possibility of catalyst leaking or diluting in the reaction mixture cannot be discarded. In order to probe that no residual photocatalyst remained in the reaction mixture, after the photocatalytic pretreatment, we analyzed this mixture by atomic absorption spectrometry, using an AA-7000 spectrophotometer provided by Shimadzu (Shimadzu Corporations, Tokyo, Japan). Results revealed a Cu concentration lower than 2 ppm in the sample. This low-level indicates a very low leaching of 10%Cu/TiO_2 catalyst during photocatalytic pretreatment.

Anaerobic codigestion tests

Preliminary tests for the optimization of catalyst/ cattle manure ratio

To optimize the photocatalytically pretreated coffee pulp/ cattle manure ratio in the digesters, we performed some simple tests in small stainless steel vessels with different weight ratios of photocatalytically pretreated coffee pulp and cattle manure. The volume of the test vessels was 0.1 L, with a working volume of about 0.05 L. About 1 g of the photocatalytically pretreated coffee pulp and cattle manure mixture with different (0:1, 0.5:1, 1:1, 2:1, and 1:0) weight ratios were incorporated into the test vessels along with 0.025 L of water. The test digesters were equipped with a tap connected to an external tube, which enabled the sampling of the exhausted biogas. These digesters were placed outdoor under direct solar radiation for 15 days, after which the produced biogas was characterized by its CH_4 content (vol %) through gas chromatography. The results are reported in Table 1. In this table, it can be seen that the methanation capacity is best in the mixture containing 40 wt% catalytically pretreated coffee pulp, 40 wt% cattle manure, and 20 wt% water. The preliminary experiments described above permitted to determine the optimum coffee pulp/cattle manure ratio, assuring the biological processes during anaerobic codigestion take place while utilizing photocatalytically pretreated coffee pulp as nutrient matter.

Batch digestion tests were performed using 2 L capacity stainless steel tanks as digestion reactors. Two digesters, one containing the mixture of cattle manure and photocatalytically pretreated coffee pulp (called digester A) and another (called digester B) containing the mixture of cattle manure and coffee pulp pretreated without the photocatalyst were prepared for the codigestion tests. The digesters were not pretreated with any methanogenic inoculum. A typical setup of the codigestion experiment using such digester is presented schematically in Figure 2. The mixtures contained 40 wt% pretreated or untreated coffee pulp, 40 wt% cattle manure, and 20 wt% water. After pouring the mixtures, the digesters were tightly closed with rubber septum, and left outdoor under direct

Table 1. Methane concentration in the biogas produced in the mixture of photocatalytically pretreated coffee pulp/cattle manure (20 wt % water) mixture after 15 days of digestion.

Pretreated coffee pulp/cattle manure	Methane concentration (%)
0/1.0	3
0.5/1.0	7
1.0/1.0	15
2.0/1.0	6
1.0/0	2.5

solar radiation (for heating) for 30 days. It is important to note that the stainless steel tanks do not permit the solar radiation to pass through the reaction mixture. Therefore, no photocatalytic process takes place inside the tank, even if the photocatalyst in traces remains in the reaction mixture. Under these circumstances, there would be no generation of oxidizing species such as surface-bound hydroxyl radical ($\cdot OH$), superoxide ($\cdot O_2^-$), hydroperoxy radical ($HO_2\cdot$), and free holes, and thus the growth of anaerobic bacteria remains unaffected.

As in the case of photocatalytic pretreatment, the average illumination intensity was about 1000 W m^{-2}, and the maximum temperature of the reaction mixture varied in between 35 and 45°C. To provide mixing of the reactants in the containers, the reactors were agitated manually with a crank handle for about 1 min twice a day.

Analytical methods

The contents of cellulose, hemicellulose, and lignin of the coffee pulp were analyzed every 5 days for 30 days of photocatalytic pretreatment, following the method proposed by Van Soest et al. [48]. The values of the pH, total solids (TS), and the volatile solids (VS) in the reactors were determined before (initial values) and after 30 days of codigestion (final values), according to the Mexican standard method [49]. The initial and final parameters are reported in Table 2.

After 30 days of codigestion, the generated biogas (100 mL min^{-1}) was passed through an activated carbon column and collected into an analysis gas cell coupled to a Bruker (Vertex 70) FTIR gas spectrometer for composition analysis. The analysis of the biogas composition was performed using the QAsoft-quantitative analysis software to determine the presence and quantity of the volatile compounds in the flow. All the measurements were performed in triplicate at room temperature.

Results and Discussion

Catalyst characterization

Figure 3 shows the XRD patterns of TiO_2, and 10%Cu/ TiO_2 samples calcined at 800°C for 4 h. While the pure TiO_2 sample revealed its crystallinity in anatase phase, the annealed 10%Cu/TiO_2 sample reveled its rutile phase without considerable change in crystallinity. The sample also revealed a weak diffraction peak associated with CuO in monoclinic phase. Appearance of CuO peak in the 10%Cu/TiO_2 composite indicates the presence of Cu dispersed in TiO_2 powder. It must be noted that all the peaks associated with rutile phase of TiO_2 in 10%Cu/ TiO_2 sample suffered a small shift toward lower angle,

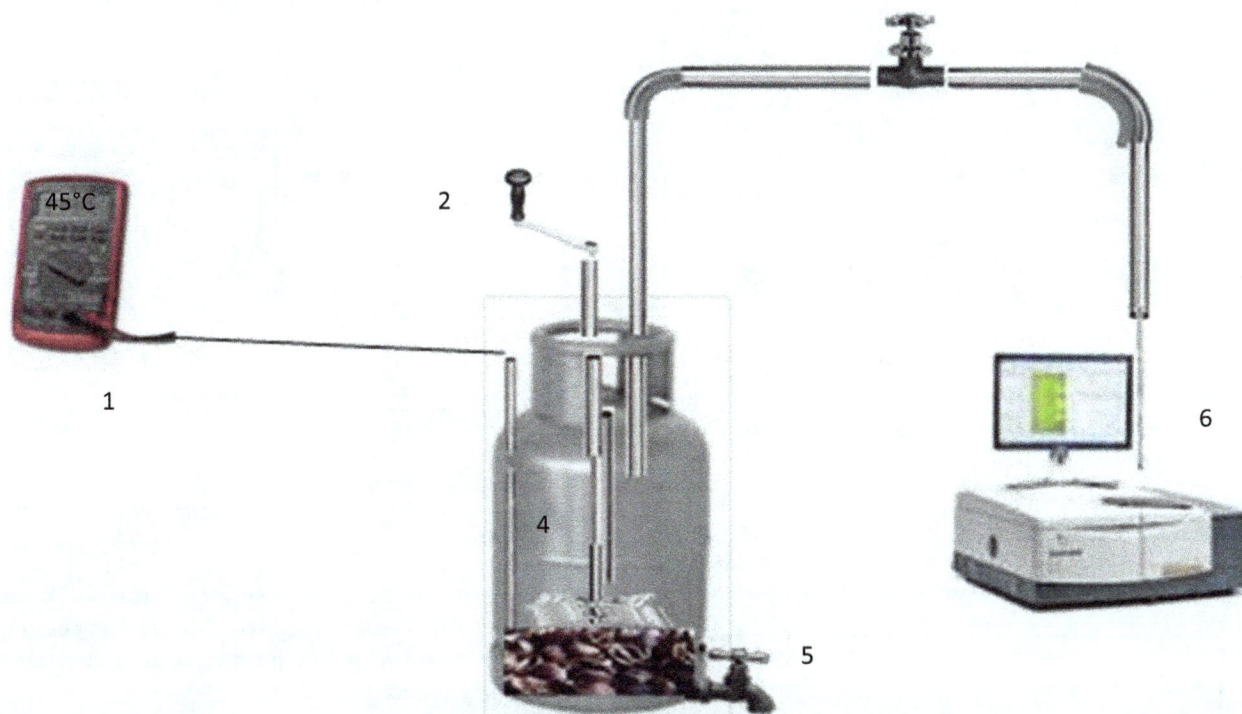

Figure 2. Schematic illustration of the used codigestion setup (1): thermocouple; (2): mixing crank handle; (3): activated carbon; (4): digestion reactor; (5): sampling valve for pH measurement; (6): FTIR spectrometer.

Table 2. Composition of coffee pulp after photocatalytic pretreatment.

	In fresh coffee pulp (%)	In the product of reactor A (%)	In the product of reactor B (%)
Lignin	6.88	4.05	6.88
Cellulose	43.98	29.45	43.56
Hemicellulose	27.80	11.86	27.74

demonstrating a lattice expansion due to Cu incorporation.

In Figure 4, the UV–Vis absorption spectra of the TiO_2 and 10%Cu/TiO_2 samples are presented. Absorption spectrum of the 10%Cu/TiO_2 sample revealed three clearly distinguishable absorption regions. In the first region in between 220 and 310 nm, there appeared a broad absorption signal peaked at about 235 nm, which can be assigned to the charge-transfer transition of the ligand O^{2-}, to isolated metal center Cu^{2+} and the d–d transition of CuO particles [50–52].

The broad absorption signal in the 350–400 nm spectral range, peaked around 370 nm indicates the presence of Cu_2O layers on the particle surface [51]. The absorption in the visible region (400–800 nm) can be attributed to the absorption of Cu_2O and CuO on the TiO_2 surface. On the other hand, the absorption spectrum of the pure TiO_2 sample is featureless, except a strong absorption

edge near about 385 nm. The band gap calculated by extrapolating the absorbance near the absorption edge to zero revealed the band gap energy values of about 3.04 and 2.77 eV for the pure TiO_2 and 10%Cu/TiO_2 samples, respectively. The estimated band gap energy values clearly indicate a red shift (toward visible region) of band gap energy of TiO_2 on Cu incorporation. The BET estimated specific surface area and optical band gap energy values of the pure and Cu incorporated calcined TiO_2 samples are presented in Table 3.

Photocatalytic pretreatment of coffee pulp

Figure 5 shows the evolution of lignin, cellulose, and hemicellulose contents present in the fresh coffee pulp, as a function of the reaction time in reactor A (with 10%Cu/TiO_2 photocatalyst), and in reactor B (without the photocatalyst). Table 2 shows the estimated values of lignin, cellulose, and, hemicellulose present in the fresh coffee pulp after 30 days of photocatalytic pretreatment in reactor A, and in reactor B. From the data presented in Figure 5 and Table 2, we can conclude that:

- The contents of lignin, cellulose, and hemicellulose remain almost same on irradiating the pulp in reactor B (in absence of the photocatalyst).

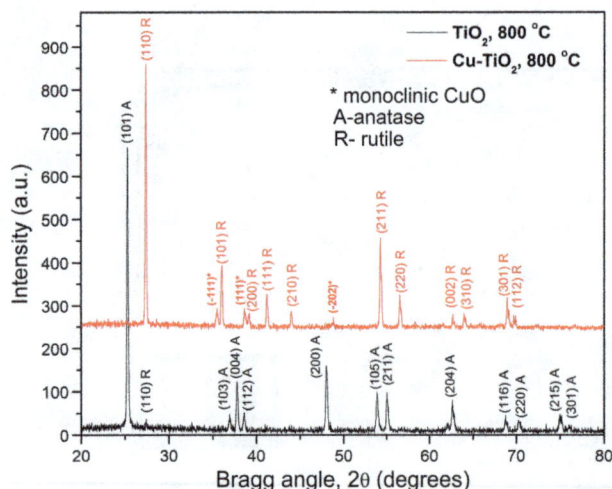

Figure 3. XRD patterns of powder TiO_2 and $10\%Cu/TiO_2$ samples after calcination in air at 800°C for 4 h. For clarity, the XRD pattern of the $10\%Cu/TiO_2$ samples was vertically shifted. XRD, X-ray diffraction.

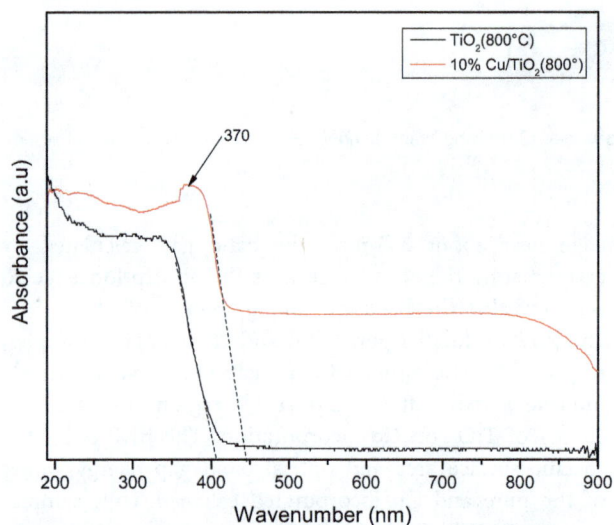

Figure 4. UV–Vis absorbance spectra of TiO_2 and $10\%Cu/TiO_2$ samples after calcination at 800°C for 4 h in air.

Table 3. Catalysts characterization data.

Sample	Specific surface area (m² g⁻¹)	Band gap energy, E_g (eV)
TiO_2	10.02	3.04
$10\%Cu/TiO_2$	12.10	2.77

- After the photocatalytic pretreatment, the contents of lignin, cellulose, and hemicellulose in the coffee pulp decreased considerably.
- Lignin, cellulose, and hemicelluloses photodegradation rate is high during the first 10 days of solar irradiation pretreatment. After this time, only a small variation in

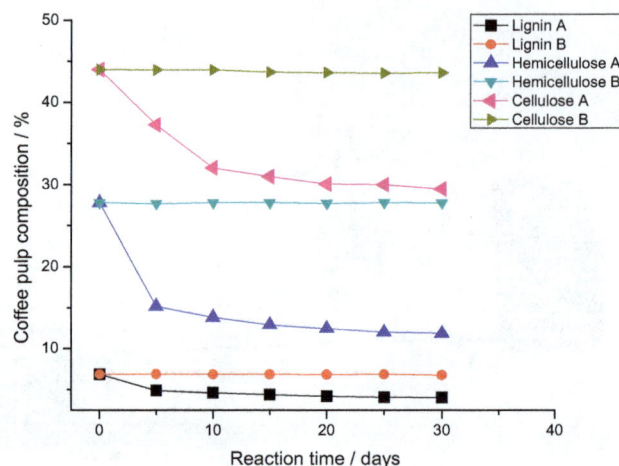

Figure 5. Evolution of the composition of coffee pulp as a function of reaction time.

the contents of these components in reactor A was detected. This result suggests that for application convenience, coffee pulp's photocatalytic pretreatment time can be reduced.

The results obtained from the pretreatments in two reactors (reactor A and reactor B) clearly indicate that the pretreatment without the photocatalyst has negligible effect on the degradation of lignocellulosic biomass in comparison with the pretreatment in presence of 10% Cu/TiO_2 photocatalyst. As the UV–Vis absorption spectrum of the 10% Cu/TiO_2 catalyst (Figure 4) revealed the presence of Cu^{2+} and Cu^{1+} ions and an extended TiO_2 absorption in the visible region, the photocatalytic degradation of lignocellulosic biomass can be explained considering the following facts:

- The $10\%Cu/TiO_2$ sample acts as a very efficient photocatalyst under solar irradiation due to its characteristic absorptions near UV region and extended absorption in the visible spectral region.
- Electron trapping by the copper ions present in the TiO_2 matrix probably inhibits the recombination process of photogenerated electrons and holes [53–56].
- The $10\%Cu/TiO_2$ photocatalyst probably generated a large amount of hydroxyl radicals ($\cdot OH$), superoxides ($\cdot O_2^-$), hydroperoxy radicals ($HO_2\cdot$), or free holes on its surface as proposed by Linsebigler et al. [45]. These species should have enhanced the lignocellulose degradation rate in a similar way as proposed by Machado et al. [40]. These authors found that the hydroxylation of aromatic structures in lignin occurs during photochemical process in the presence of TiO_2, which is known to generate hydroxyl radicals under UV–Vis irradiation. The incorporation of hydroxyl groups into the benzene rings leads to the creation of radical sites

allowing the action of superoxide radicals, and resulting in a structural fragmentation of lignin. On the other hand, Ma et al. have observed a rapid photocatalytic lignin degradation using TiO_2 and Pt/TiO_2 [45].

Our results indicate that a photocatalytic pretreatment might be an effective way for breaking the complex lignocellulosic biomass matrix, changing chemical components and physical structures of biomass like coffee pulp. As the lignocellulosic biomasses are very difficult to be degraded biologically, a photocatalytic pretreatment of biomass can generate biodegradable components, which would be easier to digest for microorganisms.

Anaerobic codigestion process

In order to investigate the effect of photocatalytic pretreatment on the subsequent biogas production, the coffee pulps pretreated in the reactors A and B were anaerobically codigested with cattle manure.

As can be seen from Table 4, the initial and final pH values measured for the two digesters remained constant, whether the coffee pulp was photocatalytically pretreated or not. The percentage decrease in TS and VS in the mixtures of the two digesters after 30 days of solar irradiation is considerably notable. The decrease is huge for the digester A where we sued the photocatalytically pretreated coffee pulp in comparison with the same in the digester B fed with untreated coffee pulp for the codigestion by cattle manure.

These results indicate that the effect of photocatalytic pretreatment of coffee pulp results in an enhanced degradation of TS and VS through the codigestion process. The result of the codigestion process is seen to be closely related with the results of photocatalytic pretreatment (Table 2), suggesting a possible correlation between the amount of degraded lignin, cellulose, and hemicelluloses in the fed charge of the digester and the produced biogas.

Biogas production

In Table 5, the results of quantitative analysis of the produced biogas in the digesters A and B are presented in comparative manner. From the obtained results, we can observe that after only 1 month of codigestion: (1)

valuable combustion gases like methane, propane, etc. are generated from the coffee pulp–cattle manure mixture through codigestion; (2) the amounts of generated combustion gases are considerably higher for the digester containing photocatalytically pretreated coffee pulp (digester A) than the amounts generated from the digester B, where untreated coffee pulp was used.

Table 6 shows the combustion enthalpies (ΔH_c) of the combustion gases detected in 100 mL of the biogas generated in digesters A and B. These values were calculated from the volume (V) of each component measured in 100 mL of biogas (Table 5) and on basis of the standard combustion enthalpies (ΔH_c°) reported in the literature using the following relation[57]:

$$\Delta H_c = \frac{V \times \Delta H_c^\circ}{22,400}$$

As can be seen from the Table 6, the total combustion enthalpy (ΔH_c) of the biogas generated in digester A is almost three times to the total combustion enthalpy of the biogas generated in the digester B ([−2.59]/[−0.93] = 2.78), indicating the necessity of utilizing the photocatalytic pretreatment of coffee pulp before its digestion for obtaining higher amounts of biogas, that is,

Table 4. Initial and final compositions determined in the digesters.

Digester	pH Initial	pH Final	Total solids (%) Initial	Total solids (%) Final	Volatile solids (%) Initial	Volatile solids (%) Final
A	7.1	7.1	14.26	5.45	92.25	10.86
B	7.2	7.0	14.02	10.15	91.09	89.03

Table 5. Amounts of different components in 100 mL of biogas generated from the digester A (containing photocatalytically pretreated coffee pulp) and the digester B (containing photocatalytically untreated coffee pulp).

Compound	Estimated volume (mL/100 mL of biogas) Digester A	Digester B
Butane	3.26	1.13
1-Butene	0.00	0.26
Cis-2-butene	0.00	1.15
Cyclopropane	6.41	2.17
Ethane	0.06	0.00
Isobutane	2.38	0.71
Methane	21.09	17.79
Propane	23.78	8.44
2-Methyl-1-butene	0.00	2.08
3-Methyl-1-butene	7.56	2.51
2-Pentene	10.61	1.97
Cyclohexane	0.48	0.07
Cyclohexene	0.05	0.01
Dodecane	1.03	0.26
n-Hexane	6.85	0.35
1-Hexene	6.83	2.07
3-Methyl pentane	2.62	0.92
n-Octane	0.05	0.22
Carbon dioxide	11.03	43.25
Hydrogen sulfide	2.06	0.92
Other gases	5.68	15.79
Total	100	100

Table 6. Standard combustion enthalpies (ΔH_c°) and combustion enthalpies (ΔH_c) of the component gases in 100 mL of the biogas generated in the digester A (containing photocatalytically pretreated coffee pulp) and digester B (containing photocatalytically untreated coffee pulp).

Compound	ΔH_c° kcal mol^{-1}	ΔH_c kcal/100 mL biogas	
		Digester A	Digester B
Butane	−687.982	−0.10	−0.03
1-Butene	−649.757	0	0
Cis-2-butene	−648.115	0	−0.03
Cyclopropane	−499.52	−0.14	−0.04
Ethane	−372.820	0	0
Isobutane	−686.342	−0.07	−0.02
Methane	−212.798	−0.20	−0.17
Propane	−530.605	−0.56	−0.19
2-Methyl-1-butene	−803.17	0	−0.07
3-Methyl-1-butene	−804.93	−0.27	−0.09
2-Pentene	−805.34	−0.38	−0.07
Cyclohexane	−944.79	−0.02	−0.003
Cyclohexene	−907.38	−0.00	−0.001
Dodecane	−1947.23	−0.09	−0.02
n-Hexane	−1002.57	−0.30	−0.01
1-Hexene	−973.42	−0.29	−0.08
3-Methyl pentane	−1001.51	−0.11	−0.04
n-Octane	−1317.45	−0.002	−0.01
Total		−2.59	−0.93

for higher economic benefits. The equivalent heat of combustion of the biogas generated in digester A is approximately 2427 kJ/mol, which is much higher than the heat of combustion of methane (889 kJ/mol), the principal component of usual biogas.

The results presented in the Tables 2 and 4–6 suggest that a photocatalytic pretreatment of the coffee pulp before the codigestion process might have accelerated its subsequent anaerobic biodegradation.

Anaerobic degradation of biomass is considered to follow a sequence of four steps: hydrolysis, acidogenesis, acetogenesis, and methanogenesis. During the hydrolysis phase, the water-insoluble compounds like cellulose, proteins, and fats get broken into monomers (water-soluble fragments) by exoenzimes (hydrolase). The conversions of carbohydrates into simple sugars, lipids into fatty acids, and proteins into amino acids take place in the hydrolysis phase [58]. However, the degradation of lignocellulose and lignin present in biomass is slow and incomplete [59].

As we can see from Table 2, after the photocatalytic pretreatment, the contents of lignin, hemicelluloses, and cellulose in the coffee pulp decrease, probably due to the cleavage of chemical bonds of these components during the photocatalytic pretreatment process. The presence of lignin in lignocellulosic biomass leads to a protective

barrier to the biomass and provides resistance to any chemical and biological degradation, preventing the plant cell destruction by fungi, bacteria, or enzymes. For the generation of biogas from coffee pulp through codigestion, the cellulose and hemicellulose must be broken to their corresponding monomers, suitable for their digestions through biological route (by microorganisms). Therefore, the photocatalytic pretreatment of coffee pulp with 10%Cu/TiO$_2$ under solar radiation in presence of oxygen helps to cleavage the impermeable/resistant layer of lignin, and breaks the containing long-chain cellulose and hemicellulose into their corresponding monomers. The enhanced codigestion of the photocatalytically pretreated coffee pulp and cattle manure mixture is due to the presence of a reduced amount of lignin and increased availability of lignocellulosic monomers in the pretreated coffee pulp.

It is worth noting (Tables 5 and 6) that besides methane, high amounts of propane, butane, and other high molecular weight combustible hydrocarbons are generated during the codigestion of photocatalytically pretreated coffee waste with cattle manure. The presence of these high molecular weight products could be due to the following factors:

- The variations in temperature inside the digester.
- An inhibition effect due to traces of 10%Cu/TiO$_2$ (remaining from the photocatalytic pretreatment) in the anaerobic digester.
- The presence of additional products generated from the photocatalytic hydroxylation of the aromatic structures of lignin, hemicelluloses, and cellulose.

As the temperature variation was similar inside both the digesters (A and B), the contribution of the first factor can be discarded. However, the composition of the biogas generated from digester A is different from the digester B. The contribution of the second factor should also be ruled out, as the used photocatalyst (10% Cu/TiO$_2$) is not active inside the stainless steel digester (as discussed in the section Anaerobic codigestion tests). Now, it is possible that during the photocatalytic pretreatment of lignin, hemicelluloses, and cellulose, several derivatives of their aromatic structures are produced in high concentration through photocatalytic hydroxylation, which might also act as bacterial nutrients. The excessive increase in bacterial nutrients might also cause a reduction in consumption of higher carbon content organic molecules by the anaerobic bacteria in the digestion process.

The results presented in Table 6 indicate that the biogas produced by our process can generate higher amount of energy during its combustion than the conventional biogases rich in methane. However, we are still optimizing the catalysis process and expecting even a higher biogas output. The optimized yield and the viability data of the process will be communicated soon.

Conclusions

We have demonstrated that a photocatalytic pretreatment of waste coffee pulp using 10%Cu/TiO$_2$ photocatalyst under solar radiation and air as oxidizing agent can enhance the solubility and biodegradability of its lignocellulosic components. This improved biodegradability also enhances the production of biogas from this waste biomass through anaerobic digestion. The process described in this study is technically viable, produces biogas of high combustion energy, and thus promising for higher economic benefits. Moreover, the process has potential for larger-scale production of energy-rich biogas from natural wastes like coffee pulp.

Acknowledgments

The authors acknowledge VIEP & DITCo, BUAP (Grants # VIEP 45/NAT/2014, DITCo-BUAP/INOVA/3 and DIT-Co-BUAP/INOVA/19), and CONACyT, Mexico (Grant # CB-2010/151767) for their financial supports.

Conflict of Interest

None declared.

References

1. Simpson-Holley, M., A. Higson, and G. Evans. 2007. Bring on the biorefinery. J. Chem. Eng. 163:46–59.
2. Fantozzi, F., and C. Buratti. 2009. Biogas production from different substrates in an experimental continuously stirred tank reactor anaerobic digester. Bioresour. Technol. 100:5783–5789.
3. Abbasi, T., and S. A. Abbasi. 2010. Biomass energy and the environmental impacts associated with its utilization. Renew. Sustain. Energy Rev. 14:919–937.
4. Fantozzi, F., and C. Buratti. 2011. Anaerobic digestion of mechanically treated OFMSW: experimental data on biogas/methane production and residues characterization. Bioresour. Technol. 102:8885–8892.
5. Boullaqui, H., Y. Touhami, R. B. Cheikh, and M. Hamdi. 2005. Bioreactor performance in anaerobic digestion of fruit and vegetables waste. Process Biochem. 40:989–995.
6. Gallert, C., A. Henning, and J. Winter. 2003. Scale-up of anaerobic digestion of the biowaste fraction from domestic wastes. Water Resour. 37:1433–1441.
7. Abarca-Guerrero, L., G. Maas, and W. Hogland. 2013. Solid waste management challenges for cities in developing countries. Waste Manag. 33:220–232.
8. Lopes, W. S., V. D. Leite, and S. Prasad. 2004. Influence of inoculum on performance of anaerobic reactors for treating municipal solid waste. Bioresour. Technol. 94:261–266.
9. Brand, D., A. Pandey, S. Roussos, and C. R. Soccol. 2000. Biological detoxification of coffee husk by filamentous fungi using a solid state fermentation system. Enzyme Microb. Technol. 26:127–130.
10. Pandey, A., C. R. Soccol, P. Nigam, D. Brand, R. Mohan, and S. Roussos. 2000. Biotechnological potential of coffee pulp and coffee husk for bioprocesses. J. Biochem. Eng 6:153–158.
11. Deublein, D., and A. Steinhauser. 2008. Biogas from waste and renewable sources: an introduction. WILEY-VCH Verlag GmbH & Co. KGaA, Weinheim.
12. Mussatto, S. I., S. E. M. Ercilla, S. Martins, and A. T. Jose. 2011. Production, composition, and application of coffee and its industrial residues. Food Bioprocess Technol. 4:661–672.
13. Franca, A. S., I. S. Oliviera, and M. E. Ferreira. 2009. Kinetics and equilibrium studies of methylene blue adsorption by spent coffee grounds. Desalination 249:267–272.
14. Murthy, P. S., and M. M. Naidu. 2012. Resources. Conservation and Recycling 66:45–58.
15. Adams, M. R., and J. Dougan. 1981. Biological management of coffee processing wastes. Trop. Sci. 23:177–185.
16. Boopathy, R., and M. Mariappan. 1984. Coffee pulp: a potential source of energy. J. Coffee Res. 14:108–110.
17. Boopathy, R., M. Mariappan, and B. B. Sunderasan. 1986. The carbon to nitrogen ratio and methane production of coffee pulp. J. Coffee Res. 16:47–50.
18. Fan, L., A. T. Soccol, A. Pandey, and C. R. Soccol. 2003. Cultivation of pleurotus mushroom on Brazilian coffee husk and its effect on caffeine and tannic acid. Micologia Aplicada International 15:15–21.
19. Belgiorno, V., D. Panza, L. Russo, V. Amodio, and A. Cesaro. 2011. Alternative stabilization options of mechanically sorted organic fraction from municipal solid wastes prior to landfill disposal. Int. J. Environ. Eng. 3:318–335.
20. Neves, L., R. Oliveira, and M. M. Alves. 2006. Anaerobic co-digestion of coffee waste and sewage sludge. Waste Manag. 26:176–181.
21. Li, Y. Y., and T. Noike. 1992. Upgrading of anaerobic digestion of waste activated sludge by thermal pretreatment. Water Sci. Technol. 26:857–866.
22. Climent, M., I. Ferrer, M. Baeza, A. Artola, F. Vazquez, and X. Font. 2007. Effects of thermal and mechanical pretreatments on secondary sludge on biogas production under thermophylic conditions. Chem. Eng. J. 133:335–342.
23. Komenmoto, K., Y. G. Lim, Y. Onoue, C. Niwa, and T. Toda. 2007. Effects of temperature on VFA's and biogas

production in anaerobic solubilization of food waste. Waste Manag. 29:2950–2955.

24. Heo, N., S. Park, and H. Kang. 2009. Solubilization of waste activated sludge by alkaline pretreatment and biochemical methane potential (BMP) test for anaerobic codigestion of municipal organic waste. Water Sci. Technol. 48:1505–1509.

25. Zhang, G., J. Yang, H. Liu, and J. Zhang. 2009. Sludge ozonation: disintegration, supernatant changes and mechanisms. Bioresour. Technol. 100:1505–1509.

26. Masse, L., D. I. Masse, and K. J. Kennedy. 2003. Effect of hydrolysis pretreatment on fat degradation during anaerobic digestion of slaughterhouse wastewater. Process Biochem. 38:1365–1372.

27. Mendes, A. A., E. B. Pereira, and H. F. de Castro. 2006. Effect of enzymatic hydrolysis of lipids-rich wastewater on the anaerobic biodigestion. Biochem. Eng. J. 32:185–190.

28. Xu, G., S. Chen, J. Shi, S. Wang, and G. Zhu. 2010. Combination of ultrasound and ozone for improving solubilization and anaerobic biodegradability of waste activated sludge. J. Hazard. Mater. 180:340–346.

29. Fdez-Guelfo, L., C. Alvarez-Gallego, B. Sales, and L. I. Romero. 2011. The use of thermochemical and biological pretreatments to enhance organic matter hydrolysis and solubilization from organic fraction of municipal solid waste. Chem. Eng. J. 168:249–254.

30. Liu, J., Y. Wu, and N. Xu. 1995. Effects of ammonia bicarbonate treatment on kinetics of fiber digestion, nutrient digestibility and nitrogen utilization of rice straw by sheep. Anim. Feed Sci. Technol. 52:131–139.

31. Patel, M. M., and R. M. Bhatt. 1992. Optimization of the alkaline peroxide pretreatment for the delignification of rice straw and its applications. J. Chem. Technol. Biotechnol. 53:253–263.

32. Sankat, C., and B. Lauckner. 1991. The effect of ammonia treatment on the digestibility of rice straw under various process conditions. Can. Agr. Eng. 33:309–314.

33. Driscoll, M., A. Stipanovic, W. Winter, K. Cheng, M. Manning, J. Spese, et al. 2009. Electron beam irradiation of cellulose. Radiat. Phys. Chem. 78:539–542.

34. Ng, F. M. F., and K. N. Yu. 2007. X-ray irradiation induced degradation of cellulose nitrate. Mater. Chem. Phys. 100:38–40.

35. Discoll, M., A. Stimanovic, W. Winter, K. Cheng, M. Manning, J. Spiese, et al. 2009. Electron beam irradiation of cellulose. Radiat. Phys. Chem. 78:539–542.

36. Ghanbari, F., T. Ghoorchi, P. Shawrang, H. Monsouri, and N. M. Torbati-Nejad. 2012. Comparison of electron beam and gamma ray irradiations effects on ruminal crude protein and amino acid degradation kinetics and in vitro digestibility of cottonseed meal. Radiat. Phys. Chem. 81:672–678.

37. Uyguner, C. S., and M. Bekbolet. 2010. TiO_2 assisted photocatalytic degradation of humic acids: effect of copper ions. Water Sci. Technol. 61:2581–2590.

38. Machado, A. E. H., A. M. Furuyama, S. Z. Falone, R. Ruggerio, D. S. Perez, and A. Castellan. 2000. Photocatalytic degradation of lignin and lignin models, using titanium dioxide: the role of the hydroxyl radical. Chemosphere 40:115–124.

39. Selli, E., D. Baglio, L. Montanarella, and G. Bidoglio. 1999. Role of humic acids in the TiO_2-photocatalyzed degradation of tetrachloroethene in water. Water Resour. 33:1827–1936.

40. Tanaka, K., C. R. Calanag, and T. Hisanaga. 1999. Photocatalyzed degradation of lignin on TiO_2. J Mol. Catal. A Chem. 38:287–294.

41. Dunn, W. W., Y. Aikawa, and A. J. Bard. 1981. Characterization of particulate titanium dioxide photocatalysts by photoelectrophoretic and electrochemical measurements. J. Am. Chem. Soc. 100:3456–3459.

42. Peterson, M. W., J. A. Turner, and A. J. Nozic. 1991. Mechanistic studies of the photocatalytic behavior of TiO_2. Particles in a photoelectrochemical slurry cell and the relevance of photodetoxification reactions. J. Phys. Chem. 95:221–225.

43. Fox, M. A., and M. T. Dulay. 1993. Heterogeneous photocatalysis. Chem. Rev. 93:341–357.

44. Linsebigler, A. L., G. Lu, and J. T. Yates Jr. 1995. Photocatalysis on TiO_2 surfaces: principles, mechanisms, and selected results. Chem. Rev. 95:735–741.

45. Ma, Y. S., C. N. Chang, Y. P. Chiang, H. F. Sung, and A. C. Chao. 2008. Photocatalytic degradation of lignin using Pt/TiO_2 as the catalyst. Chemosphere 71:998–1004.

46. Gaya, U. I., and A. H. Abdullah. 2010. Heterogeneous photocatalytic degradation of organic contaminants over titanium dioxide: a review of fundamentals, progress and problems. J. Photochem. Photobiol. C-9:1–12.

47. Huang, G. L., S. C. Zhang, T. G. Xu, and Y. F. Zhu. 2008. Fluorination of $ZnWO_4$ photocatalyst and influence on the degradation mechanism for 4-chlorophenol. Environ. Sci. Technol. 42:8516–8521.

48. Van Soest, P., J. Robertson, and B. Lewis. 1991. Symposium: carbohydrate methodology, metabolism, and nutritional implications in dairy cattle. J. Dairy Sci. 74:3583–3597.

49. Mexican Official Standard: NOM-AA-34-1976, ISO 14000.

50. Cordoba, G., R. Arroyo, J. L. G. Fierro, and M. Viniegra. 1996. Study of Xerogel-Glass Transition of CuO/SiO_2. J. Solid State Chem. 123:93–98.

51. Boccuzzi, F., S. Coluccia, G. Martra, and N. Ravasio. 1999. Cu/SiO_2 and Cu/SiO-$_2TiO_2$ catalysts: 1. TEM, DR UV–vis-NIR, and FTIR characterization. J. Catal. 184:316–322.

52. Velu, S., K. Suzyki, M. Akazaki, M. P. Kapoor, T. Osaki, and F. Ahashi. 2000. Oxidative steam reforming of methanol over CuZnAl(Zr)-oxide catalysts for the selective production of hydrogen for fuel cells: catalyst

characterization and performance evaluation. J. Catal. 194:373–379.

53. Xin, B., P. Wang, D. Ding, J. Liu, Z. Ren, and H. Fu. 2008. Effect of surface species on Cu-TiO$_2$ photocatalytic activity. Appl. Surf. Sci. 254:2569–2574.

54. Foster, N., R. Noble, and C. Koval. 1993. Reversible photoreductive deposition and oxidative dissolution of copper ions in titanium dioxide, aqueous suspensions. Environ. Sci. Technol. 27:350–356.

55. Ward, M., and A. Bard. 1982. Photocurrent enhancement via trapping of photogenerated electrons of TiO, particles. J. Phys. Chem. 86:3599–3605.

56. Butler, E. C., and A. P. Davis. 1993. Photocatalytic oxidation in aqueous titanium dioxide suspensions: the influence of dissolved transition metals. J. Photochem. Photobiol., A 70:273–283.

57. Perry RH, R. H. and D. W. Gree. 1999. Perry's Chemical Engineers' handbook. Mc Graw Hill Companies Inc, McGraw Hill, New York, USA.

58. Gerardi, M. H. 2003. The microbiology of anaerobic digesters, waste water microbiology series. John Wiley and Sons Inc, Hoboken, NJ.

59. Deublein, D., and A. Steinhauser. 2008. Biogas from waste and renewable sources: an introduction. WILEY-VCH Verlag GmbH and Co. KGaA, Weinheim.

6

Butanol and ethanol production from lignocellulosic feedstock: biomass pretreatment and bioconversion

Sonil Nanda[1], Ajay K. Dalai[2] & Janusz A. Kozinski[1]

[1]Lassonde School of Engineering, York University, Ontario, Canada
[2]Department of Chemical and Biological Engineering, University of Saskatchewan, Saskatchewan, Canada

Keywords

Butanol, dilute acid pretreatment, enzymatic hydrolysis, ethanol, fermentation, lignocellulosic biomass

Abstract

Lignocellulosic feedstock has tremendous potential to sustain the renewable production of biofuels such as ethanol and butanol. Although lignocellulosic biomass is a storehouse of energy in the form of cellulose and hemicellulose, yet lignin acts as a barrier against their hydrolysis. A dilute acid pretreatment disintegrates the biomass complex and allows cellulolytic enzymes to hydrolyze cellulose and hemicelluloses in releasing fermentable sugars. The current study investigates the effect of different H_2SO_4 doses (0–2.5%) on the three lignocellulosic feedstock material, especially pinewood, timothy grass, and wheat straw at 121°C for 1 h. Furthermore, the pretreated feedstock was subjected to enzymatic hydrolysis using cellulase, β-glucosidase, and xylanase at 45°C for 72 h. The biomass hydrolysates containing monomeric sugars (glucose and xylose) were fermented using *Saccharomyces cerevisiae* and *Clostridium beijerinckii* for ethanol and butanol production, respectively. A comparative evaluation for the concentrations of ethanol and butanol, residual sugars as well as byproducts such as acetone, acetate, and butyrate from biomass hydrolysates was performed. Pinewood hydrolysate revealed high ethanol (24.1 g/L) and butanol (11.6 g/L) concentrations due to greater sugar content. In contrast to ethanol fermentation by *S. cerevisiae*, butanol fermentation by *C. beijerinckii* demonstrated low butanol levels in the hydrolysates due to butanol toxicity toward clostridia.

Introduction

Lignocellulosic residues are inexpensive and attractive renewable resources for the production of next generation biofuels. With their vast availability, they are considered to be a suitable alternative to the diminishing fossil fuels. The inedible plant material including residues from agriculture (e.g., corn stover, wheat straw, rice husk); energy crops (e.g., switchgrass, timothy grass), and forest refuse (e.g., pinewood, spruce) are produced in abundance every year [1, 2]. Since, much of these biomasses are thrown away, turning these surplus discarded plant materials into biofuels is of great appeal and importance. The global energy consumption in 2008 being 533 EJ is projected for increase to 653 EJ by 2020 and 812 EJ by 2030 [3]. It is acceptable that a major proportion of this future energy supply (250–500 EJ per year by 2050) will be contributed by lignocellulosic biomass [4].

The major components of lignocellulosic biomass are cellulose, hemicellulose, and lignin. Cellulose is composed of glucose polymers which are largely insoluble and exist in crystalline microfibrils making the sugars difficult to extract [5]. Hemicellulose which comprises various pentose and hexose sugars is attached to the cellulose microfibrils. On the other hand, lignin is a phenyl propane polymer which forms a complex network cross-linking the cellulose and hemicellulose together. In order to break the lignocellulosic framework and extract the fermentable sugars for bioconversion to higher fuel alcohols, a pretreatment method prior to conversion is required. A number of pretreatment methods available for biomass are dilute acid, alkali, hot water, ammonia fiber explosion, carbon dioxide explosion, and organic solvent [6].

An ideal pretreatment should have the following properties: (1) disintegrate the lignin and hemicellulose complex with cellulose, (2) improve the sugar yield as a result of hydrolysis of cellulose and hemicellulose, and (3) prevent excessive degradation or loss of carbohydrate. Last but not the least, it should be cost-effective. Dilute sulfuric acid pretreatment solubilizes the hemicelluloses and thereby disrupts the lignocellulosic composite linked by covalent bonds, hydrogen bonds and van der Waals forces [7]. The pretreatment process is used to overcome the recalcitrance of lignocellulose, increase enzyme efficiency, and improve the yields of monomeric sugars.

During the past few decades, there has been a significant interest in the production of fermentation derived fuel such as ethanol from agriculture-based substrates. However, due to the controversies of food versus fuel there has been a shift in the ethanol substrates from food grains to crop residues [8]. On the other hand, there have been a few attempts to produce alternative alcohol fuels from lignocellulosic residues. One such fuel is butanol that has superior fuel properties than ethanol and can be efficiently produced from lignocellulosic feedstock. A few bacteria known to produce butanol through ABE (acetone-butanol-ethanol) fermentation are *Clostridium acetobutylicum*, *C. beijerinckii*, *C. saccharoperbutylacetonicum* and *C. saccharobutylicum* [9].

Butanol-producing *Clostridium* is advantageous over the long-established ethanol-producing *Saccharomyces cerevisiae* in efficiently metabolizing both pentose and hexose sugars. *Saccharomyces cerevisiae* is inefficient in metabolizing pentose (e.g., xylose); however it is known to utilize hexose (e.g., glucose) proficiently [10]. In contrast, *Clostridium* is capable of metabolizing both pentose and hexose [11]. *Clostridium* utilizes xylose by directly converting it into xylulose with the enzyme xylose isomerase. Engineered *S. cerevisiae* has a tendency to utilize xylulose, although it requires xylose to be reduced to xylitol and xylulose by xylose reductase and xylitol dehydrogenase, respectively [12].

As a fuel, butanol can be used in pure form or blended in gasoline in any concentration unlike ethanol that can be blended only up to 85%; however, a higher blend would necessitate motor engine modification [13]. In addition, the energy content of butanol being 29.2 MJ/L is 30% higher than that of ethanol (21.2 MJ/L) and much closer to gasoline (32.5 MJ/L). Also, butanol's low vapor pressure, hydroscopic nature, less volatility and less flammability facilitates its blending and supply in existing gasoline channels and pipelines [11].

Despite many advantages, production of butanol has some drawbacks such as its low final titer levels, culture toxicity due to low butanol tolerance by clostridia, and purification issues [14]. This paper emphasizes on bioproduction of ethanol and butanol from wood (pinewood) and herbaceous-based (timothy grass, wheat straw) lignocellulosic feedstock. These biomasses are least explored in terms of their butanol and ethanol-producing potentials. In addition, a dilute sulfuric acid pretreatment followed by enzymatic hydrolysis of these lignocellulosic residues was developed for higher sugar yields. Furthermore, the biomass hydrolysates with cellulosic and hemicellulosic components were fermented for the production of ethanol and butanol.

Materials and Methods

Lignocellulosic biomass

The biomass samples used in this study were pinewood, PW (*Pinus banksiana*), timothy grass, TG (*Phleum pratense*), and wheat straw, WS (*Triticum aestivum*). The biomasses were collected from Saskatchewan, Canada in 2011. After collection, the biomass samples were air-dried and pulverized using a Wiley mill (Wiley Mill No. 1.706.648, Arthur H. Thomas Co., Philadelphia, PA, USA) to pass through a sieve screen of 1.18 mm. The crushed biomass samples were stocked in clean glass jars at room temperature and used as necessary.

Compositional analysis

The compositional analysis (cellulose-hemicellulose-lignin) of the feedstock samples was performed using a two-step standard NREL method. In the first step, the feedstock was subjected to a Soxhlet extraction procedure using a sequential application of water, ethanol, and hexane for the extraction of any inorganic materials and nonstructural sugars from the samples such as chlorophyll, waxes, sterols, and lipids [15]. The solvents after the extraction were removed in a rotary evaporator under reduced pressure and the residual biomass was air-dried for sugar determination. The extracts after evaporation were recovered and weighed.

In the next step, the air-dried residual biomasses were hydrolyzed with 72% H_2SO_4 at 30°C for 1 h followed by 4% H_2SO_4 at 121°C for 1 h [16]. The hydrolyzed sugars (e.g., arabinose, cellobiose, glucose and xylose) were quantified through an Agilent 1100 Series HPLC (Agilent Technologies, Waldbronn, Germany) equipped with refractive index detector. The analysis was performed using an Aminex HPX 87H ion-exclusion column (BioRad, Hercules, CA, USA) with a Cation H micro-guard cartridge (BioRad). The mobile phase used was 5 mmol/L H_2SO_4 with a flow rate of 0.6 mL/min and column temperature of 55°C. After a complete conversion, the composition of glucose and cellobiose was considered as cellulose and that of xylose and arabinose as hemicellulose [17]. The sugar standards used in the experiments were purchased from Sigma-Aldrich, Oakville, Canada.

Acid and enzymatic hydrolysis

For dilute acid pretreatment, 10 g of biomass (dry basis) was mixed well with 100 mL of 0–2.5% v/v H_2SO_4 to distilled water in 250 mL conical flask. The solid loading ratio or the amount of dry feedstock loading was 1:10 w/v biomass/water. The flasks were cotton plugged and covered with aluminum foil prior to autoclaving at 121°C for 1 h in a Primus PSS5-C-MSSD steam sterilizer (Primus Sterilizer Company LLC, Omaha, NE). The flasks were weighed before and after autoclaving to account for any loss of water. The difference in water loss was supplemented by adding sterilized distilled water to the flasks. After autoclaving, the hydrolysates were cooled to room temperature and neutralized to pH 5.0 with 10 mol/L NaOH [18].

About 6 mL/L each of cellulase, β-glucosidase, and xylanase was added to the neutralized biomass hydrolysates and incubated at 45°C for 72 h with agitation at 80 rpm in an incubator shaker [18]. The enzymes cellulase (brand name: Celluclast 1.5 L; enzyme activity: ≥700 EGU/g), β-glucosidase (brand name: Novozyme 188; enzyme activity ≥250 CBU/g) and xylanase (brand name: Xylanase 1; activity: ≥40 units/mg protein) were obtained from Sigma-Aldrich, Canada. After 72 h of enzymatic hydrolysis, the mixture was centrifuged at 3300g to remove the biomass residues. The supernatant liquid containing hydrolyzed sugars were recovered and filter sterilized by passing through a 0.2 μm filter disc (Fisher Scientific, Ottawa, Canada). The sterilized biomass hydrolysates were stored in sterilized screw-capped Pyrex bottles at 4°C prior to the fermentation experiments. The sugar (e.g., glucose and xylose) levels in the hydrolysates were quantified through HPLC using an Aminex HPX 87H column with similar operating conditions mentioned in the section Compositional analysis. The hydrolysate with greater sugar yields was used for ethanol and butanol fermentation.

The percent saccharification was calculated using the following equation [19].

$$\text{Saccharification (\%)} = \frac{\text{Glucose (g/L)} \times 0.9}{\text{Substrate (g/L)}} \times 100 \quad (1)$$

Ethanol fermentation

Saccharomyces cerevisiae ATCC 96581 (Cedarlane, Burlington, Canada) was used for ethanol fermentation. The inoculum for fermentation was maintained in sterilized ATCC 1245 YPD broth medium containing yeast extract (10 g/L), peptone (20 g/L) and dextrose (20 g/L) for 24 h at 30°C, pH 5.6 with 150 rpm.

For ethanol fermentation, 100 mL of biomass hydrolysate was supplemented with yeast extract (1.5 g/L), peptone (1 g/L), ammonium sulfate (1 g/L), dipotassium phosphate (0.5 g/L), magnesium sulfate (0.5 g/L), and manganese(II) sulfate (0.5 g/L) at pH 5.6 and autoclaved at 121°C for 30 min [20]. After cooling to room temperature, about 6 mL of actively growing *S. cerevisiae* was added to the sterilized media and incubated for 60 h at 30°C and 150 rpm. Ethanol production from glucose as the control substrate was studied at levels varying from 50–150 g/L. The specified concentrations of glucose solutions were supplemented with the above-mentioned nutrients. All the chemicals were purchased from Sigma-Aldrich, Canada.

About 1.5 mL of samples was drawn every 12 h for ethanol and residual sugar determinations in an HPLC with conditions previously mentioned. Prior to all measurements, the liquid samples were filtered through 0.2 μm filter disks to remove any sediment.

Butanol fermentation

Clostridium beijerinckii B-592 (NRRL, Peoria, IL, USA) was used for butanol fermentation. The dried clostridia spores were rejuvenated in a sterilized ATCC 2107 modified reinforced clostridial broth medium (Cedarlane, Burlington, Canada) and incubated in an anaerobic chamber (New Brunswick Galaxy 170R CO_2 incubator; Eppendorf, Mississauga, Canada). About 6 mL of actively growing *C. beijerinckii* culture was transferred to 100 mL of freshly prepared pre-sterilized ATCC 2107 media in 125 mL screw-capped Pyrex bottle for inoculum development. The ATCC 2107 modified reinforced clostridial broth contained peptone (10 g/L), beef extract (10 g/L), yeast extract (3 g/L), dextrose (5 g/L), NaCl (5 g/L), starch (1 g/L), L-cysteine HCl (0.5 g/L), sodium acetate (3 g/L), and 0.025% resazurin (4 mL/L). The inoculum was allowed to grow for 16–18 h at 35°C prior to inoculation into the fermentation media.

Butanol production was studied with glucose as the control substrate at levels varying from 50–150 g/L. To 100 mL of the specified concentration of glucose, 2.5 mL of 40 g/L yeast extract was added and sterilized at 121°C for 15 min followed by cooling to room temperature. After cooling, 6 mL of *C. beijerinckii* (grown in ATCC 2107) culture was added to the medium and incubated in the anaerobic chamber for 72 h at 35°C.

Similarly, 100 mL of filter sterilized biomass hydrolysate (pH 6.5) was transferred to 250 mL pre-sterilized screw-capped Pyrex bottle. To the hydrolysate, 2.5 mL of 40 g/L sterilized yeast extract solution was added followed by inoculation with 6 mL of the actively growing *C. beijerinckii* (in ATCC 2107 broth) as described above. During all fermentations, 1.5 mL of liquid samples was taken every 12 h and filtered with 0.2 μm filter disks. The filtered samples were quantified for acetone, butanol, ethanol, acetic acid, butyric acid, and residual sugar in an HPLC (Aminex HPX 87H column) with conditions previously mentioned.

All the analyses mentioned above such as compositional analysis, acid/enzymatic hydrolysis, and ethanol and butanol fermentations were performed in triplicates with standard deviation less than 5%. The product yields (g/g) were calculated as gram quantities of ethanol or butanol produced per gram of glucose or sugar utilized. The productivity (g/L/h) was calculated as the maximum ABE concentration (g/L) divided by the particular fermentation time (h).

Results and Discussion

The lignocellulosic biomasses were pretreated using 0–2.5% H_2SO_4 followed by the addition of a mixture of cellulolytic enzymes. The dilute acid and enzymatic hydrolysis resulted in the breakdown of lignocellulosic network for the release of pentose and hexose sugars. Prior to this pretreatment, the compositional analysis of PW, TG and WS was performed to quantify the lignocellulosic components along with extractives and ash.

About 90% of dry matter in lignocellulosic biomass comprises cellulose, hemicellulose, and lignin, whereas the remaining consists of extractives and ash [21]. The exhaustive extraction process of feedstock samples using water, ethanol, and hexane led to the removal of extractives. Table 1 gives the compositional analysis of the three lignocellulosic feedstock. The yield of extractives in PW, TG, and WS was 15.7, 16.5, and 19.2 wt%, respectively. The high amount of total extractives in WS indicated higher amount of water, ethanol, and hexane soluble components such as terpenes, terpenoids, tannins, resins, fats, waxes, lipids, proteins, and organic acids. PW, TG and WS represented 1.5, 1.3, and 1.1 wt% of ash content, respectively.

On the other hand, the cellulose concentration in the three feedstock components ranged between 34.2 and 39.1 wt%,

with WS demonstrating highest levels. The hemicellulose levels were highest (30.1 wt%) in TG and lowest (23.6 wt %) in PW. Similar results on high hemicellulose contents (26–32 wt%) have been reported from reed canary grass and switchgrass [22]. This was due to the fact that fast-growing plants are abundant in hemicellulose that aids in conducting and concentrating tissue for mineralized solutions rich in sulphates, chlorides, nitrates and silicic acid in plants [23]. The holocellulosic (cellulose and hemicellulose) composition in the three feedstock samples was in the range of 62.4–64.3 wt% indicating higher structural carbohydrate (sugar) levels.

PW showed utmost levels of lignin (20.4 wt%) compared to those of herbaceous biomasses (16.3 and 18.1 wt %), which is in accordance with various authors [24, 25]. As lignin concentrates between primary and secondary cell walls, it generally tends to occur in high amounts in woods due to their tightly bound fibrous texture compared to the loosely bound fibers in herbaceous plants [26]. Lignin acts as an organic polymer to bind the cellulose and hemicellulose together forming a complex network that makes the biomass recalcitrance to acids and enzymes for releasing the fermentable sugars. For this purpose, the biomasses were subjected to dilute H_2SO_4 pretreatment and enzymatic (i.e., cellulase, β-glucosidase and xylanase) hydrolysis.

Figure 1 illustrates the sugar yields from PW, TG, and WS as a result of different H_2SO_4 concentrations ranging from 0% to 2.5% and enzymatic hydrolysis (denoted as "E" in the figure). Enzymatic hydrolysis resulted in higher sugar yields within all feedstocks than the dilute acid pretreatment. Respectively, enzymatic hydrolysis resulted in an increase of 66.5%, 65.7%, and 60% total sugar (glucose and xylose) yield in PW, TG, and WS. Among all feedstock components, PW showed maximum fermentable sugar release of 68.5 g/L at 2% H_2SO_4 and enzymatic hydrolysis. On the other hand, 1.5% H_2SO_4 and enzymatic hydrolysis resulted in 57.4 and 63.6 g/L of sugars from TG and WS, respectively. In terms of the highest percent saccharification (Fig. 2), the following sequence was observed: PW (29.6%) > TG (24%) > WS (23.4%).

As a result of dilute H_2SO_4 pretreatment for feedstock, there was a significant increase in the xylose yields compared to that of glucose. In contrast to glucose, xylose yield was amplified by 37.6%, 60.3%, and 54.7% in dilute

Table 1. Compositional analysis (in weight percent) of lignocellulosic feedstock.

Biomass	Cellulose	Hemicellulose	Lignin	Extractives	Ash
Pinewood	38.8 ± 1.4	23.6 ± 0.8	20.4 ± 1.0	15.7 ± 0.6	1.5 ± 0.2
Timothy grass	34.2 ± 1.2	30.1 ± 1.0	18.1 ± 0.7	16.5 ± 0.8	1.1 ± 0.4
Wheat straw	39.1 ± 0.8	24.1 ± 0.6	16.3 ± 1.2	19.2 ± 0.8	1.3 ± 0.1

(A)

(B)

(C)

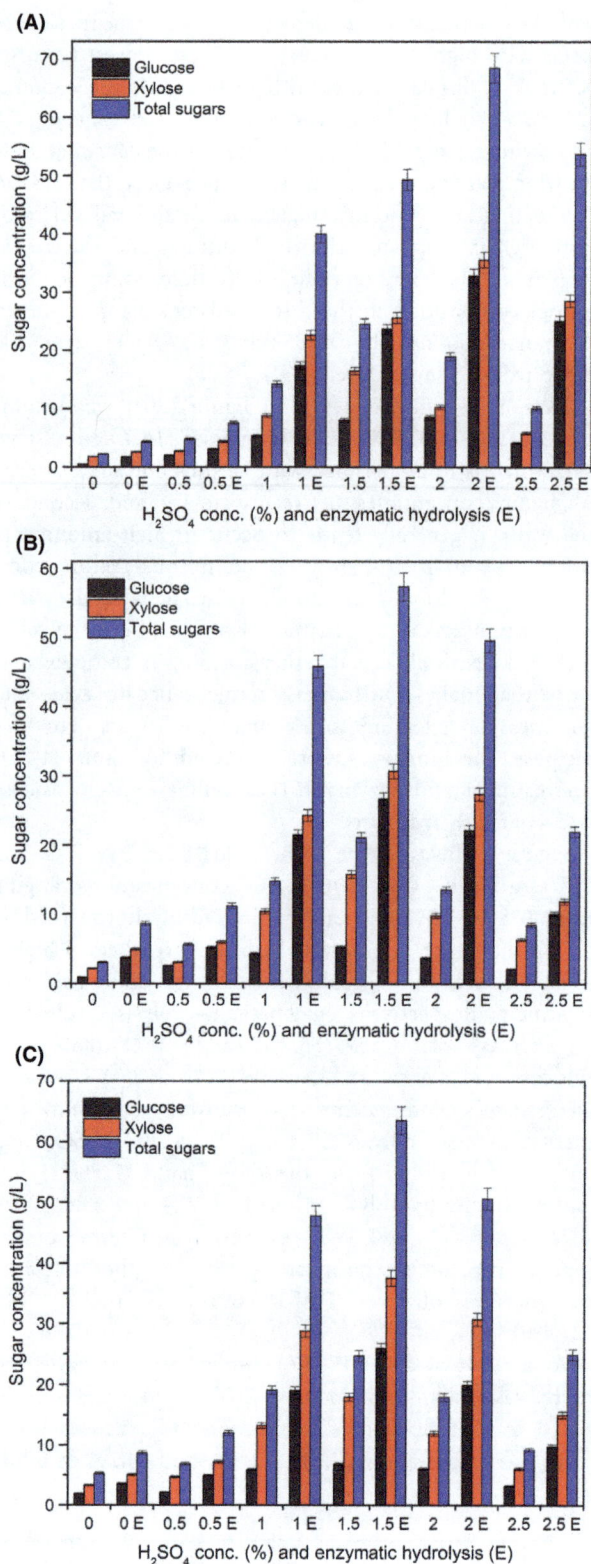

Figure 1. Sugar release from (A) pinewood, (B) timothy grass, and (C) wheat straw at 0–2.5% H_2SO_4 concentrations and enzymatic hydrolysis.

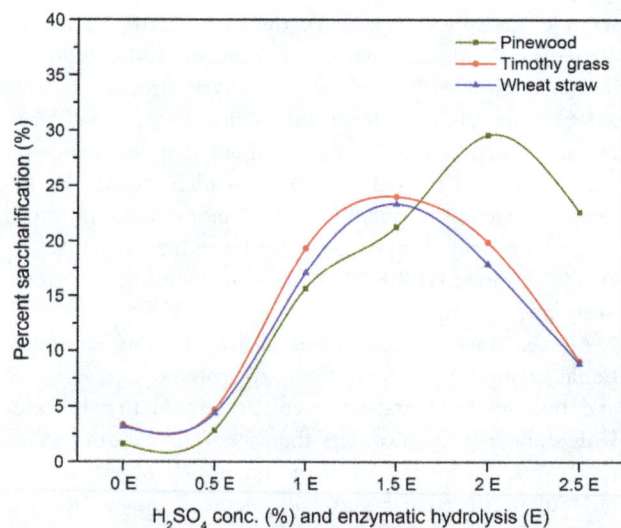

Figure 2. Percent saccharification of biomass after H_2SO_4 and enzymatic hydrolysis.

acid hydrolysates of PW, TG, and WS, respectively. This was due to the fact that hemicelluloses (primarily xylose) are easily hydrolysable in dilute acid than cellulose (glucose) because of their amorphous nature and lower degree of polymerization [5]. Compared to dilute acid pretreatment, enzymatic hydrolysis resulted in an increase in glucose concentration (g/L) by 72.3%, 78.8%, and 68.9% in PW, TG, and WS, respectively. Correspondingly, xylose levels (g/L) increased by 61.4%, 54.7%, and 54% in the enzymatic hydrolysates of PW, TG, and WS.

This escalation of glucose and xylose levels in enzymatic hydrolysis was due to the activity of cellulase, β-glucosidase, and xylanase. Cellulases (e.g., endoglucanases and cellobiohydrolases) hydrolyze the β-(1,4)-glycosidic linkages of cellulose releasing cellobiose molecules [27]. Cellobiose comprises cellulose chains with repeat units of D-glucose established through β-(1,4)-glycosidic linkages. Further, β-glucosidases break down cellobiose molecules releasing two glucose subunits. On the other hand, xylanases hydrolyze the β-(1,4) bond in xylan backbone, producing short xylo-oligomers and xylose [28]. Xylan is a polysaccharide carbohydrate made from xylose units.

However, dilute H_2SO_4 pretreatment is necessary prior to enzymatic hydrolysis as it could lead to high reaction rates and significantly improve cellulose hydrolysis [29]. The removal of majority of hemicelluloses (xylose) enhances cellulose (glucose) recovery by exposing the cellulose fibers to enzymes for catalysis. The highest total sugar yield from enzymatic hydrolysis of PW (68.5 g/L) was found at 2% H_2SO_4, whereas for TG (57.4 g/L) and WS (63.6 g/L) it was 1.5% H_2SO_4. A comparative analysis by Wyman et al. [30] suggests that high glucose (41.4 g/L) and xylose (22.3 g/L) fractions were recovered from poplar wood at 2% H_2SO_4 concentration. The digestibility of cellulose in biomass is influenced by the physico-chemical, structural, and compositional factors [6]. The woody and fibrous nature of PW along with its high amount of lignin (20.4 wt%) led to the requirement of a relatively higher H_2SO_4 concentration than that of TG and WS. Unlike PW, herbaceous biomasses such as TG and WS were found to be porous and fragile with lesser amount of lignin (16.3–18.1 wt%) necessitating lower H_2SO_4 concentration for hydrolysis.

A gradual decrease in glucose and xylose yields was found with an increase in H_2SO_4 concentrations. There was a notable decrease in the total sugar yields of 21.5%, 61.7%, and 60.8% in the enzymatic hydrolysis of PW, TG and WS, respectively at 2.5% H_2SO_4 concentration. It has been reported that high-severity in acid levels result in the undesired conversion of sugars to furfurals and hydroxymethylfurfural [6, 31]. Recent investigations reveal that xylose degrades into insoluble compounds called pseudo-lignin at high dilute acid levels which can significantly retard cellulose digestibility [32, 33]. The deposition of pseudo-lignin on cellulose fibers (in biomass) could block the surface binding sites for acids/enzymes and may result in lower cellulose hydrolysis. A similar effect is presumed to occur in this study at high H_2SO_4 levels.

The effectiveness of biomass pretreatment could be improved by avoiding undesired sugar degradation that may lead to pseudo-lignin formation [34]. Furthermore, these sugar degradation components tend to inhibit the fermentation. Overliming is a conditioning method to reduce the toxicity of pretreated hydrolysates caused by furans (e.g., furfurals and hydroxymethylfurfural) and other phenolics [35]. It is an additional step after enzymatic hydrolysis that uses $Ca(OH)_2$ to increase the alkalinity of the hydrolysate followed by heating and neutralization [18, 36].

The evaluation of ethanol yields from *S. cerevisiae* ATCC 96581 was performed on glucose medium at different substrate levels varying from 50 to 150 g/L (Fig. 3). The final ethanol concentrations of 19.3, 31.5 and 47.1 g/L were obtained from *S. cerevisiae* at 50, 100 and 150 g/L glucose levels, respectively in 60 h of fermentation. In other words, the final ethanol yields were 0.39, 0.32 and 0.31 g/g in 50, 100 and 150 g glucose media (Table 2). Table 2 shows the yield and productivity of ethanol from different glucose media and biomass hydrolysates. Among the three glucose doses, maximum ethanol concentration of 48.3 g/L (productivity: 1.34 g/L/h) was obtained from 150 g/L glucose medium in 36 h. However, the highest ethanol levels in 50 and 100 g/L glucose levels were 20.3 and 32.9 g/L. High sugar and ethanol contents in the

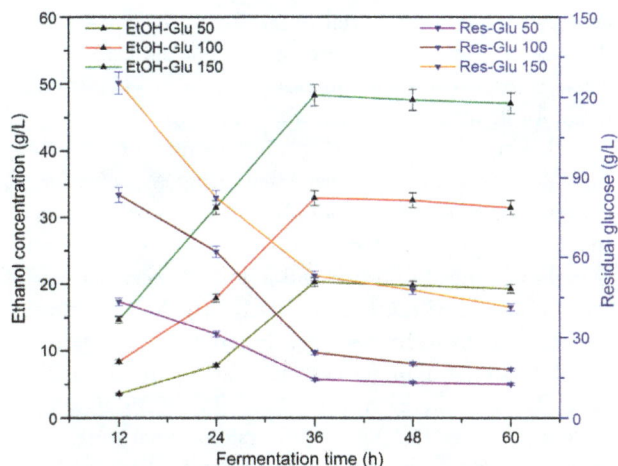

Figure 3. Ethanol production by *Saccharomyces cerevisiae* ATCC 96581 from 50, 100, and 150 g/L glucose substrates.

Table 2. Yield and productivity of ethanol from glucose media and biomass hydrolysates.

Fermentation media	Ethanol yield (g/g)					Productivity (g/L per hour)
	12 h	24 h	36 h	48 h	60 h	
Glucose (50 g/L)	0.07	0.16	0.41	0.4	0.39	0.56
Glucose (100 g/L)	0.08	0.18	0.33	0.33	0.32	0.91
Glucose (150 g/L)	0.1	0.21	0.32	0.32	0.31	1.34
Pinewood	0.07	0.15	0.35	0.32	0.31	0.67
Timothy grass	0.05	0.16	0.39	0.35	0.33	0.63
Wheat straw	0.05	0.13	0.36	0.34	0.33	0.64

The yield and productivity values presented are average of triplicate measurements with the standard error less than ±5%.

fermentation broth inhibit growth and the rate of product formation by yeast. The specific growth rate (per hour) for yeast is found to decrease from 0.61 at 50 g/L glucose to 0.51 at 140 g/L glucose [37]. The trend of glucose conversion increased from Glu-50 g/L (~74%) to Glu-100 g/L (~80%), and decreased in Glu-150 g/L (~68%). This was due to the fact that fermentation process by yeast is inhibited at glucose concentrations above 100 g/L [37, 38]. As a result of substrate inhibition, the residual glucose was relatively high (41.4 g/L) in 150 g/L glucose medium at 60 h of fermentation. In the same way, the unutilized glucose content in 50 and 100 g/L glucose media were 12.6 and 18.1 g/L, respectively.

Similarly, ethanol at a titer level of 90 g/L is completely inhibitory (no growth) to *S. cerevisiae*, although the inhibition is initiated at 24 g/L ethanol in a batch fermentation experiment. Ethanol at higher concentrations can alter the composition, structure, and function of the microbial cell membranes as well as inhibit cell division and decrease cell viability, thus reducing metabolic activity [39]. However, cell organelles such as mitochondria and vacuoles along with cellular metabolism such as sugar transport systems are found to play vital roles in ethanol tolerance mechanism for yeasts.

Following the glucose control fermentations, the biomass hydrolysates were fermented using *S. cerevisiae* to evaluate their ethanol yields. The bioconversion experiments were performed on PW, TG and WS hydrolysates with initial sugar (glucose and xylose) concentrations of 68.5, 57.4, and 63.6 g/L, respectively (Fig. 4). The ethanol levels in all hydrolysates were found to be high (22.6–24.1 g/L) at 36 h of fermentation at 30°C. The highest ethanol yields were in the order: TG (0.39 g/g) > WS (0.36 g/g) > PW (0.35 g/g). The residual total sugar levels at 60 h of fermentation were found to be 13.6, 11.5 and 14.4 g/L for PW, TG, and WS, respectively.

Two significant phases were observed in the ethanol fermentation for glucose and biomass hydrolysates. The first phase was characterized with a marked increase in ethanol levels between 12 and 36 h, whereas the second phase showed a significant decrease in the sugar levels within the same period. This could be explained through the yeast metabolism. *Saccharomyces cerevisiae* started metabolizing the sugars for ethanol production at 12–36 h of fermentation and as the level reached up to 24.1 g/L in hydrolysates and 25.5 g/L in glucose media, an inhibition mechanism restricted the yeast to further multiply and produce ethanol.

Figure 5 illustrates the trend of butanol production from glucose substrates with concentrations ranging from 50–150 g/L. Butanol fermentation by *Clostridium* spp. is characterized by the production of two major classes of products, namely solvents (acetone, ethanol and butanol) and organic acids (acetic acid and butyric acid) [14]. However, some fractions of gas components with CO_2 and H_2 are also produced. The yield of acetone, butanol and ethanol along with acetic and butyric acid was recorded up to 72 h of fermentation. Maximum butanol concentration of 11.9 g/L (yield: 0.12 g/g) was found in 100 g/L glucose media at 60 h of fermentation (Fig. 5B). Similarly, the levels of acetone (6.1 g/L), ethanol (1.9 g/L) were high in 100 g/L glucose media at 60 h. Acetic acid (4.5 g/L) and butyric acid (2.5 g/L) levels were greater in 100 g/L glucose media at 72 h of the fermentation. The total amount of ABE produced was high (17.9 g/L) in 100 g/L glucose medium compared to those of 50 and 150 g/L glucose media.

With an increase in the substrate (glucose) level, there was an increase in the residual glucose concentration. In particular, the residual glucose contents in 50, 100 and 150 g/L glucose media were 18.2, 28.3 and 53.8 g/L, respectively at 72 h (Fig. 5). This large amount of remaining glucose in the fermentation media was due to the substrate inhibition and butanol toxicity. Glucose

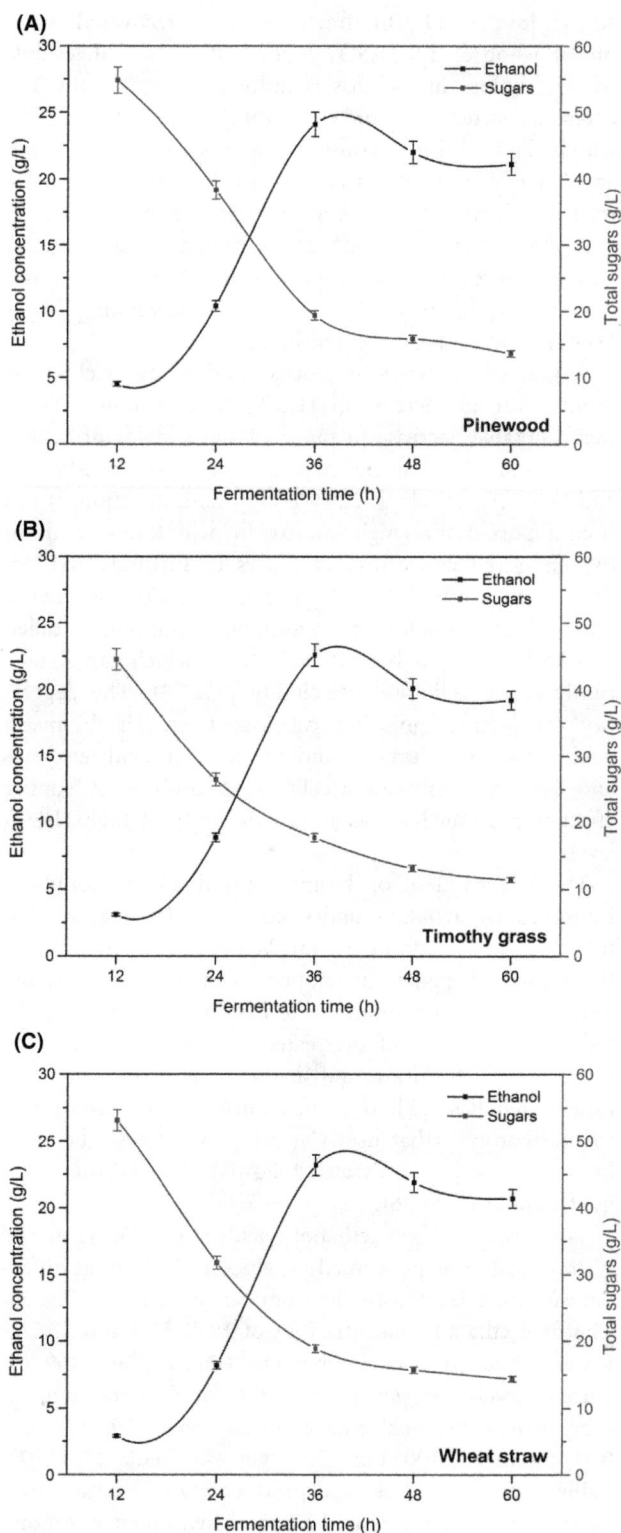

Figure 4. Ethanol production by *Saccharomyces cerevisiae* ATCC 96581 from (A) pinewood, (B) timothy grass, and (C) wheat straw hydrolysates.

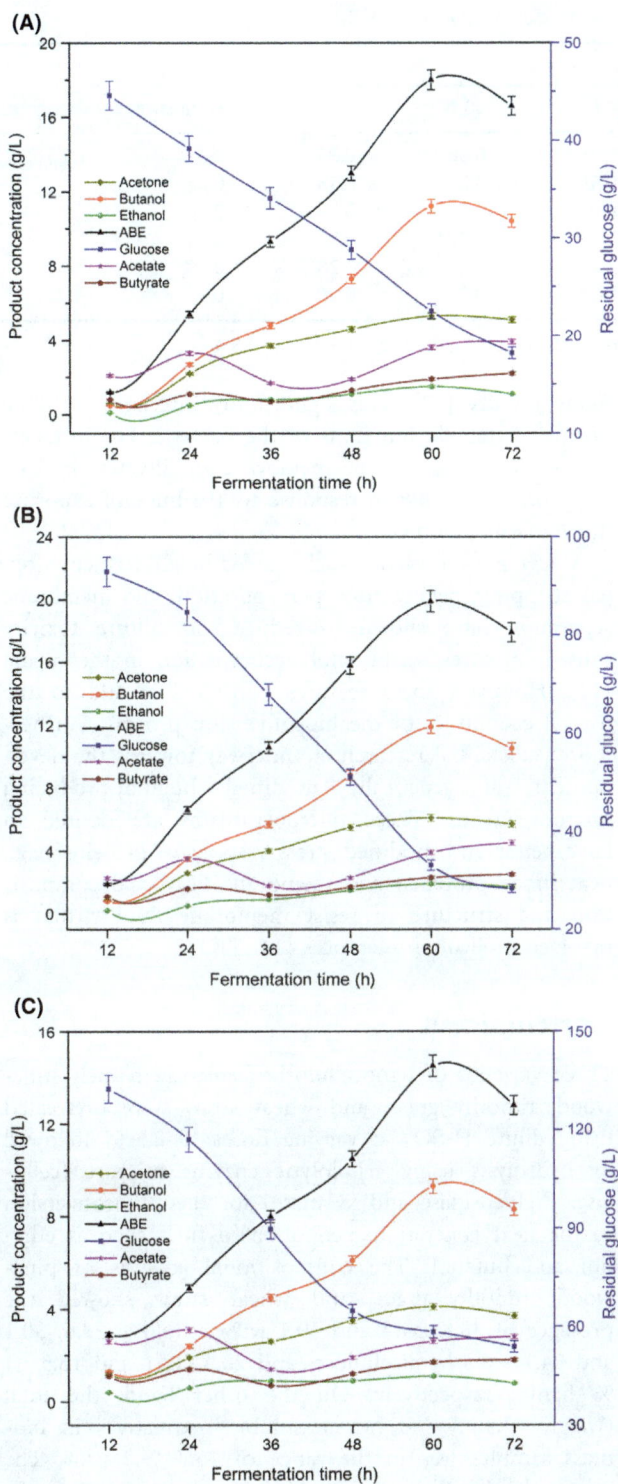

Figure 5. Butanol production by *Clostridium beijerinckii* B-592 from (A) 50, (B) 100, and (C) 150 g/L glucose substrates.

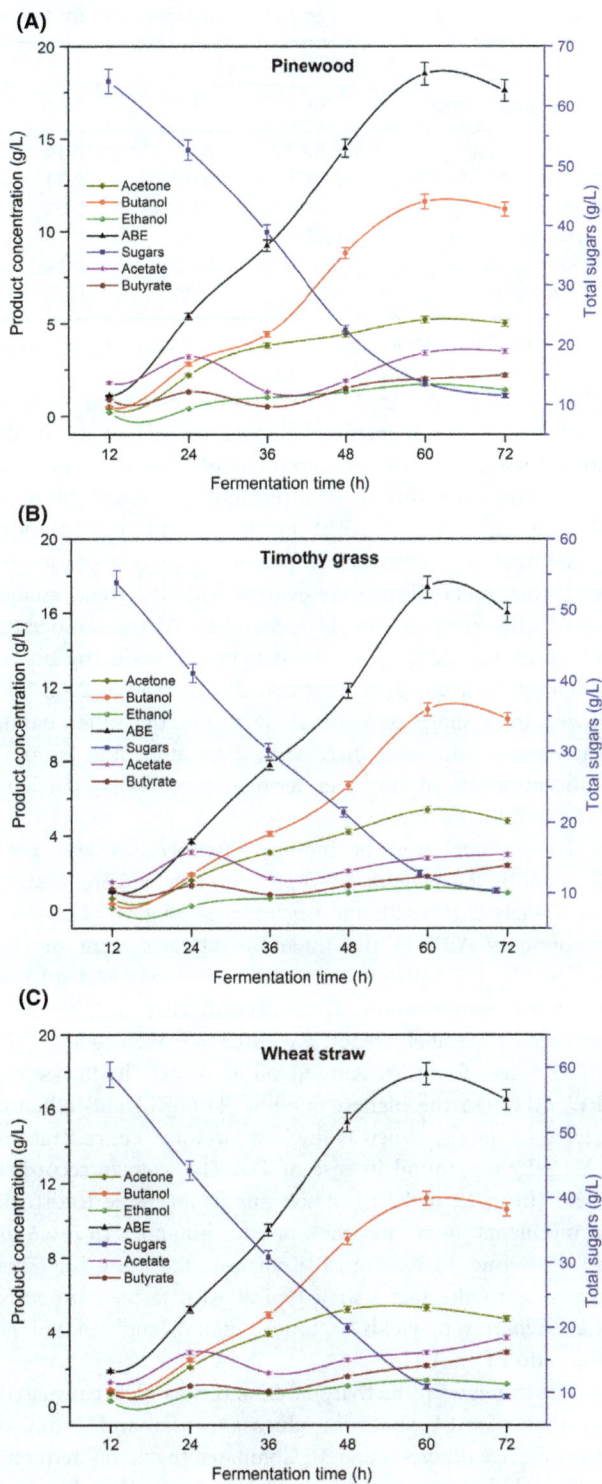

Figure 6. Butanol production by *Clostridium beijerinckii* B-592 from (A) pinewood, (B) timothy grass, and (C) wheat straw hydrolysates.

concentration of 161.7 g/L has been found to cause substrate inhibition for clostridia in butanol fermentation, resulting in a lag phase of ~40 h [40].

A typical ABE batch fermentation by clostridia occurs in two phases, especially acidogenic phase and solventogenic phase [9]. The acidogenic phase is the initial growth

Table 3. Yield and productivity of acetone-butanol-ethanol (ABE) from glucose media and biomass hydrolysates.

Fermentation media	ABE yield (g/g)						Productivity (g/L per hour)
	12 h	24 h	36 h	48 h	60 h	72 h	
Glucose (50 g/L)	0.02	0.11	0.19	0.26	0.36	0.33	0.3
Glucose (100 g/L)	0.02	0.07	0.11	0.16	0.2	0.18	0.33
Glucose (150 g/L)	0.02	0.03	0.05	0.07	0.1	0.09	0.24
Pinewood	0.02	0.08	0.13	0.21	0.27	0.26	0.31
Timothy grass	0.02	0.06	0.14	0.21	0.3	0.28	0.29
Wheat straw	0.03	0.08	0.15	0.24	0.28	0.26	0.3

The yield and productivity values presented are average of triplicate measurements with the standard error less than ±5%.

phase that results in H_2, CO_2, acetic acid and butyric acid. Due to the formation of acids, the pH of the medium lowers and the bacteria enters stationary growth stage. This shifts the bacterial metabolism toward the production of solvents (ABE) in the second fermentation phase that is, solventogenic phase. These two phases of clostridial metabolism were evident from the trend analysis of ABE fermentation (Figs. 5 and 6). As the acidogenic phase occurs early in the fermentation, producing acids, the levels of acetic and butyric acid were high at 24 h followed by a sharp decrease at 36 h. On the other hand, there was a noticeable increase in the acetone and butanol concentrations in the later fermentation hours, particularly at 60 h.

The fermentation of biomass hydrolysates was performed by *C. beijerinckii* B-592 to have a comparative yield analysis for ABE and organic acid (Fig. 6). The total amount of ABE in the three hydrolysates were in the order: PW (18.5 g/L) > WS (17.9 g/L) > TG (17.4 g/L) at 60 h of fermentation. The productivity (g/L/h) also showed a parallel trend: PW (0.31) > WS (0.3) > TG (0.29) (see Table 3). Among all feedstock hydrolysates, PW exhibited the highest butanol (11.6 g/L) and ethanol (1.7 g/L) levels, whereas highest acetone concentration (5.4 g/L) was found in case of TG. The average recorded yield (from 12 to 72 h) of solvents in the three feedstock components was in the order: butanol (5.7–6.6 g/L) > acetone (3.3–3.7 g/L) > ethanol (0.6–1.0 g/L). This was due to the fact that a typical ABE fermentation by *Clostridium* spp. yields acetone, butanol, and ethanol in the ratio of 3:6:1 [14].

The average productivity of ethanol was high compared to that of ABE, both in glucose media and biomass hydrolysates (Tables 2 and 3). Compared to ethanol fermentation, ABE fermentation resulted in lower final butanol levels, which was because of butanol toxicity. Butanol at a level of 12 g/L is inhibitory to *C. beijerinkii* [41]. Moreover, a maximum butanol production of 19.6 g/L has been reported by *C. beijerinckii* BA101 [39]. Butanol as a solvent enters in to the bacterial cytoplasmic membranes and changes the membrane structures resulting in mem-

brane fluidity [42]. This significantly interferes with the normal metabolic functions of the bacteria. For instance, a 20–30% increase in the membrane fluidity was noticed in *C. acetobutylicum* in response to 1% butanol exposure during fermentation.

A few advancements such as fed-batch fermentation, gas stripping, perstraction, pervaporation, and membrane separation have shown to reduce the culture toxicity caused by excessive butanol accumulation in the media [14]. However, these features considerably add to the overall economics of the bioconversion process. Furthermore, substantial research is underway towards the development of genetically modified butanol-producing microorganisms. These microorganisms are desired to have better and modified stress responsive proteins (e.g., heat shock proteins) and membrane fatty acid composition and structure to resist membrane fluidity that is involved in alcohol tolerances [14, 39].

Conclusions

Three varieties of lignocellulosic feedstock, namely pinewood, timothy grass, and wheat straw were pretreated using dilute H_2SO_4 at varying doses (0–2.5%) followed by hydrolysis using cellulolytic enzyme mixture (cellulase, β-glucosidase and xylanase) for their bioconversion to the next generation alcohol-based fuels such as ethanol and butanol. The compositional analysis of pinewood, timothy grass, and wheat straw showed the presence of 38.8, 34.2 and 39.1 wt% cellulose; 23.6, 30.1 and 24.1 wt% hemicellulose; and 20.4, 18.1 and 16.3 wt% lignin, respectively. On the other hand, the total (water, ethanol and hexane-soluble) extractives in biomass samples was in the range of 15.7–19.2 wt%. The utmost levels of glucose (32.9 g/L) and xylose (35.6 g/L) were recovered from pinewood with 2% H_2SO_4 and enzymatic hydrolysis. In contrast, 1.5% H_2SO_4 pretreatment with enzymatic hydrolysis resulted in 26.7 and 26 g/L glucose and 30.7 and 37.6 g/L xylose yields from timothy grass and wheat straw hydrolysates, respectively. Pinewood (29.6%) showed high levels of saccharification, followed by timothy grass (24%) and wheat straw (23.4%).

Maximum ethanol concentrations of 20.3, 32.9 and 48.3 g/L were obtained using *S. cerevisiae* ATCC 96581 from 50, 100 and 150 g/L glucose levels, respectively in 36 h of fermentation. Among the feedstock hydrolysates, pinewood demonstrated high ethanol levels (24.1 g/L) followed by wheat straw (23.2 g/L) and timothy grass (22.6 g/L). ABE fermentation using *C. beijerinckii* B-592 led to uppermost butanol concentrations of 11.2, 11.9, and 9.3 g/L from 50, 100, and 150 g/L glucose substrate in 60 h. The butanol levels from the biomass hydrolysates decreased as: pinewood (11.6 g/L) > wheat straw (11.2 g/L) > timothy grass (10.8 g/L). Furthermore, the total ABE levels from pinewood, timothy grass, and wheat straw hydrolysates were found to be 18.5, 17.4, and 17.9 g/L, respectively for 60 h of fermentation.

Acknowledgment

The authors thank Natural Sciences and Engineering Research Council of Canada (NSERC) and Canada Research Chair (CRC) program for the financial support toward this biomass conversion research.

Conflict of Interest

None declared.

References

1. Sims, R. E. H., W. Mabee, J. N. Saddler, and M. Taylor. 2010. An overview of second generation biofuel technologies. Bioresour. Technol. 101:1570–1580.

2. Lynd, L. R., C. E. Wyman, and T. U. Gerngross. 1999. Biocommodity engineering. Biotechnol. Prog. 15:777–793.

3. United States Energy Information Administration, USEIA. 2011. International energy outlook 2011. Available at: http://www.eia.gov/forecasts/ieo/pdf/0484(2011).pdf (accessed 03 January 2012).

4. Berndes, G., M. Hoogwijk, and R. van den Broek. 2003. The contribution of biomass in the future global energy system: a review of 17 studies. Biomass Bioenergy 25:1–28.

5. Hu, F., and A. Ragauskas. 2012. Pretreatment and lignocellulosic chemistry. Bioenergy Res. 5:1043–1066.

6. Kumar, P., D. M. Barrett, M. J. Delwiche, and P. Stroeve. 2009. Methods for pretreatment of lignocellulosic biomass for efficient hydrolysis and biofuel production. Ind. Eng. Chem. Res. 48:3713–3729.

7. Karimi, K., M. Shafiei, and R. Kumar. 2013. Progress in physical and chemical pretreatment of lignocellulosic biomass. Pp. 53–96 *in* V. K. Gupta and M. G. Tuohy, eds. Biofuel technologies. Springer, Berlin, Heidelberg.

8. Graham-Rowe, D. 2011. Beyond food versus fuel. Nature 474:S6–S8.

9. Jones, D. T., and D. R. Woods. 1986. Acetone-butanol fermentation revisited. Microbiol. Rev. 50:484–524.

10. Ha, S. J., J. M. Galazka, S. R. Kim, J. H. Choi, X. Yang, J. H. Seo, et al. 2011. Engineered *Saccharomyces cerevisiae* capable of simultaneous cellobiose and xylose fermentation. PNAS 108:504–509.

11. Qureshi, N., and T. C. Ezeji. 2008. Butanol, 'a superior biofuel' production from agricultural residues (renewable biomass): recent progress in technology. Biofuels Bioprod. Bioref. 2:319–330.

12. Walfridsson, M., J. Hallborn, M. Penttila, S. Keranen, and B. Hahn-Hagerdal. 1995. Xylose-metabolizing *Saccharomyces cerevisiae* strains overexpressing the TKL1 and TAL1 genes encoding the pentose phosphate pathway enzymes transketolase and transaldolase. Appl. Environ. Microbiol. 61:4184–4190.

13. Durre, P. 2007. Biobutanol: an attractive biofuel. Biotechnol. J. 2:1525–1534.

14. Zheng, Y. N., L. Z. Li, M. Xian, Y. J. Ma, J. M. Yang, X. Xu, et al. 2009. Problems with the microbial production of butanol. J. Ind. Microbiol. Biotechnol. 36:1127–1138.

15. Sluiter, A., R. Ruiz, C. Scarlata, J. Sluiter, and D. Templeton. 2008. Determination of extractives in biomass. Technical report NREL/TP-510-42619. National Renewable Energy Laboratory (NREL), Colorado.

16. Sluiter, A., B. Hames, R. Ruiz, C. Scarlata, J. Sluiter, and D. Templeton. 2008. Determination of sugars, byproducts, and degradation products in liquid fraction process samples. Technical report NREL/TP-510-42623. National Renewable Energy Laboratory (NREL), Colorado.

17. Lenihan, P., A. Orozco, E. O'Neill, M. N. M. Ahmad, D. W. Rooney, and G. M. Walker. 2010. Dilute acid hydrolysis of lignocellulosic biomass. Chem. Eng. J. 156:395–403.

18. Qureshi, N., B. C. Saha, B. Dien, R. E. Hector, and M. A. Cotta. 2010. Production of butanol (a biofuel) from agricultural residues. Part I – use of barley straw hydrolysate. Biomass Bioenergy 34:559–565.

19. Araujo, A., and J. D'Souza. 1986. Enzymatic saccharification of pretreated rice straw and biomass production. Biotechnol. Bioeng. 18:1503–1509.

20. Govumoni, S. P., S. Koti, S. Y. Kothagouni, S. Venkateshwar, and V. R. Linga. 2013. Evaluation of pretreatment methods for enzymatic saccharification of wheat straw for bioethanol production. Carbohydr. Polym. 91:646–650.

21. Balat, M. 2011. Production of bioethanol from lignocellulosic materials via the biochemical pathway: a review. Energy Convers. Manage. 52:858–875.

22. Bridgeman, T. G., L. I. Darvell, J. M. Jones, P. T. Williams, R. Fahmi, A. V. Bridgwater, et al. 2007.

Influence of particle size on the analytical and chemical properties of two energy crops. Fuel 86:60–72.

23. Vassilev, S. V., D. Baxter, L. K. Andersen, C. G. Vassileva, and T. J. Morgan. 2012. An overview of the organic and inorganic phase composition of biomass. Fuel 94:1–33.

24. Naik, S., V. V. Goud, P. K. Rout, K. Jacobson, and A. K. Dalai. 2010. Characterization of Canadian biomass for alternative renewable biofuel. Renew. Energy 35:1624–1631.

25. Shen, D. K., S. Gu, K. H. Luo, A. V. Bridgwater, and M. X. Fang. 2009. Kinetic study on thermal decomposition of woods in oxidative environment. Fuel 88:1024–1030.

26. McKendry, P. 2002. Energy production from biomass (Part 1): overview of biomass. Bioresour. Technol. 83:37–46.

27. Perez, J., J. Munoz-Dorado, T. de la Rubia, and J. Martinez. 2002. Biodegradation and biological treatments of cellulose, hemicellulose and lignin: an overview. Int. Microbiol. 5:53–63.

28. Shallom, D., and Y. Shoham. 2003. Microbial hemicellulases. Curr. Opin. Microbiol. 6:219–228.

29. Nanda, S., J. Mohammad, S. N. Reddy, J. A. Kozinski, and A. K. Dalai. 2014. Pathways of lignocellulosic biomass conversion to renewable fuels. Biomass Convers. Bioref. 4:157–191.

30. Wyman, C. E., B. E. Dale, R. T. Elander, M. Holtzapple, M. R. Ladisch, Y. Y. Lee, et al. 2009. Comparative sugar recovery and fermentation data following pretreatment of poplar wood by leading technologies. Biotechnol. Prog. 25:333–339.

31. Rosatell, A. A., S. P. Simeonov, R. F. M. Frade, and C. A. M. Afonso. 2011. 5-Hydroxymethylfurfural (HMF) as a building block platform: biological properties, synthesis and synthetic applications. Green Chem. 13:754–793.

32. Kumar, R., F. Hu, P. Sannigrahi, S. Jung, A. J. Ragauskas, and C. E. Wyman. 2013. Carbohydrate

derived-pseudo-lignin can retard cellulose biological conversion. Biotechnol. Bioeng. 110:737–753.

33. Hu, F., S. Jung, and A. Ragauskas. 2012. Pseudo-lignin formation and its impact on enzymatic hydrolysis. Bioresour. Technol. 117:7–12.

34. Hu, F., and A. Ragauskas. 2014. Suppression of pseudo-lignin formation under dilute acid pretreatment conditions. RSC Adv. 4:4317–4323.

35. Chi, Z., M. Rover, E. Jun, M. Deaton, P. Johnston, R. C. Brown, et al. 2013. Overliming detoxification of pyrolytic sugar syrup for direct fermentation of levoglucosan to ethanol. Bioresour. Technol. 150:220–227.

36. Mohagheghi, A., M. Ruth, and D. J. Schell. 2006. Conditioning hemicellulose hydrolysates for fermentation: effects of overliming pH on sugar and ethanol yields. Process Biochem. 41:1806–1811.

37. Ghose, T. K., and T. D. Tyagi. 1979. Rapid ethanol fermentation of cellulose hydrolysate. II. Product and substrate inhibition and optimization of fermentor design. Biotechnol. Bioeng. 21:1401–1420.

38. Moulin, G., H. Boze, and P. Galzy. 1980. Inhibition of alcoholic fermentation by substrate and ethanol. Biotechnol. Bioeng. 22:2375–2381.

39. Liu, S., and N. Qureshi. 2009. How microbes tolerate ethanol and butanol. New Biotechnol. 26:117–121.

40. Ezeji, T. C., N. Qureshi, and H. P. Blaschek. 2004. Acetone butanol ethanol (ABE) production from concentrated substrate: reduction in substrate inhibition by fed-batch technique and product inhibition by gas stripping. Appl. Microbiol. Biotechnol. 63:653–658.

41. Westhuizen, A. V. D., D. T. Jones, and D. R. Woods. 1982. Autolytic activity and butanol tolerance of *Clostridium acetobutylicum*. Appl. Environ. Microbiol. 44:1277–1281.

42. Durre, P. 2008. Fermentative butanol production bulk chemical and biofuel. Ann. N. Y. Acad. Sci. 1125:353–362.

7

Oxyfuel combustion in rotary kiln lime production

Matias Eriksson[1,2,3], Bodil Hökfors[1,4] & Rainer Backman[1]

[1]Umeå University, SE-901 87 Umeå, Sweden
[2]NorFraKalk AS, Kometveien 1, 7650 Verdal, Norway
[3]Nordkalk Oy Ab, Skräbbölevägen 18, 21600 Pargas, Finland
[4]Cementa AB, Årstaängsvägen 25, SE-11743 Stockholm, Sweden

Keywords

Carbon capture and storage, energy efficiency, lime, oxyfuel

Abstract

The purpose of this article is to study the impact of oxyfuel combustion applied to a rotary kiln producing lime. Aspects of interest are product quality, energy efficiency, stack gas composition, carbon dioxide emissions, and possible benefits related to carbon dioxide capture. The method used is based on multicomponent chemical equilibrium calculations to predict process conditions. A generic model of a rotary kiln for lime production was validated against operational data and literature. This predicting simulation tool is used to calculate chemical compositions for different recirculation cases. The results show that an oxyfuel process could produce a high-quality lime product. The new process would operate at a lower specific energy consumption thus having also a reduced specific carbon dioxide emission per ton of product ratio. Through some processing, the stack gas from the new process could be suitable for carbon dioxide transport and storage or utilization. The main conclusion of this paper is that lime production with an oxyfuel process is feasible but still needs further study.

Introduction

It is the purpose of this article to study the impact of oxyfuel combustion applied to a rotary kiln- producing lime. Aspects of interest are product quality, energy efficiency, stack gas composition, carbon dioxide emissions, and possible benefits related to carbon dioxide capture.

Oxyfuel combustion is combustion of a fuel with pure oxygen or a mixture of oxygen, water, and carbon dioxide [1] in contrast to conventional combustion which is done with air. This oxyfuel technology study involves combustion with pure oxygen and flue gas recirculation.

Lime (calcium oxide, CaO) is produced by calcination of limestone, containing a high concentration of calcium carbonate ($CaCO_3$). Limestone is an abundant natural raw material. Lime is used for environmental purposes, e.g., waste neutralization or flue gas desulphurization, and in industrial processes, e.g., for formation of metallurgical slags or for production of paper pigments.

The method used is based on multicomponent chemical equilibrium calculations to predict process conditions. A predicting simulation tool is used to model processes and calculate chemical compositions for different cases. A generic model of a rotary kiln for lime production was validated against operational data and literature and used to examine different oxyfuel recirculation cases.

Oxyfuel technology applied to lime production would cause several major changes to the lime plant. The implications of all these changes, e.g., air separation unit, new piping, and fans for gas recirculation, are not discussed in this work.

Conventional lime production in rotary kilns

Lime is produced by calcination of calcium carbonates in industrial kilns. The mineral calcite containing the calcium carbonates is the main component in naturally abundant limestone. The limestone is quarried or mined, mechanically pretreated and delivered to the lime plant. One of the most common kiln types is the rotary kiln.

Calcination is an endothermic reaction requiring heat to evolve gaseous carbon dioxide from the calcite to form lime [2].

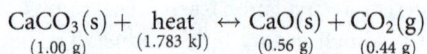

$$\underset{(1.00\,g)}{CaCO_3(s)} + \underset{(1.783\,kJ)}{heat} \leftrightarrow \underset{(0.56\,g)}{CaO(s)} + \underset{(0.44\,g)}{CO_2(g)}$$

The calcination starts between 800°C and 900°C and the operational solid temperature usually reaches 1000–1200°C. The calcination temperature is dependent on the partial pressure of carbon dioxide in the kiln. A simplified illustration of a rotary kiln for lime production is presented in Figure 1. The upper setup shows the conventional air-fired process used as a reference case in this work. The lower setup shows the oxyfuel process with added air separation unit and flue gas recirculation are used for simulation cases in this work.

The limestone is fed in the upper end. The feed is usually a 10–40 mm fraction of the limestone produced in the quarry. A small inclination of the kiln enables the material to move downward during rotation. The rotation is slow

and the residence time is usually in the range of 5–10 h. It is a countercurrent process. When the limestone enters the kiln, it is dried and heated by the flue gases flowing in the opposite direction. At the lower end, the fuel is fed into the kiln through a burner. When exiting the kiln through the lower end, the limestone has released its content of carbon dioxide and the majority of the material is calcined into calcium oxide. Residual content of carbon dioxide in the product is usually in the range of 0.1–2.0 mass percent, depending on customer application. The mass balance shows that at full calcination, up to 44% of the feed is released as carbon dioxide gas during the process. After the kiln the lime product enters a cooler. The cooling is done with air to suitable product temperature. The cooling procedure is fast so that the product remains as lime and carbonation, which is uptake of carbon dioxide from the atmosphere, is minimal.

Oxyfuel combustion in rotary kiln lime production

Oxyfuel combustion aims to concentrate carbon dioxide in the flue gases enough for utilization or storage. Replacing air (79% nitrogen and 21% oxygen) with pure oxygen can decrease the volume of flue gases from the process. This also increases the carbon dioxide concentration since no nitrogen is added to the system. Combustion in elevated oxygen concentrations increases the flame temperature. This increase in flame temperature changes the heat load of the kiln combustion area. Thus, flame temperature needs to be controlled. This is achieved by recirculating the flue gas back to the lower end of the kiln. This work studies the effect of recirculation levels 50%, 60%, 70%, and 80%. The recirculated flue gas has a high carbon dioxide concentration. The increased carbon dioxide concentration increases the calcination temperature of the raw material. This will change the heat balances of the

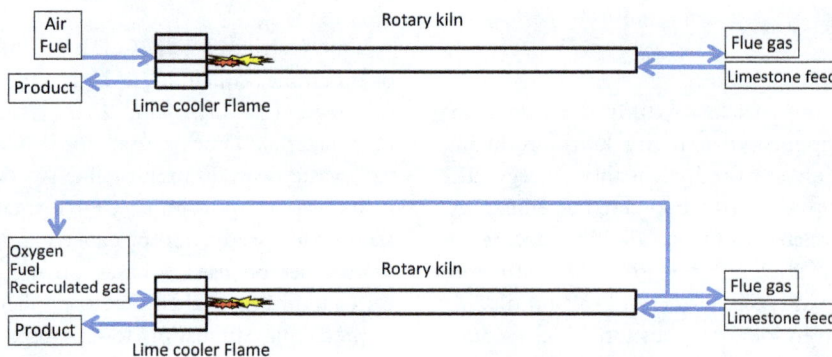

Figure 1. Simplified illustration of conventional and oxyfuel rotary kiln technology for lime production.

kiln. The basic recirculation scenario applied to a lime kiln is illustrated in Figure 1.

Carbon dioxide emissions from lime production

Within the European Union Emission Trade Schemes (EU ETS) 2008–2012 and 2013–2020, the lime process has two types of carbon dioxide emissions. *Combustion emissions* relate to combustion of carbon-based fuels. The carbon dioxide released from the raw material during production is referred to as *process emissions*. The process emissions are the main source of carbon dioxide emissions. They constitute 60–75% of the total emissions. Their portion varies with kiln type and quality requirements of fuels, raw materials, and products. The combustion emissions are defined as fossil and nonfossil. The process emissions are defined as fossil. All fossil carbon dioxide emissions are monitored and reported and the corresponding amount of emission allowances are surrendered annually. The nonfossil carbon dioxide emissions are monitored and reported but no emission allowances need to be surrendered. The product benchmark in the EU ETS 2013–2020 is 0.954 ton carbon dioxide per ton product. EU operational data show that lime plants with low specific carbon dioxide emissions usually are low-carbon fuel firing shaft kilns, around 1.0 ton carbon dioxide per ton product. At around 1.4 ton carbon dioxide per ton of product, the rotary kiln technology is not as emission efficient as the shaft kiln technology. The advantage of the shaft kiln lies in the high-energy efficiency of the technology.

The energy consumption of the kiln as fuel for calcination, studied in this work, varies mainly with kiln technology, process setup, raw material properties, and product properties. The variations are large. Shaft kiln operations are reported at 3.2–4.9 GJ per ton product and rotary kilns at 5.1–9.2 GJ per ton product [3].

The benefits of the rotary kiln are, e.g., higher utilization of quarried raw material, different product quality, and higher fuel flexibility.

The combustion emissions from lime production can be reduced by [1, 4]:

- Increased energy efficiency
- Fuel switch to low-carbon fuels
- Fuel switch to nonfossil fuels
- Precombustion capture of carbon dioxide with storage or utilization
- Postcombustion capture of carbon dioxide with storage or utilization
- Capture of carbon dioxide through oxyfuel combustion with storage or utilization

The process emissions from lime production can be reduced by:

- Postcombustion capture of carbon dioxide with storage or utilization
- Capture of carbon dioxide through oxyfuel combustion with storage or utilization

The only two technologies that can allow a near-zero carbon dioxide emission lime plant are postcombustion capture and capture through oxyfuel combustion. Both technologies require storage or utilization of carbon dioxide to achieve the emissions' reduction potential. Postcombustion capture of carbon dioxide with application on lime production was reviewed by Eriksson and Backman in 2009 [4] and this paper studies oxyfuel combustion in lime production.

Literature review

There are no known lime kilns operating in an oxyfuel mode. No publications on oxyfuel lime kilns have been found.

There are publications on oxygen enrichment in lime production [5, 6]. Oxygen enrichment usually aim at e.g., higher production capacity, increased fuel efficiency, or increased fuel flexibility, but as such the technology still operates with high nitrogen levels in the process and has no relevance as carbon dioxide capture. Further work is needed to optimize the oxygen feed to the lime kiln in an oxyfuel process but this is outside the scope of this study. The pulp lime recovery kiln operates at similar though different conditions. Some oxygen enrichment publications on lime recovery kilns in the pulp industry are available [7, 8].

There are publications on oxyfuel in other industrial processes, e.g., cement production [9–12], iron and steel production [13, 14], and power production [15, 16]. The technology has been demonstrated, e.g., for power production, but there are no full-scale plants in operation. Of these three processes, cement production is the closest to lime production but still significantly different. The results published on oxyfuel in cement production cannot be transferred as such to lime production.

Method

Through chemical thermodynamic modeling utilizing Gibbs free energy minimization, one can predict the chemical composition of a mixture of species at a given temperature and pressure. A predictive simulation tool has been developed by the Energy Technology and Thermal Process Chemistry Research Group at Umeå University in cooperation with Nordkalk Oy and Cementa AB.

Predictive simulation tool

The predictive simulation tool utilized in this paper is a combination of commercially available software: Aspen-Plus™ V7.1 [17] and ChemApp™ [18]. In this work, Aspen Plus™ is used for drawing the process scheme and iteratively solving the mass and energy balances. ChemApp™ is used to calculate the equilibrium chemistry. In addition, FactSage V6.1 [19, 20] is used as a source of thermochemical data for the system. The reasons for choosing ChemApp™ over the Aspen Plus™ are several. ChemApp™ makes it possible to utilize user-defined component data and the thermodynamic data from FACT. ChemApp™ also makes it possible to calculate solution phases not available in AspenPlus™. The predictive simulation tool has been validated for an operational lime plant and an operational cement plant [21–24]. The system components are aluminum, carbon, calcium, chloride, iron, hydrogen, potassium, magnesium, nitrogen, sodium, oxygen, phosphorus, sulfur, silica, titanium, and zinc. There is one gas phase and two pure liquids, 173 pure solid phases, seven solid solution phases, and two liquid solution phases. The system is further described by Hökfors et al. [22].

The model produces a vast amount of data. Not all components are critical when interpreting the results. Therefore, only data for selected compounds are included in the discussion. The selection criterion varies and is stated in the discussion. Some presented data are statistically normalized so that the selected data comprise 100 percentages.

This selection method is used to support the purpose of this article. A more comprehensive study on each specific retrofit or green field project is of course needed.

Generic lime kiln model

The generic lime kiln model is simplified from a more extensive model of an operational lime plant of Nordkalk Oy. The full plant model has been validated against the operational data. The validated full plant model was reduced to a more generic rotary kiln lime process setup and equipped with a flue gas circulation option. The generic model reference case was validated against operational data. This generic model is designed to be easier and faster to operate than the full plant model. The result should also be easier to export and apply to other lime plants. The generic model utilizes the same compounds present in the database as the full model. The full model is further described by Hökfors et al. [22].

In this work, eight different recirculation cases are compared with the validated reference case.

All of the chemical equilibria are calculated by defining the pressures and the enthalpy changes in the system. The model allows for setting either the equilibrium temperatures or the enthalpy changes. When setting enthalpy changes, one can observe changes in temperature related to different fuels or process setups. This means that the model is controlled by energy losses. The simulated temperatures are calculated at a constant energy-loss profile. The energy-loss profile is calculated from kiln surface temperature measurements as part of the model validation. The total energy loss is set at 5.5 MJ/s. In the model the rotary kiln is described by three ChemApp™ equilibrium blocks. The Aspen Plus™ schematic of the generic model with recirculation can be seen in Figure 2. The main streams are indicated as streams 1–4 in Figure 2. Stream 1 is the input data of fuel and stream 2 is the input data for limestone. The chemical composition of the input streams is presented in Table 1. The simulation results, the output data, are streams 3 and 4. Stream 3 is the condensed phase, the lime product, and stream 4 the noncondensed phase, the stack gas. The recirculation gas has the identical equilibrium composition and properties as the stack gas. The composition of the model output streams, the simulation results, are presented and discussed below.

Results and Discussion

To study the impact of process changes nine cases were simulated. The simulations are referred to as cases A–I. Reference case A is the conventional lime process. The cases B–I are oxyfuel processes varied by energy input and flue gas circulation. The simulation matrix is seen in Table 2. The results show that we have achieved to simulate oxyfuel combustion. In Figure 3, cases B to I show that composition of the gas recirculated for combustion with pure oxygen is mainly oxygen, carbon dioxide, and water, which is the expected stack gas composition from oxyfuel combustion. The recirculation gas is identical to stream 4 in Figure 2, the stack gas. Case A is the conventional air combustion process with high nitrogen.

The reference case modeled in this study has an energy consumption of ~7 GJ/t lime produced. The energy consumption for oxygen production is reported at 400 kJ/mole of oxygen, as O_2 [9]. This corresponds to <0.5% of the energy as fuel consumed for calcination. The energy consumption of oxygen production is not significant and is, therefore, excluded from this study. It can be noted that oxygen production consumes energy as electricity which is more expensive than the low-grade fuels used for calcination, so of course, a more detailed study will be required to establish impact of oxyfuel on the cost of production.

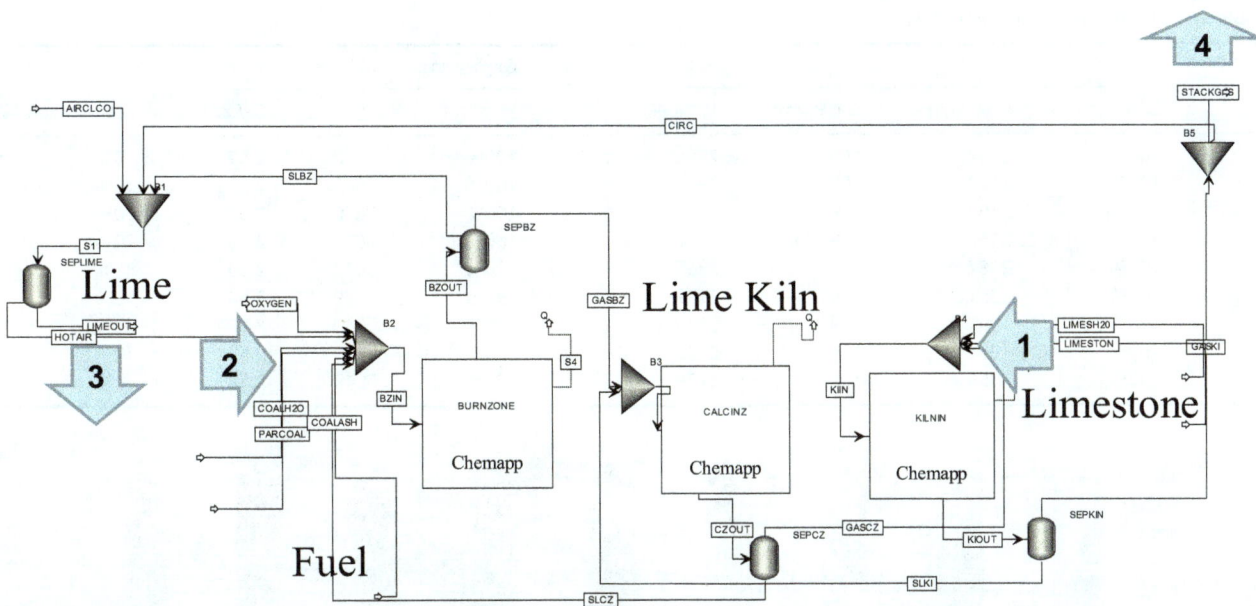

Figure 2. ASPEN schematic of generic lime model.

Table 1. Chemical composition of model input streams.

Fuel		Ash		Limestone	
C	75.84	CaO	6.90	$CaCO_3$	97.41
H	5.00	SiO_2	39.00	SiO_2	1.25
O	8.80	Al_2O_3	43.90	Al_2O_3	0.58
N	1.20	Fe_2O_3	7.90	Fe_2O_3	0.14
S	0.76	K_2O	2.30	$FeSO_4$	0.33
Cl	0.00			K_2O	0.24
Ash	8.40			$CaCl_2$	0.05
	100.00		100.00		100.00

Influence on product quality, energy efficiency, and implications to the production process

The main component of the product is calcium oxide. The main quality value is the amount calcium oxide readily available for reaction in industrial processes. This is here described as free calcium oxide. In the product calcium oxide is also found as e.g., carbonates, oxide melts, and salts. Stream 3 in Figure 2 is the output data for the condensed phase, the lime product. The selected product compositions with mole fractions equal to or >0.01 can be seen in Figure 4.

The calculated free lime is here used as product quality indicator. The results show that the amount of free calcium oxide in the product ranges from 0.894 to 0.952 mole fraction, meaning that there is a loss in product quality in some of the cases. This is a wide range taking into consideration the identical raw material and fuel.

Being the single most important compound in the product, calcium was investigated in more detail. The distribution of calcium species in Figure 5 shows why several cases suffer high quality losses. The data are selected at equal to or >0.01 mole fraction. In cases B and C, high temperatures increase the oxide melt phase increasing the amount of calcium oxide dissolved. Calcium also forms solid solutions with e.g., silica as di-calcium silicates and tri-calcium silicates. These types of solutions lower the amount of free calcium oxide reducing product quality. In cases H and I, changes in the gas phase, presented later in Figure 9, and the low temperature results in high amounts of calcium as carbonate.

None of the cases increased product quality. The cases D, E, F, and G show equal free calcium oxide levels as in the reference case A. These are high-quality products.

Energy efficiency is an important operational result. To investigate whether any significant energy efficiency improvements can be achieved, cases F–I were run with only 90% energy input.

Of the five high-quality products two; F and G, used 10% less energy. Although requiring further study this is a significant result. A 10% reduction in energy consumption while maintaining product quality reduces carbon dioxide emissions and production costs.

The results show that the heat balances of the kiln have moved in a beneficial direction. Recirculating hot flue gases (from preheating zone to burn zone) will conserve energy compared to the air-fired conventional reference case A. In case F, the temperatures rise over the length of

Table 2. Simulation matrix for cases A–I.

Case	Fuel in				Oxygen in			Recirculation %
	Energy %	Coal kg/h	Ash kg/h	Tot fuel in kg/h	% of ref	kg/h	O_2/coal	
A	100%	2167	199	2366	100%	5610	2.59	0%
B	100%	2167	199	2366	100%	5610	2.59	50%
C	100%	2167	199	2366	100%	5610	2.59	60%
D	100%	2167	199	2366	100%	5610	2.59	70%
E	100%	2167	199	2366	100%	5610	2.59	80%
F	90%	1951	179	2129	90%	5049	2.59	50%
G	90%	1951	179	2129	90%	5049	2.59	60%
H	90%	1951	179	2129	90%	5049	2.59	70%
I	90%	1951	179	2129	90%	5049	2.59	80%

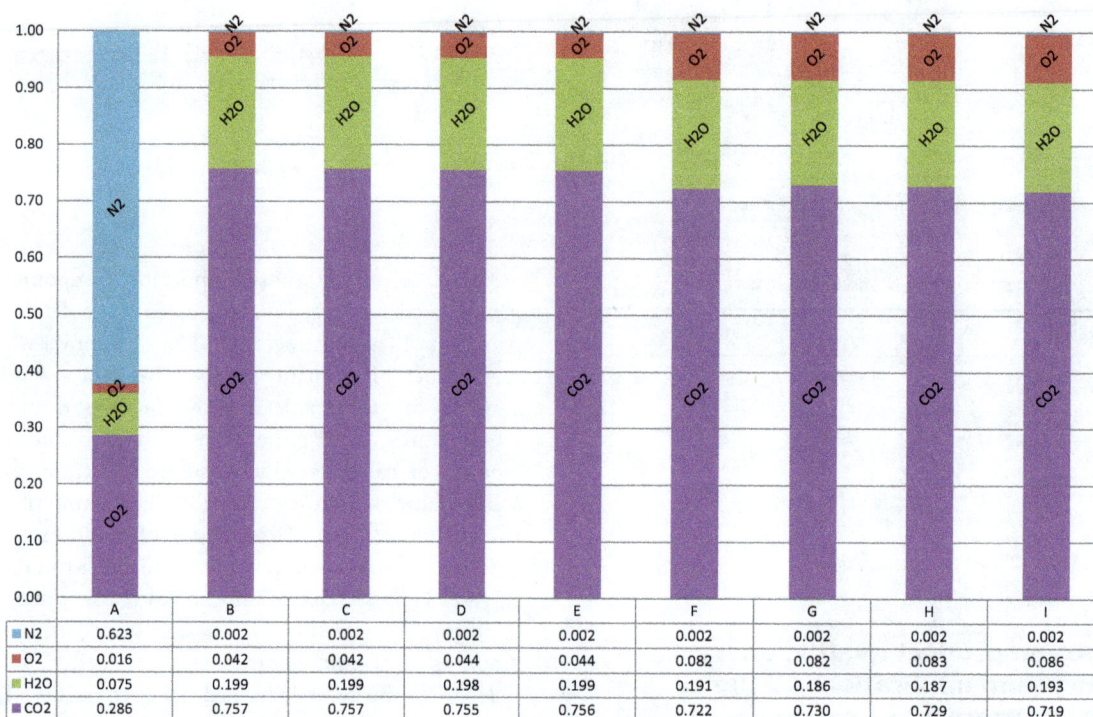

	A	B	C	D	E	F	G	H	I
N2	0.623	0.002	0.002	0.002	0.002	0.002	0.002	0.002	0.002
O2	0.016	0.042	0.042	0.044	0.044	0.082	0.082	0.083	0.086
H2O	0.075	0.199	0.199	0.198	0.199	0.191	0.186	0.187	0.193
CO2	0.286	0.757	0.757	0.755	0.756	0.722	0.730	0.729	0.719

Figure 3. Gas compositions in mole fraction for the recirculation gases in cases B–I and case A kiln exit gas.

the kiln although energy input is lowered by 10%. The temperature profiles can be seen in Table 3.

The high temperature of the process causes melting of the product. The total weight percent melt is seen in Figure 6. The conventional lime kiln is not equipped to handle such high temperatures and large amount of melt. Case B and C do not achieve full product quality because of the high amount of calcium oxide in the melt.

Influence on stack gas composition and implications to the production process

The results show significant changes in stack gas composition. Stream 4 in Figure 2 is the output data for the non-condensed phase. Figure 7 show selected concentrations equal to or higher than 0.00001 mole fractions of all compounds in the stack gas. The selection criterion is set so that hydrochloride is present in the gas. The hydrochloride concentration in the flue gases is environmentally and technically nonbeneficial. Some components are present at concentrations below the selection criteria and are, therefore, excluded from this study. The results show 28.6% carbon dioxide in the reference case. With circulation the carbon dioxide concentration increases to a maximum of 75.7% at full fuel load and 73.0% at reduced fuel load. Nitrogen levels decrease from 62.3% to 0.2%. The water content of the stack gas is increased from 7.5% to 19.9% when the circulation is introduced. The changes

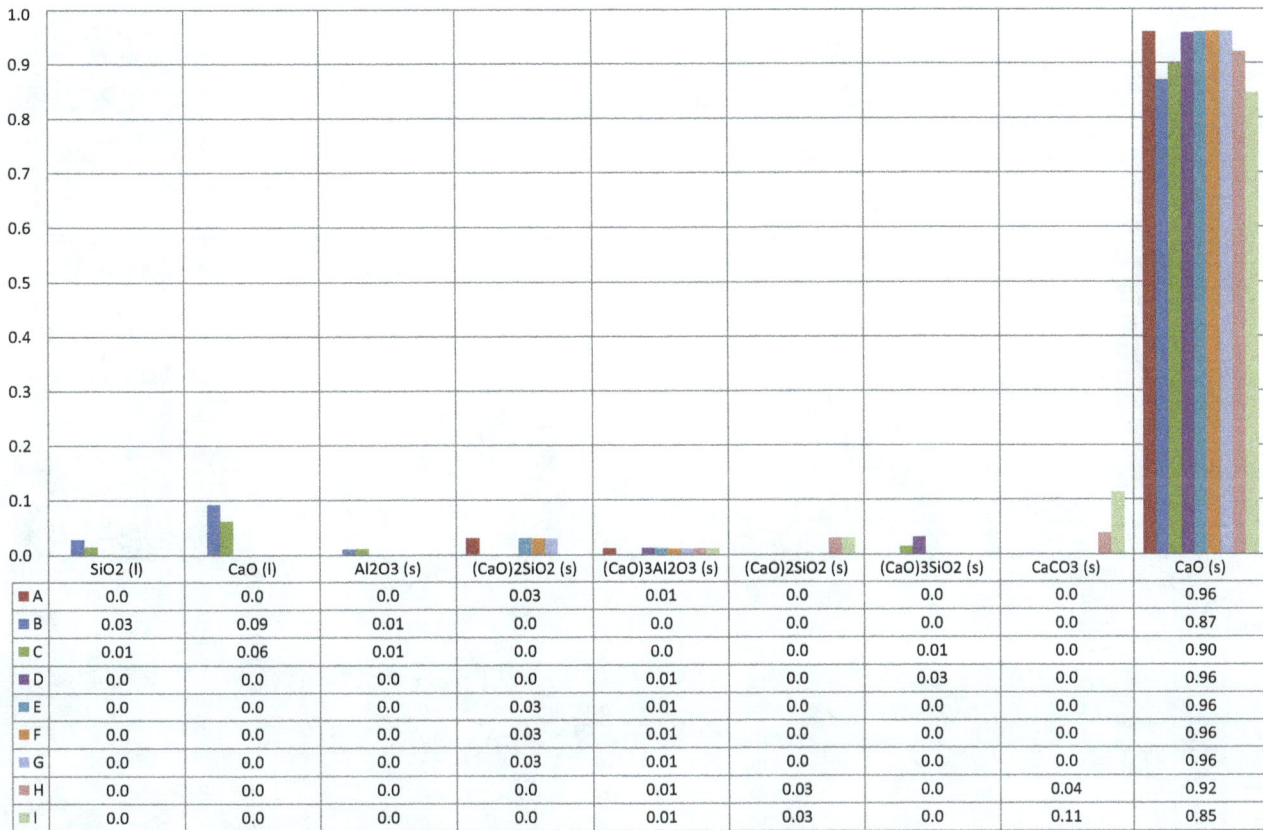

	SiO2 (l)	CaO (l)	Al2O3 (s)	(CaO)2SiO2 (s)	(CaO)3Al2O3 (s)	(CaO)2SiO2 (s)	(CaO)3SiO2 (s)	CaCO3 (s)	CaO (s)
A	0.0	0.0	0.0	0.03	0.01	0.0	0.0	0.0	0.96
B	0.03	0.09	0.01	0.0	0.0	0.0	0.0	0.0	0.87
C	0.01	0.06	0.01	0.0	0.0	0.0	0.01	0.0	0.90
D	0.0	0.0	0.0	0.0	0.01	0.0	0.03	0.0	0.96
E	0.0	0.0	0.0	0.03	0.01	0.0	0.0	0.0	0.96
F	0.0	0.0	0.0	0.03	0.01	0.0	0.0	0.0	0.96
G	0.0	0.0	0.0	0.03	0.01	0.0	0.0	0.0	0.96
H	0.0	0.0	0.0	0.0	0.01	0.03	0.0	0.04	0.92
I	0.0	0.0	0.0	0.0	0.01	0.03	0.0	0.11	0.85

Figure 4. Mole fractions for all components in product.

in level of circulation and fuel load show only small variations on the water content of the stack gas. The oxygen levels increase with recirculation.

Chloride enters the system with the limestone raw material feed. As seen in Figure 8, the circulation levels and fuel loads influence the levels and compounds of chloride present in the kiln exit gas. All in all the chloride levels remain low, peaking at 0.054% for circulation case E. In the full fuel recycle scenarios, B–F, the chloride containing compounds increase with the recirculation level. Even if the overall chloride levels are very low the results show a distinct accumulation of chlorides that might be significant for operations. Increasing chloride concentrations with increasing recirculation rates can be described as an accumulative phenomenon. The accumulative effect of chloride is not as significant at reduced fuel level. The chloride is present as hydrochloride and potassium chloride. The distribution is seen in Figure 9. Hydrogen and potassium varies widely between the cases. This might also be significant for operations. The variations in chloride component distribution can be described as a function of temperature. At high temperatures, potassium evaporates increasing potassium concentration in the gas phase.

Influence on carbon dioxide emissions and implications on Carbon Capture and Storage

The use of oxyfuel combustion enables high carbon dioxide concentrations in the lime kiln flue gas. Very high carbon dioxide content opens up opportunities for carbon utilization and storage processes. Recirculation of hot flue gases offers a potential to reduce the operational fuel energy per ton of product ratio. If an oxyfuel lime plant is operated with carbon neutral fuels, e.g., biofuels, and equipped with carbon dioxide storage or utilization facilities, it also allows for "below zero emission" operations storing separated biogenic carbon away from the atmosphere. With sustainable nonfossil fuel production, this could enable a reduction of carbon dioxide in the atmosphere.

Even though carbon dioxide capture and storage are not employed in this model, the results show decreased carbon dioxide emissions in the oxyfuel cases in comparison to the reference case. The ton carbon dioxide per ton product ratio for the cases is shown in Figure 10. The reference case A has a ratio of 1.34 ton carbon dioxide per ton product. The lowest emitting case with acceptable product quality is F at 1.23 ton carbon dioxide per ton product. Cases H and

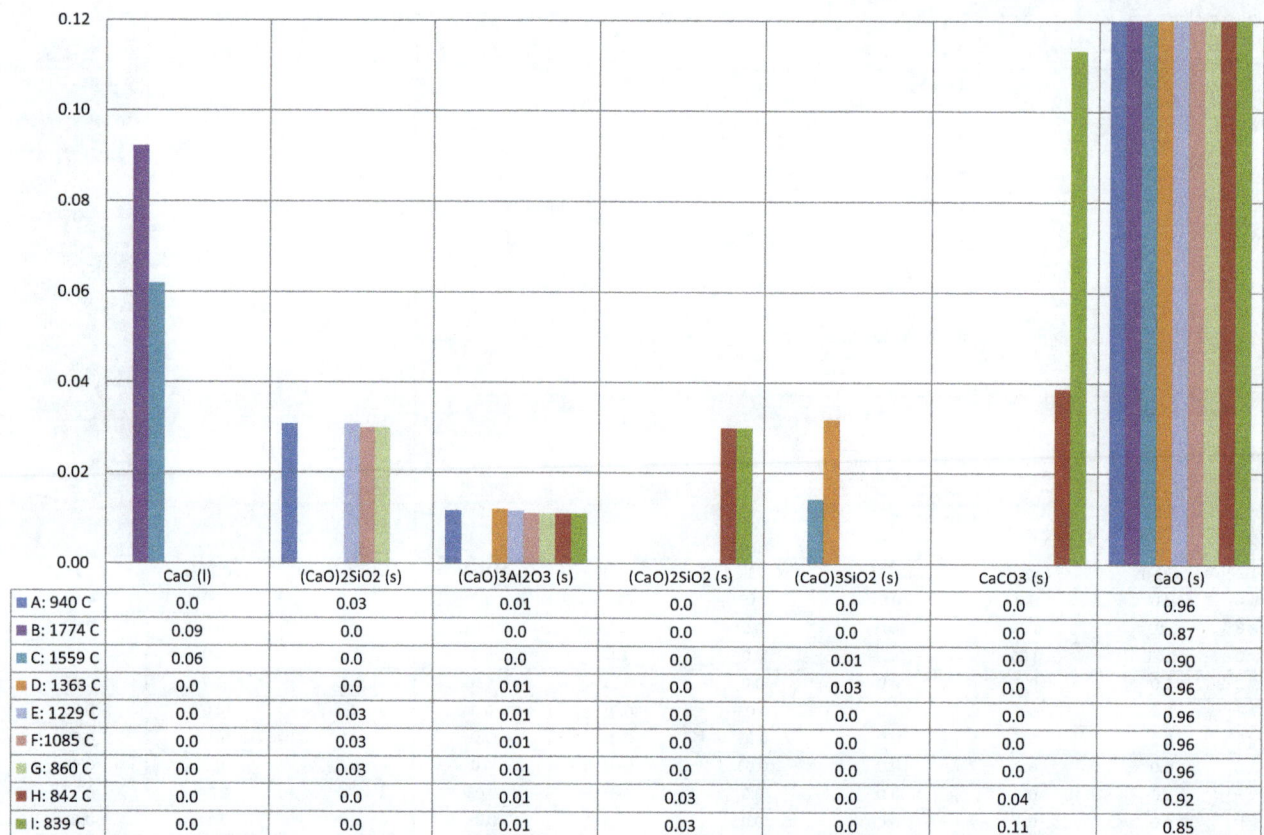

Case	CaO (l)	(CaO)2SiO2 (s)	(CaO)3Al2O3 (s)	(CaO)2SiO2 (s)	(CaO)3SiO2 (s)	CaCO3 (s)	CaO (s)
A: 940 C	0.0	0.03	0.01	0.0	0.0	0.0	0.96
B: 1774 C	0.09	0.0	0.0	0.0	0.0	0.0	0.87
C: 1559 C	0.06	0.0	0.0	0.0	0.01	0.0	0.90
D: 1363 C	0.0	0.0	0.01	0.0	0.03	0.0	0.96
E: 1229 C	0.0	0.03	0.01	0.0	0.0	0.0	0.96
F: 1085 C	0.0	0.03	0.01	0.0	0.0	0.0	0.96
G: 860 C	0.0	0.03	0.01	0.0	0.0	0.0	0.96
H: 842 C	0.0	0.0	0.01	0.03	0.0	0.04	0.92
I: 839 C	0.0	0.0	0.01	0.03	0.0	0.11	0.85

Figure 5. Distribution of calcium species in the product in mole fraction and maximum equilibrium temperature in °C.

Table 3. Temperature profile of kiln for different cases.

Case	Energy input	Recirculation	Burnzone [C]	Calcination zone [C]	Preheating zone [C]
A	100%	0%	940	784	532
B	100%	50%	1774	979	660
C	100%	60%	1559	987	709
D	100%	70%	1363	967	742
E	100%	80%	1229	936	783
F	90%	50%	1085	842	556
G	90%	60%	860	753	538
H	90%	70%	842	760	572
I	90%	80%	839	796	647

I have low-quality products containing high amounts of calcium carbonate. The product benchmark for free allocation of emission allowances within the European Union Emissions trade scheme, ETS PMB, as decided by the European Commission [25] is seen in the last column. The benchmark is set at 0.954 on carbon dioxide per ton lime product. The Commission has reserved the right to cut free allocation to comprise only a certain, yet undefined, quota of the benchmark. It is clearly shown that the free alloca-

tion is not sufficient to cover rotary kiln lime production completely at the given fuel mixtures.

From a carbon capture view, the dry stack gas composition is of interest. Drying of the gas can be easily achieved. The result in Figure 11 shows high carbon dioxide concentrations in the dry gas. Reference case A has a carbon dioxide concentration of 31.0%. Introducing oxyfuel combustion increases the concentration to the range between 89.1% and 94.5% the rest being mainly oxygen. These levels seem attractive for compression and transportation. The feasibility of reducing the oxygen concentration, e.g., by injection of carbon carrying fuels, resulting in even higher carbon dioxide purity in the processed gas, should be further assessed.

Conclusions

Oxyfuel combustion in rotary kiln lime production was studied. A validated predictive simulation tool has been applied to lime production.

The results show that an oxyfuel process could produce a high-quality lime product at reduced specific energy consumption. The new process would have a lower car-

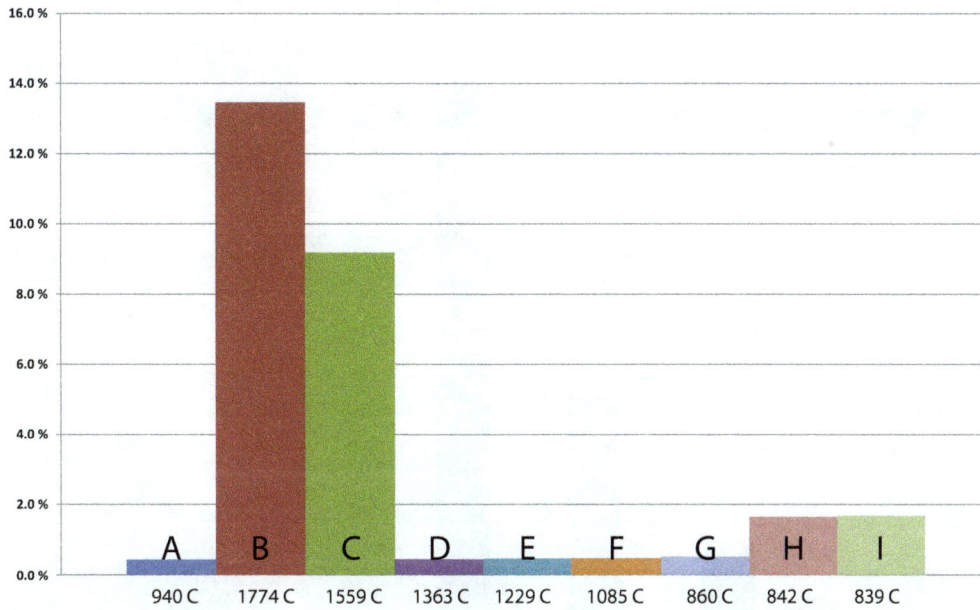

Figure 6. Weight percent melt in product and equilibrium temperature.

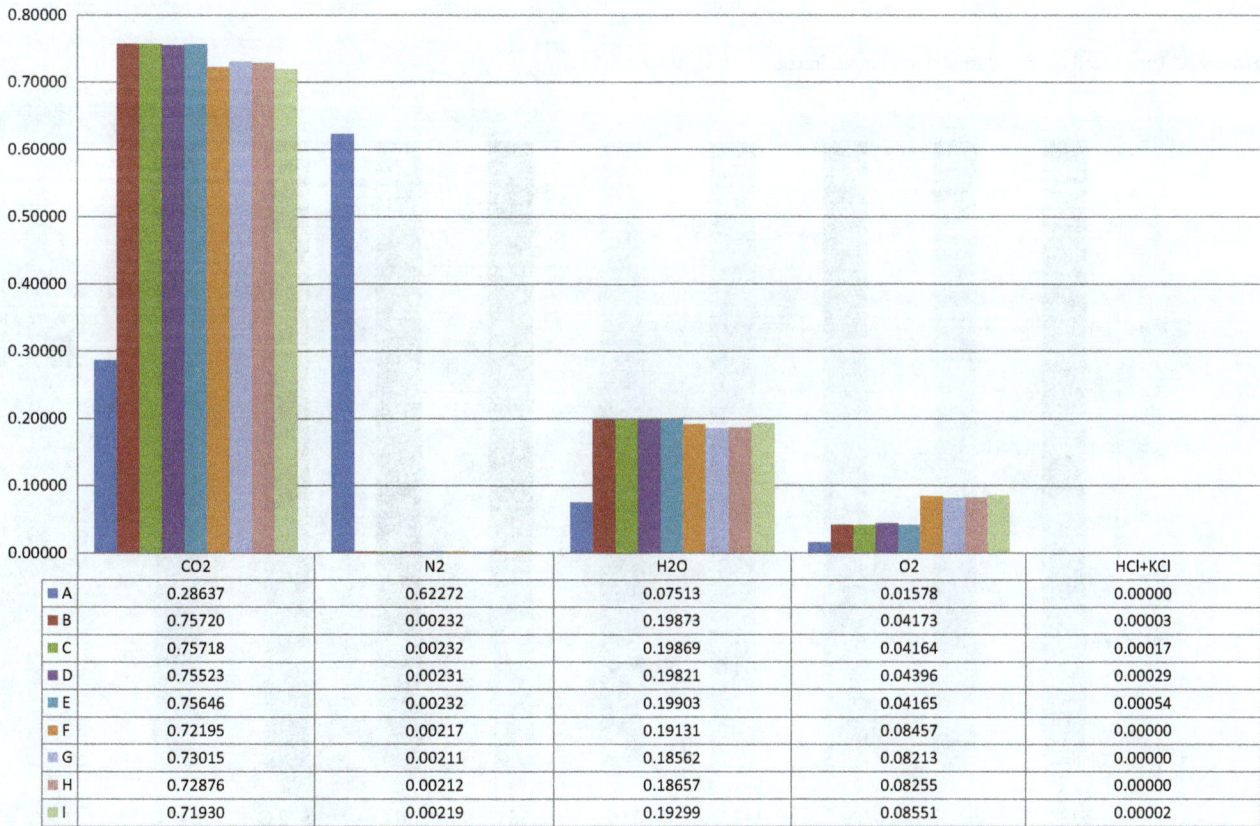

	CO2	N2	H2O	O2	HCl+KCl
A	0.28637	0.62272	0.07513	0.01578	0.00000
B	0.75720	0.00232	0.19873	0.04173	0.00003
C	0.75718	0.00232	0.19869	0.04164	0.00017
D	0.75523	0.00231	0.19821	0.04396	0.00029
E	0.75646	0.00232	0.19903	0.04165	0.00054
F	0.72195	0.00217	0.19131	0.08457	0.00000
G	0.73015	0.00211	0.18562	0.08213	0.00000
H	0.72876	0.00212	0.18657	0.08255	0.00000
I	0.71930	0.00219	0.19299	0.08551	0.00002

Figure 7. Stack gas compositions in mole fraction.

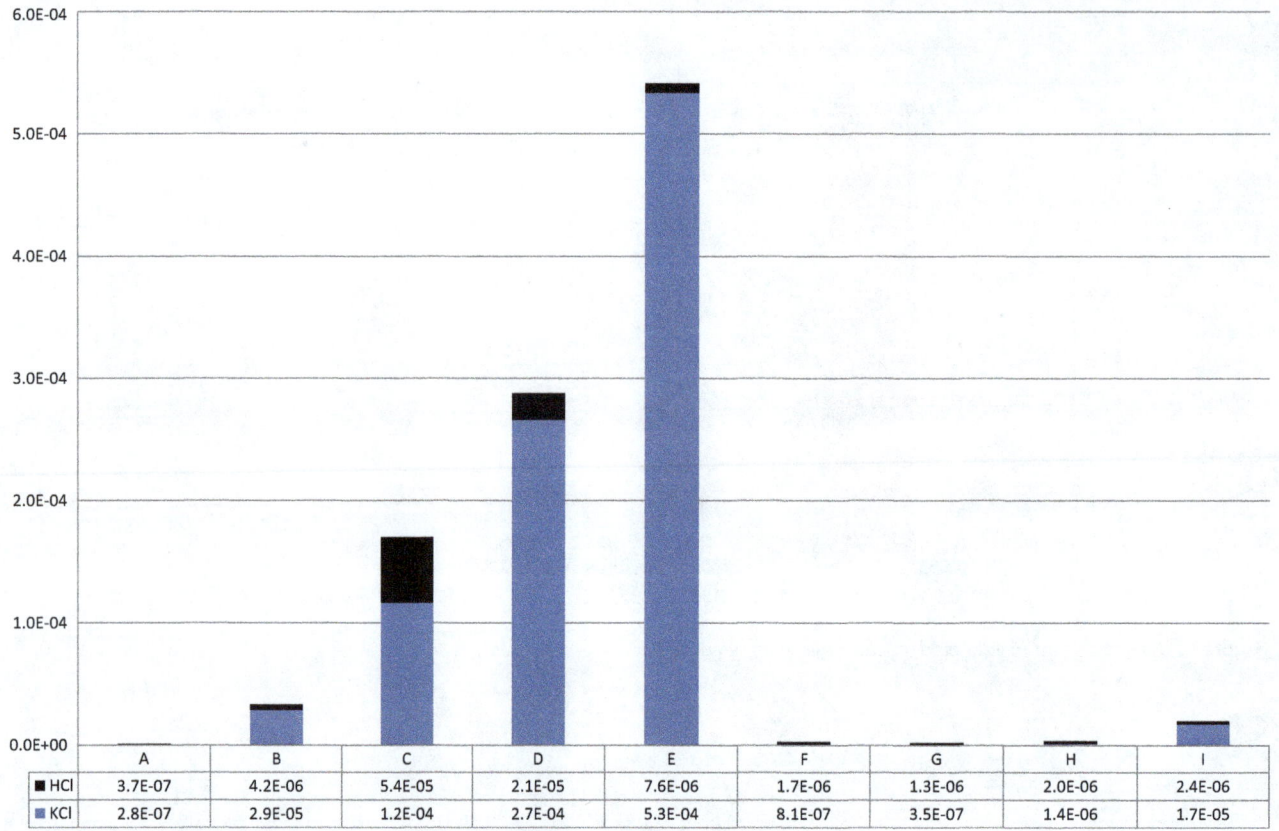

	A	B	C	D	E	F	G	H	I
■ HCl	3.7E-07	4.2E-06	5.4E-05	2.1E-05	7.6E-06	1.7E-06	1.3E-06	2.0E-06	2.4E-06
■ KCl	2.8E-07	2.9E-05	1.2E-04	2.7E-04	5.3E-04	8.1E-07	3.5E-07	1.4E-06	1.7E-05

Figure 8. Total chlorides and distribution in mole fraction of total stack gas.

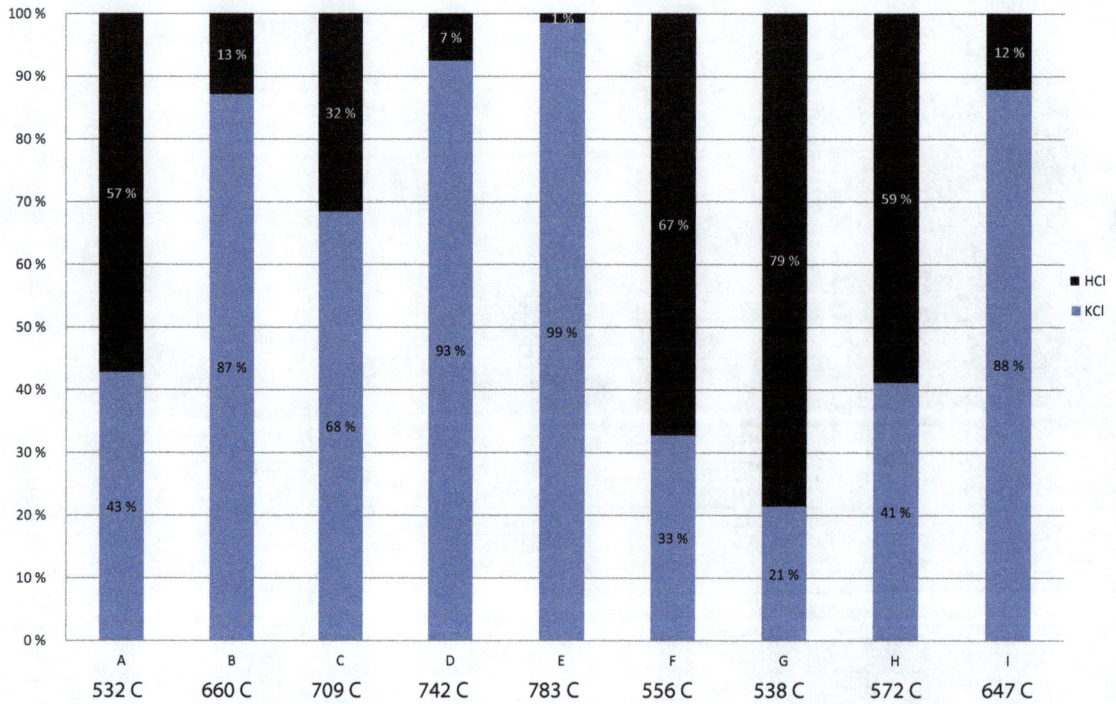

Figure 9. Distribution of chloride compounds in stack gas.

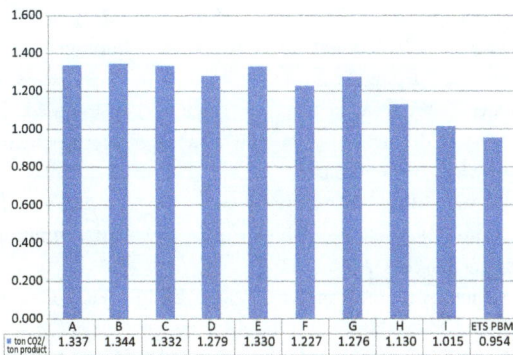

Figure 10. Ton carbon dioxide per ton of product ratio.

	A	B	C	D	E	F	G	H	I	ETS PBM
ton CO2/ ton product	1.337	1.344	1.332	1.279	1.330	1.227	1.276	1.130	1.015	0.954

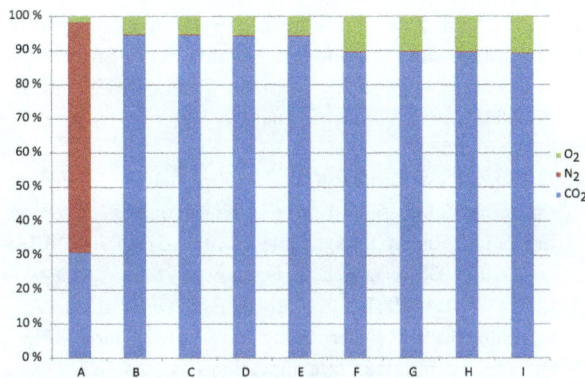

Figure 11. Dry stack gas concentrations.

bon dioxide emission per ton of product. The results show that to avoid a low-quality product, the process energy balances the need to be optimized. If the kiln runs too warm we will experience melting of oxide and salt phases in the product, and if the kiln runs too cold there is not enough energy to complete the calcination.

If dried the flue gas from the new process could be suitable for carbon dioxide transport and storage or utilization. The new process would have different energy and material balances compared to a conventional process. To exemplify the method the chloride balances were studied. The results show accumulation of chloride and variation in chloride species in the oxyfuel process.

In this work recirculation cases F and G, at 50% and 60% recirculation, show full product quality with reduced fuel consumption. Both cases show low chloride accumulation and high carbon dioxide content in the flue gas. Having lower specific carbon dioxide emissions than G, F at 1.227 ton carbon dioxide per ton product is considered the best case simulated in this work.

It is the conclusion that lime production with an oxyfuel process seems feasible but still requires further studies in order to reach a technical application level. A more comprehensive study on each specific retrofit or green field project is of course needed.

Acknowledgments

The authors would like to acknowledge Kjell Dahlberg at Nordkalk Oy Ab, Bo-Erik Eriksson, Thomas Lind, Anders Lyberg, Stefan Sandelin, and Erik Viggh at Cementa, Jan Bida, and Marianne Thomaeus at MinFo (Swedish Mineral Processing Research Association). The Swedish Energy Agency (No. 2006-06679. Project 30527-1) and the National (Swedish) Strategic Research Program Bio4Energy for financial support.

Conflict of Interest

None declared.

References

1. IPCC. 2005. Capture of CO_2. P. 442 *in* B. Metz, O. Davidson, H. C. de Coninck, M. Loos, and L. A. Meyer, eds. IPCC special report on carbon dioxide capture and storage. Prepared by Working Group III of the Intergovernmental Panel on Climate Change. Cambridge University Press, Cambridge, United Kingdom; New York, NY.

2. Oates, J. O. H. 1998. Lime and limestone: chemistry and technology, production and use. Wiley-VCH, Weinheim, ISBN 3-527-29527-5.

3. Industrial Emissions Directive 2010/75/EU, European IPPC Bureau, Best Available Techniques (BAT) Reference Document for the Production of Cement, Lime and Magnesium Oxide, JOINT RESEARCH CENTRE, Institute for Prospective Technological Studies Sustainable Production and Consumption Unit, 2013.

4. Eriksson, M. 2009. Post combustion CO_2 capture in the Swedish lime and cement industries. Licentiate thesis, Umeå university, ETPC Report 09-05, ISSN 1653-0551, ISBN 978-91-7264-922-4.

5. Wrampe, P., H. C. Rolseth. 1976. The effect of oxygen upon the rotary Kiln's production and fuel efficiency: theory and practice. IEEE Trans. Ind. Appl. IA-12:568–573.

6. Kudrina, A. P., V. N. Andryushchenko, and L. P. Kushchenko. 1979. Use of oxygen for limestone burning in rotary Kilns. Heat Eng. 19:353–357.

7. Garrido, G. F., A. S. Perkins, and J. R. Ayton. 1982. Upgrading lime recovery with oxygen enrichment. Pulp Paper Canada 83:T11–15, 67th Annual Meeting of Technical Section, CPPA, at Montreal 1981.

8. Watkinson, A. P., and J. K. Bricombe. 1983. Oxygen Enrichment in roatry lime kilns. Can. J. Chem. Eng. 61:842–849.

9. Zeman, F., K. Lackner. 2008. The reduced emission oxygen Kiln. The Earth Institute at Columbia University, New York, NY.

10. ECRA, European Cement Research Academy. Phase I report: 2007 Phase II report: 2009 Phase III report: 2012. CCS project, phase I-III reports. Available at http://www.ecra-online.org/226/ (accessed 28 March 2014).

11. Hökfors, B., E. Viggh, and M. Eriksson. 2013. Simulation of oxy-fuel combustion in cement clinker manufacturing. Adv. Cement Res. 25:1–8.

12. Zeman, F. 2009. Oxygen combustion in cement production. Energy Procedia 1:187–194.

13. ULCOS. Ultra low CO_2 steelmaking. Available at http://www.ulcos.org/en/press.php# (accessed 28 March 2014).

14. Danloy, G., Berthelemot, A., M. Grant, J. Borlée, D. Sert, and J. van der Stel, et al. 2009. ULCOS - pilot testing of the Low-CO2 Blast Furnace process at the experimental BF in Luleå. Revue de Métallurgie 106:1–8. doi: 10.1051/metal/2009008

15. Wilkinson, M, Boden, J., R. Panesar, and R. Allam et al. 2001. CO_2 capture via oxyfuel firing: optimisation of a retrofit design concept for a refinery power station boiler in USA First National Conference on Carbon Sequestration, Washington, DC, 15–17 May 2001.

16. Strömborg, L., Lindgren, G., J. Jacoby, R. Giering, M. Anheden, and U. Burchhardt et al. 2009. Update on Vattenfall's 30 MWth oxyfuel pilot plant in Schwarze Pumpe, Proceedings of the 9th International Conference on Greenhouse Gas Control Technologies (GHGT-9), 16–20 November 2008, Washington DC, Energy Procedia 1:581–589, Greenhouse Gas Control Technologies 9.

17. AspenPlus. Conceptual design of chemical processes, 1994–2011. Available at http://www.aspentech.com/core/aspen-plus.aspx (accessed 1 August 2011).

18. ChemApp. 2011. The Thermochemistry Library for Your Software. Available at http://www.gtt-technologies.de (accessed 1 August 2011).

19. Bale, C. W., P. Chartrand, S. Degterov, G. Eriksson, K. Hack, R. Mahfoud, et al. 2002. FactSage thermochemical software and databases. CALPHAD 26:189–228.

20. FactSage. 2011. Interactive programs for computational thermochemistry. Available at http://www.gtt-technologies.de (accessed 1 August 2011).

21. Wilhelmson Hökfors, B., E. Viggh, and R. Backman. 2008. A predictive chemistry model for the cement process, Zement-kalk-gips, 61:60–70, no 7, [Note(s): 60-70 [8 p.]].

22. Hökfors, B., M. Eriksson, and R. Backman. 2012. Improved process modeling for a lime rotary kiln using equilibrium chemistry, J. Eng. Technol. 29:8–18.

23. Eriksson, M. 2009. An industrial perspective on modelling of a rotary kiln for lump lime production in Proceedings at the NAFEMS NORDIC Seminar: Multi-Disciplinary Simulation in Engineering analysis, Helsinki, Finland, 21–22 April, 2009.

24. Hökfors, B., M. Eriksson, and E. Viggh. 2013. Modelling the cement process and cement clinker quality, Advances in Cement Research. Available at http://dx.doi.org/10.1680/adcr. 13.00050 (accessed 22 September 2014).

25. EU EEC 2011/278/EU, COMMISSION DECISION, 27 April 2011. 2011. Determining transitional Union-wide rules for harmonized free allocation of emission allowances pursuant to Article 10a of Directive 2003/87/EC of the European Parliament and of the Council, notified under document C(2011) 2772, 2011.

8

A bridge to nowhere: methane emissions and the greenhouse gas footprint of natural gas

Robert W. Howarth

Department of Ecology & Evolutionary Biology, Cornell University, Ithaca, New York 14853

Keywords

Greenhouse gas footprint, methane emissions, natural gas, shale gas

Abstract

In April 2011, we published the first peer-reviewed analysis of the greenhouse gas footprint (GHG) of shale gas, concluding that the climate impact of shale gas may be worse than that of other fossil fuels such as coal and oil because of methane emissions. We noted the poor quality of publicly available data to support our analysis and called for further research. Our paper spurred a large increase in research and analysis, including several new studies that have better measured methane emissions from natural gas systems. Here, I review this new research in the context of our 2011 paper and the fifth assessment from the Intergovernmental Panel on Climate Change released in 2013. The best data available now indicate that our estimates of methane emission from both shale gas and conventional natural gas were relatively robust. Using these new, best available data and a 20-year time period for comparing the warming potential of methane to carbon dioxide, the conclusion stands that both shale gas and conventional natural gas have a larger GHG than do coal or oil, for any possible use of natural gas and particularly for the primary uses of residential and commercial heating. The 20-year time period is appropriate because of the urgent need to reduce methane emissions over the coming 15–35 years.

Introduction

Natural gas is often promoted as a bridge fuel that will allow society to continue to use fossil energy over the coming decades while emitting fewer greenhouse gases than from using other fossil fuels such as coal and oil. While it is true that less carbon dioxide is emitted per unit energy released when burning natural gas compared to coal or oil, natural gas is composed largely of methane, which itself is an extremely potent greenhouse gas. Methane is far more effective at trapping heat in the atmosphere than is carbon dioxide, and so even small rates of methane emission can have a large influence on the greenhouse gas footprints (GHGs) of natural gas use.

Increasingly in the United States, conventional sources of natural gas are being depleted, and shale gas (natural gas obtained from shale formations using high-volume hydraulic fracturing and precision horizontal drilling) is rapidly growing in importance: shale gas contributed only 3% of United States natural gas production in 2005, rising to 35% by 2012 and predicted to grow to almost 50% by 2035 [1]. The gas held in tight sandstone formations is another form of unconventional gas, also increasingly obtained through high-volume hydraulic fracturing and is growing in importance. In 2012, gas extracted from shale and tight-sands combined made up 60% of total natural gas production, and this is predicted to increase to 70% by 2035 [1]. To date, shale gas has been almost entirely a North American phenomenon, and largely a U.S. one, but many expect shale gas to grow in global importance as well.

In 2009, I and two colleagues at Cornell University, Renee Santoro and Tony Ingraffea, took on as a research challenge the determination of the GHG of unconventional gas, particularly shale gas, including emissions of methane. At that time, there were no papers in the peer-reviewed literature on this topic, and there were relatively few papers even on the contribution of methane to the GHG of conventional natural gas [2–4]. At the end of 2009, the U.S. Environmental Protection Agency (EPA) still did not distinguish between conventional gas and shale gas, and they estimated methane emissions for the natural gas industry using emission factors from a 1996 study conducted jointly with the industry [5]; shale gas is not mentioned in that report, which is not surprising since significant shale gas production only started in the first decade of the 2000s.

We began giving public lectures on our analysis in March 2010, and these attracted media attention. One of our points was that it seemed likely that complete life cycle methane emissions from shale gas (from well development and hydraulic fracturing through delivery of gas to consumers) were greater than from conventional natural gas. Another preliminary conclusion was that the EPA methane emission estimates (as they were reported in 2009 and before, based on [5]) seemed at least two- to three-fold too low. In response to public attention from our lectures, the EPA began to reanalyze their methane emissions [6], and in late 2010, EPA began to release updated and far higher estimates of methane emissions

from the natural gas production segment [7]. In April 2011, we published our first paper on the role of methane in the GHG of shale gas [8]. We concluded that (1) the amount and quality of available data on methane emissions from the natural gas industry were poor; (2) methane emissions from shale gas were likely 50% greater than from conventional natural gas; and (3) these methane emissions contributed significantly to a large GHG for both shale gas and conventional gas, particularly when analyzed over the timescale of 20-years following emission. At this shorter timescale – which is highly relevant to the concept of natural gas as a bridge or transitional fuel over the next two to three decades – shale gas appeared to have the largest greenhouse warming consequences of any fossil fuel (Fig. 1). Because our conclusion ran counter to U.S. national energy policy and had large implications for climate change, and because the underlying data were limited and of poor quality, we stressed the urgent need for better data on methane emissions from natural gas systems. This need has since been amplified by the Inspector General of the EPA [9].

Our paper received immense media coverage, as evidenced by Time Magazine naming two of the authors

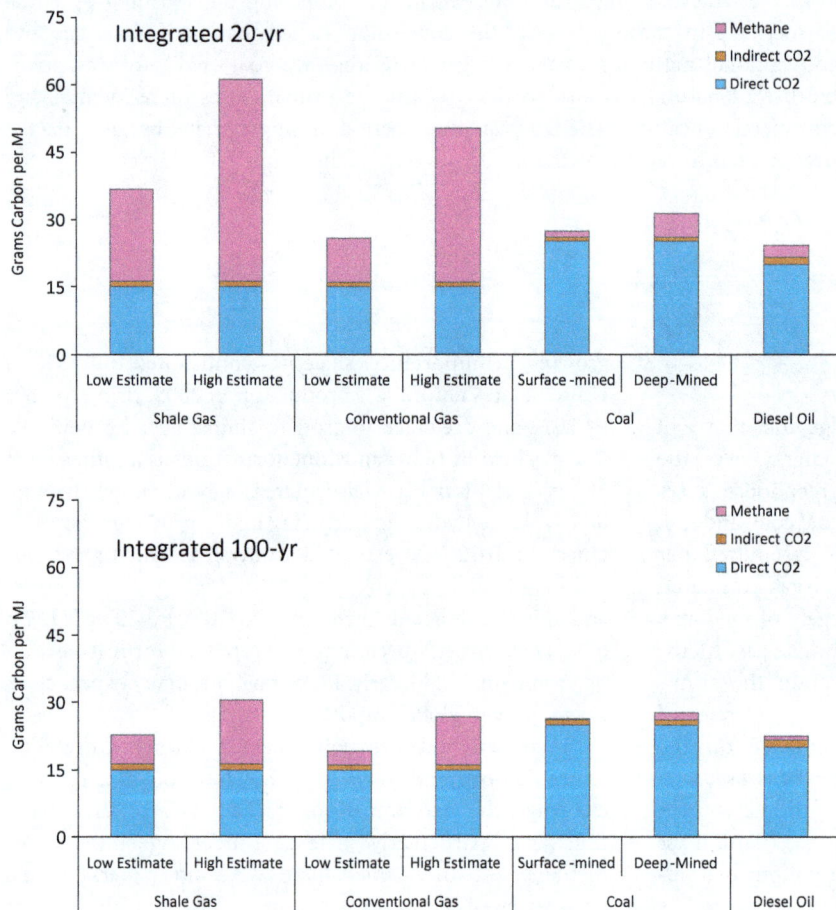

Figure 1. Comparison of the greenhouse gas footprint of shale gas, conventional natural gas, coal, and oil to generate a given quantity of heat. Two timescales for analyzing the relative warming of methane and carbon dioxide are considered: an integrated 20-year period (top) and an integrated 100-year period (bottom). For both shale gas and conventional natural gas, estimates are shown for the low- and high-end methane emission estimates from Howarth et al. [8]. For coal, estimates are given for surface-mined and deep-mined coal, since methane emissions are greater for deeper mines. Blue bars show the direct emissions of carbon dioxide during combustion of the fuels; the small red bars show the indirect carbon dioxide emissions associated with developing and using the fuels; and the magenta bars show methane emissions converted to g C of carbon dioxide equivalents using period-appropriate global warming potentials. Adapted from [8].

(Howarth and Ingraffea) "People who Mattered" to the global news in the December 2011 Person of the Year Issue [10]. The nine months after our paper was published saw a flurry of other papers on the same topic, a huge increase in the rate of publication on the topic of methane and natural gas compared to prior years and decades. While some of these offered support for our analysis, most did not and were either directly critical of our work, or without referring to our analysis reached conclusions more favorable to shale gas as a bridge fuel. Few of these papers published in the 9 months after our April 2011 paper provided new data; many simply offered different interpretations of previously presented information (as is reviewed briefly below). However, in 2012 and 2013 many new studies were published with major new insights and sources of data. In this paper, I briefly review the work on methane and natural gas published between April 2011 and February 2014, concentrating on those studies that have produced new primary data.

There are four components that are central to evaluating the role of methane in the GHG footprint of natural gas: (1) the amount of carbon dioxide that is directly emitted as the fuel is burned and indirectly emitted to obtain and use the fuel; (2) the rate of methane emission from the natural gas system (often expressed as a fraction of the lifetime production of the gas well, normalized to the amount of methane in the gas produced); (3) the global warming potential (GWP) of methane, which is the relative effect of methane compared to carbon dioxide in terms of its warming of the global climate system and is a function of the time frame considered after the emission of the methane; and (4) the efficiency of use of natural gas in the energy system. The GHG is then determined as:

GHG footprint
$$= [CO_2 \text{ emissions} + (GWP \times \text{methane emissions})]/\text{efficiency}$$

There is widespread consensus on the magnitude of the direct emissions of carbon dioxide, and the indirect emissions of carbon dioxide used to obtain and use natural gas (for example, in building and maintaining pipelines, drilling and hydraulically fracturing wells, and compressing gas), while uncertain, are also relatively small [8]. In this paper, I separately consider each of the other three factors (methane emissions, GWP, and efficiency of use) in the context of our April 2011 paper [8] and the subsequent literature.

How Much Methane is Emitted by Natural Gas Systems?

We used a full life cycle analysis in our April 2011 paper, estimating the amount of methane emitted to the atmo-

sphere as a percentage of the lifetime production of a gas well (normalized to the methane content of the natural gas), including venting and leakages at the well site but also during storage, processing, and delivery to customers. For conventional natural gas, we estimated a range of methane emissions from 1.7% to 6% (mean = 3.8%), and for shale gas a range of 3.6% to 7.9% (mean = 5.8%) [8]. We attributed the larger emissions from shale gas to venting of methane at the time that wells are completed, during the flowback period after high-volume hydraulic fracturing, consistent with the findings of the EPA 2010 report [7]. We assumed all other emissions were the same for conventional and shale gas. We estimated that downstream emissions (emissions during storage, long-distance transport of gas in high-pressure pipelines, and distribution to local customers) were 1.4–3.6% (mean = 2.5%) of the lifetime production of a well, and that the upstream emissions (at the well site and for gas processing) were in the range of 0.3–2.4% (mean = 1.4%) for conventional gas and 2.2–4.3% (mean = 3.3%) for shale gas (Table 1).

Table 1. Full life cycle-based methane emission estimates, expressed as a percentage of total methane produced in natural gas systems, separated by upstream emissions for conventional gas, upstream emissions for unconventional gas including shale gas, and downstream emissions for all natural gas. Studies are listed chronologically, and our April 2011 study is boldfaced.

	Upstream conventional gas	Upstream unconventional gas	Downstream
EPA 1996 [5]	0.2%	–	0.9%
Hayhoe et al. [2]	1.4	–	2.5
Jamarillo et al. [4]	0.2	–	0.9
Howarth et al. [8]	**1.4**	**3.3**	**2.5**
EPA [11]	1.6	3.0	0.9
Ventakesh et al. [12]	1.8	–	0.4
Jiang et al. [13]	–	2.0	0.4
Stephenson et al. [14]	0.4	0.6	0.07
Hultman et al. [15]	1.3	2.8	0.9
Burnham et al. [16]	2.0	1.3	0.6
Cathles et al. [17]	0.9	0.9	0.7

Total emissions are the sum of the upstream and downstream emissions. Studies are listed chronologically by time of publication. Dashes indicate no values provided. The full derivation of the estimates shown here is provided elsewhere [18, 19].

Although there were no prior papers on methane emissions from shale gas when our paper was published, we can compare our estimates for conventional natural gas with earlier literature (Table 1). Our mean estimates for both upstream and downstream emissions were identical to the "best estimate" of Hayhoe et al. [2], although that paper presented a wider range of estimates for both upstream and downstream. It is important to note that we used several newer sources of information not available to Hayhoe et al. [2], making the agreement all the more remarkable. The Howarth et al. [8] estimates were substantially higher than the emission factors used by the EPA through 2009 based on the 1996 joint EPA-industry study [5], which were only 1.1% for total emissions, 0.2% for upstream emissions, and 0.9% for downstream emissions. In the only other peer-reviewed paper on life cycle methane emissions from conventional gas published in the decade or two before our paper, Jamarillo et al. [4] relied on these same EPA emission factors, although new data on downstream emissions had already shown these emission factors to be too low [3].

Through late 2010 and the first half of 2011, the EPA provided a series of updates on their methane emission factors from the natural gas industry, giving estimates for shale gas for the first time as well as substantially increasing their estimates for conventional natural gas. These are discussed in detail by us elsewhere [18, 19]. Note that the EPA did not and still has not updated their estimates for downstream emissions, still using a value of 0.9% from a 1996 study [5]. For upstream emissions, the revised EPA estimates gave emission factors of 1.6% (an increase from their earlier value of 0.2%) for conventional natural gas and 3.0% for shale gas [18, 19]. Note that the EPA estimates for upstream emissions presented in 2011 [11] were 14% higher than ours for conventional gas and 10% lower than ours for shale gas. Total emissions were more divergent, due to the large difference in downstream emission estimates (Table 1).

In addition to the revised EPA emission factors, many other papers presented life cycle assessments of methane emissions from shale gas, conventional gas, or both in the immediate 9 months after April 2011 (Table 1). We and others have critiqued these publications in detail elsewhere [18–20]. Here, I will emphasize four crucial points:

1 For the upstream emissions in Table 1, all studies relied on the same type of poorly documented and highly uncertain information. These poor-quality data led us in Howarth et al. [8] to call for better measurements on methane fluxes, conducted by independent scientists. Several such studies have been published in the past 2 years, as is discussed further below, and these provide a more robust approach for estimating methane emissions.

2 At least some of the differences among values in Table 1 are due more to different assumptions about the lifetime production of a shale gas well than to differences in emissions per well [18, 20]. Note that the upstream life cycle emissions are scaled to the lifetime production of a well (normalized to the methane content of the gas produced for the estimates given in Table 1), and this was very uncertain in 2011 since shale gas development is such a new phenomenon [21]. A subsequent detailed analysis by the U.S. Geological Survey has demonstrated that the mean lifetime production of unconventional gas wells is in fact lower than any of papers in Table 1 assumed [22], meaning that upstream shale gas emissions per production of the well from all of the studies should be higher, in some cases substantially so [18, 20].

3 The downstream emissions in Table 1 are particularly uncertain, as highlighted by both Hayhoe et al. [2] and Howarth et al. [8]. Note that all of the other papers listed in Table 1 base their downstream emissions on the EPA emission factors from 1996 [5], and none are higher than those EPA estimates, even though a 2005 paper in *Nature* demonstrated higher levels of emission from long-distance pipelines in Europe [3]. Several of the papers in Table 1 have downstream emissions that are lower than the 1996 EPA values, as they are focused on electric power plants and assume that these plants are drawing on gas lines that have lower emissions than the average, which would include highly leaky low-pressure urban distribution lines [12–14, 16]. Some recent papers have noted a high incidence of leaks in natural gas distribution systems in two U.S. east coast cities [23, 24], but these new studies have yet placed an emission flux estimate on these leaks. Another study demonstrated very high methane emissions from fossil fuel sources in Los Angeles but could not distinguish between downstream natural gas emissions and other sources [25]. Given the age of gas pipelines and distribution systems in the United States, it should come as no surprise that leakage may be high [8, 18, 19]. Half of the high-pressure pipelines in the United States are older than 50 years [18], and parts of the distribution systems in many northeastern cities consist of cast-iron pipes laid down a century ago [24].

4 While one of the papers in Table 1 by Cathles and his colleagues [17], characterized our methane emission estimates as too high and "at odds with previous studies," that in fact is not the case. As noted above, both our downstream and upstream estimates for conventional gas are in excellent agreement with one of the few previous peer-reviewed studies [2]. Furthermore, our upstream emissions are in good agreement with the majority of the papers published in 9 months after

ours: for conventional gas, our mean estimate of 1.4% compares with the mean for all the other studies in Table 1 of 1.33%; if we exclude the very low estimate from Stephenson et al. [14], which was based on an analysis of what the gas industry is capable of doing rather than on any new measurements, and also the relatively low estimate from Cathles et al. [17], which was based on the assumption that the gas industry would not vent gas for economic and safety issues (see critique of this in [18]), the mean of the other four studies is 1.7, or almost twice as high as the Cathles et al. [17] estimate and 20% higher than our estimate. For shale gas, again excluding Stephenson et al. [14] and Cathles et al. [17] as well as our estimate, the other four studies in Table 1 have a mean estimate of 2.3, a value 2.5-fold greater than that from Cathles et al. [17] and 30% less than our mean estimate. From this perspective, the estimates of Cathles et al. [17] appear to be greater outliers than are ours.

Since 2012, many new papers have produced additional primary data (Fig. 2). Two of these found very high upstream methane emission rates from unconventional gas fields (relative to gross methane production), 4% for a tight-sands field in Colorado [26] and 9% for a shale gas field in Utah [27], while another found emissions from a shale gas field in Pennsylvania to be broadly consistent with the emission factors we had published in our 2011 paper [28]. All three of these studies inferred rates from atmospheric data that integrated a large number of wells at the basin scale. The new Utah data [27] are much higher than any of the estimates previously published for upstream emissions from unconventional gas fields (Fig. 2), while the measurement for the Colorado tight-

sands field [26] overlaps with our high-end estimate for upstream unconventional gas emissions in Howarth et al. [8]. The Utah and Colorado studies may not be representative of the typical methane emissions for the entire United States, in part, because they focused on regions where they expected high methane fluxes based on recent declines in air quality. But I agree with the conclusion of Brandt and his colleagues [29] that the "bottom-up" estimation approaches that we and all the other papers in Table 1 employed are inherently likely to lead to underestimates, in part, because some components of the natural gas system are not included. As one example, the recent Pennsylvania study, which quantified fluxes from discrete locations on the ground by mapping methane plumes from an airplane, found very high emissions from many wells that were still being drilled, had not yet reached the shale formation, and had not yet been hydraulically fractured [28]. These wells represented only 1% of the wells in the area but were responsible for 6–9% of the regional methane flux from all sources. One explanation is that the drill rigs encountered pockets of shallower gas and released this to the atmosphere. We, the EPA, and all of the papers in Table 1 had assumed little or no methane emissions from wells during this drilling phase.

Allen and colleagues [30] published a comprehensive study in 2013 of upstream emissions for both conventional and unconventional gas wells for several regions in the United States, using the same basic bottom-up approach as the joint EPA-industry study of 1996 used [5]. As with that earlier effort, this new study relied heavily on industry cooperation, and was funded largely by industry with coordination provided by the Environmental Defense Fund. For the United States as a whole at the

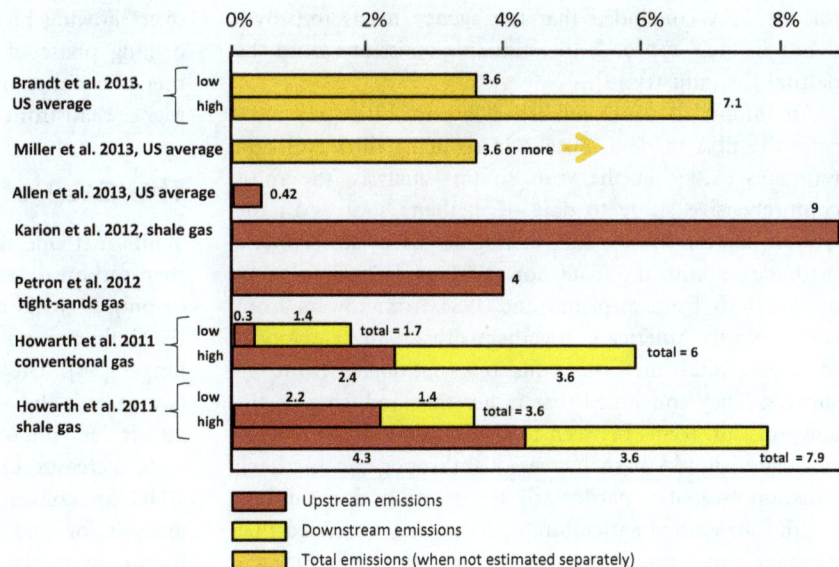

Figure 2. Comparison of recent new data on methane emissions compared to the estimates published in Howarth et al. [8]. Some of the new data are for upstream emissions, while others give only averages for natural gas systems in the United States. No new measurements for downstream emissions alone have been published since 2005 [8, 26, 27, 29, 30, 32].

time of their study, Allen et al. [30] concluded that upstream methane emissions were only 0.42% of the natural gas production by the wells (Fig. 2), a value at the low end of those seen in Table 1. Using the low-end estimates, "best-case" scenarios for upstream emissions from Howarth et al. [8] and the mix of shale gas and conventional gas produced in the United States in 2012, I estimate the U.S. national best-case emission rate would be 0.5%, or similar to that observed by Allen and colleagues. It should not be surprising that their study, in relying on industry access to their sampling points, ended up in fact measuring the best possible performance by industry.

In 2013, the EPA reduced their emission estimates for the oil and gas industry, essentially halving their upstream emissions for average natural gas systems from 1.8% to 0.88% for the year 2009 (with the mix of conventional and unconventional gas for that year) from what they had reported in 2011 and 2012; the EPA estimate for downstream emissions remained at 0.9%, giving a total national emission estimate of 1.8%. EPA took this action to decrease their emission factors for upstream emissions despite the publication in 2012 of the methane emissions from a Colorado field [26] and oral presentations at the American Geophysical Union meeting in December 2012 of the results subsequently published by Karion and colleagues [27] and Caulton and colleagues [28], all of which would have suggested higher emissions, perhaps spectacularly so. As is discussed by Karion et al. [27], the decrease in the upstream methane emissions by EPA in 2013 was driven by a non-peer-reviewed industry report [31] which argued that emissions from liquid unloading and during refracturing of unconventional wells were far lower than used in the EPA [11] assessment. At least in part in response to these changes by EPA, the Inspector General for the EPA concluded that the agency needs improvements in their approach to estimating emissions from the natural gas industry [9].

An important paper published late in 2013 [32] indicates the EPA made a mistake in reducing their emission estimates earlier in the year. In this analysis, the most comprehensive study to date of methane sources in the United States, Miller and colleagues used atmospheric methane monitoring data for 2007 and 2008 – 7710 observations from airplanes and 4984 from towers from across North America – together with an inverse model to assess total methane emissions nationally from all sources. They concluded that rather than reducing methane emission terms between their 2011 and 2013 inventories, EPA should have increased anthropogenic methane emission estimates, particularly for the oil and gas industry and for animal agriculture operations. They stated that methane emissions from the United States oil and gas

industry are very likely two-fold greater or more than indicated by the factors EPA released in 2013 [32]. This suggests that total methane emissions from the natural gas industry were at least 3.6% in 2007 and 2008 (Fig. 2).

In early 2014, Brandt and his colleagues [29] reviewed the technical literature over the past 20 years on methane emissions from natural gas systems. They concluded that "official inventories consistently underestimate actual methane emissions," but also suggested that the very high estimates from the top–down studies in Utah and Colorado [26, 27] "are unlikely to be representative of typical [natural gas] system leakage rates." In the supplemental materials for their paper, Brandt et al. [29] state that methane emissions in the United States from the natural gas industry are probably greater than the 1.8% assumed by the EPA by an additional 1.8–5.4%, implying an average rate between 3.6% and 7.1% (mean = 5.4%) [33] (Fig. 2).

This recent literature suggests to me that the emission estimates we published in Howarth et al. [8] are surprisingly robust, particularly for conventional natural gas (Fig. 2). The results from two of the recent top–down studies [26, 27] indicate our estimates for unconventional gas may have been too low. Partly in response to our work and their own reanalysis of methane emissions from shale gas wells, EPA has now promulgated new regulations that will as of January 2015 reduce methane emissions at the time of well completions, requiring capture and use of the gas instead in most cases. Some wells are exempt, and the regulation does not apply to venting of methane from oil wells, including shale oil wells, which often have associated gas. Nonetheless, the regulations are an important step in the right direction, and will certainly help, if they can be adequately enforced. Even still, though, results such as those from the Pennsylvania fly-over showing high rates of methane emission during the drilling phase of some shale gas wells [28] suggest that methane emissions from shale gas may remain at levels higher than from conventional natural gas.

The GWP of Methane

While methane is far more effective as a greenhouse gas than carbon dioxide, methane has an atmospheric lifetime of only 12 years or so, while carbon dioxide has an effective influence on atmospheric chemistry for a century or longer [34]. The time frame over which we compare the two gases is therefore critical, with methane becoming relatively less important than carbon dioxide as the time-scale increases. Of the major papers on methane and the GHG for conventional natural gas published before our analysis for shale gas, one modeled the relative radiative forcing by methane compared to carbon dioxide continuously over a 100-year time period following emission [2], and two used the global warming approach (GWP) which compares how much larger the integrated global warming from a given mass of methane is over a specified period of time compared to the same mass of carbon dioxide. Of the two that used the GWP approach, one showed both 20-year and 100-year GWP analyses [3] while another used only a 100-year GWP time frame [4]. Both used GWP values from the Intergovernmental Panel on Climate Change (IPCC) synthesis report from 1996 [35], the most reliable estimates at the time their papers were published. In subsequent reports from the IPCC in 2007 [36] and 2013 [34] and in a paper in *Science* by workers at the NASA Goddard Space Institute [37], these GWP values have been substantially increased, in part, to account for the indirect effects of methane on other radiatively active substances in the atmosphere such as ozone (Table 2).

In Howarth et al. [8], we used the GWP approach and closely followed the work of Lelieveld and colleagues [3] in presenting both integrated 20 and 100 year periods, and in giving equal credence and interpretation to both timescales. We upgraded the approach by using the most recently published values for GWP at that time [37].

Table 2. Comparison of the timescales considered in comparing the global warming consequences of methane and carbon dioxide.

Publication	Timescale considered	20-year GWP	100-year GWP
IPCC [35]	**20 and 100 years**	**56**	**21**
Hayhoe et al. [2]	0–100 years	NA	NA
Lelieveld et al. [3]	20 and 100 years	56	21
Jamarillo et al. [4]	100 years	–	21
IPCC [36]	**20 and 100 years**	**72**	**25**
Shindell et al. [37]	**20 and 100 years**	**105**	**33**
Howarth et al. [8]	20 and 100 years	105	33
Hughes [20]	20 and 100 years	105	33
Venkatesh et al. [12]	100 years	–	25
Jiang et al. [13]	100 years	–	25
Wigley [38]	0–100 years	NA	NA
Stephenson et al. [14]	100 years	–	25
Hultman et al. [15]	20 and 100 years	72, 105	25, 44
Skone et al. [39]	100 years	–	25
Burnham et al. [16]	100 years	–	25
Cathles et al. [17]	100 years	–	25
Alvarez et al. [40]	0–100 years	NA	NA
IPCC [34]	**10, 20, and 100 years**	**86**	**34**
Brandt et al. [29]	100 years	–	25

Studies are listed chronologically by time of publication. Values for the global warming potentials at 20 and 100 years given, when used in the studies. NA stands for not applicable and is shown when studies did not use the global warming potential approach. Dashes are shown for studies that did not consider the 20-year GWP. Studies that are bolded provided primary estimates on global warming potentials, while other studies are consumers of this information.

These more recent GWP values increased the relative warming of methane compared to carbon dioxide by 1.9-fold for the 20-year time period (GWP of 105 vs. 56) and by 1.6-fold for the 100-year time period (GWP of 33 vs. 21; Table 2). Our conclusion was that for the 20-year time period, shale gas had a larger GHG than coal or oil even at our low-end estimates for methane emission (Fig. 1); conventional gas also had a larger GHG than coal or oil at our mean or high-end methane emission estimates, but not at the very low-end range for methane emission (the best-case, low-emission scenario). At the 100-year timescale, the influence of methane was much diminished, yet at our high-end methane emissions, the GHG of both shale gas and conventional gas still exceeded that of coal and oil (Fig. 1).

Of nine new reports on methane and natural gas published in 9 months after our April 2011 paper [8], six only considered the 100-year time frame for GWP, two used both a 20- and 100-year time frame, and one used a continuous modeling of radiative forcing over the 0–100 time period (Table 2). Of the six papers that only examined the 100-year time frame, all used the lower GWP value of 25 from the 2007 IPCC report rather than the higher value of 33 published by Shindell and colleagues in 2009 that we had used; this higher value better accounts for the indirect effects of methane on global warming. Many of these six papers implied that the IPCC dictated a focus on the 100-year time period, which is simply not the case: the IPCC report from 2007 [36] presented both 20- and 100-year GWP values for methane. And two of these six papers criticized our inclusion of the 20-year time period as inappropriate [14, 17]. I strongly disagree with this criticism. In the time since April 2011 I have come increasingly to believe that it is essential to consider the role of methane on timescales that are much shorter than 100 years, in part, due to new science on methane and global warming presented since then [34, 41, 42], briefly summarized below.

The most recent synthesis report from the IPCC in 2013 on the physical science basis of global warming highlights the role of methane in global warming at multiple timescales, using GWP values for 10 years in addition to 20 and 100 years (GWP of 108, 86, and 34, respectively) in their analysis [34]. The report states that "there is no scientific argument for selecting 100 years compared with other choices," and that "the choice of time horizon .... depends on the relative weight assigned to the effects at different times" [34]. The IPCC further concludes that at the 10-year timescale, the current global release of methane from all anthropogenic sources exceeds (slightly) all anthropogenic carbon dioxide emissions as agents of global warming; that is, methane emissions are more important (slightly) than carbon dioxide emissions for driving the current rate of global warming. At the 20-year timescale, total global emissions of methane are equivalent to over 80% of global carbon dioxide emissions. And at the 100-year timescale, current global methane emissions are equivalent to slightly less than 30% of carbon dioxide emissions [34] (Fig. 3).

This difference in the time sensitivity of the climate system to methane and carbon dioxide is critical, and not widely appreciated by the policy community and even some climate scientists. While some note how the long-term momentum of the climate system is driven by carbon dioxide [15], the climate system is far more immediately responsive to changes in methane (and other short-lived radiatively active materials in the atmosphere, such as black carbon) [41]. The model published in 2012 by Shindell and colleagues [41] and adopted by the United Nations [42] predicts that unless emissions of methane and black carbon are reduced immediately, the Earth's average surface temperature will warm by 1.5°C by about 2030 and by 2.0°C by 2045 to 2050 whether or not carbon dioxide emissions are reduced. Reducing methane and black carbon emissions, even if carbon dioxide is not controlled, would significantly slow the rate of global warming and postpone reaching the 1.5°C and 2.0°C marks by 15–20 years. Controlling carbon dioxide as well as methane and black carbon emissions further slows the rate of global warming after 2045, through at least 2070 [41, 42] (Fig. 4).

Why should we care about this warming over the next few decades? At temperatures of 1.5–2.0°C above the

1890–1910 baseline, the risk of a fundamental change in the Earth's climate system becomes much greater [41–43], possibly leading to runaway feedbacks and even more global warming. Such a result would dwarf any possible benefit from reductions in carbon dioxide emissions over the next few decades (e.g., switching from coal to natural gas, which does reduce carbon dioxide but also increases methane emissions). One of many mechanisms for such catastrophic change is the melting of methane clathrates in the oceans or melting of permafrost in the Arctic. Hansen and his colleagues [43, 44] have suggested that warming of the Earth by 1.8°C may trigger a large and rapid increase in the release of such methane. While there is a wide range in both the magnitude and timing of projected carbon release from thawing permafrost and melting clathrates in the literature [45], warming consistently leads to greater release. This release can in turn cause a feedback of accelerated global warming [46].

To state the converse of the argument: the influence of today's emissions on global warming 200 or 300 years into the future will largely reflect carbon dioxide, and not

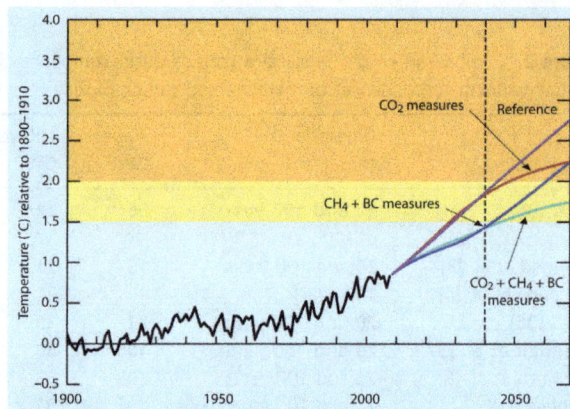

Figure 4. Observed global mean temperature from 1900 to 2009 and projected future temperature under four scenarios, relative to the mean temperature from 1890 to 1910. The scenarios include the IPCC [36] reference, reducing carbon dioxide emissions but not other greenhouse gases ("CO₂ measures"), controlling methane, and black carbon emissions but not carbon dioxide ("CH₄ + BC measures"), and reducing emissions of carbon dioxide, methane, and black carbon ("CO₂ + CH₄ + BC measures"). An increase in the temperature to 1.5–2.0°C above the 1890–1910 baseline (illustrated by the yellow bar) poses risk of passing a tipping point and moving the Earth into an alternate state for the climate system. The lower bound of this danger zone, 1.5° warming, is predicted to occur by 2030 unless stringent controls on methane and black carbon emissions are initiated immediately. Controlling methane and black carbon shows more immediate results than controlling carbon dioxide emissions, although controlling all greenhouse gas emissions is essential to keeping the planet in a safe operating space for humanity. Adapted from [42].

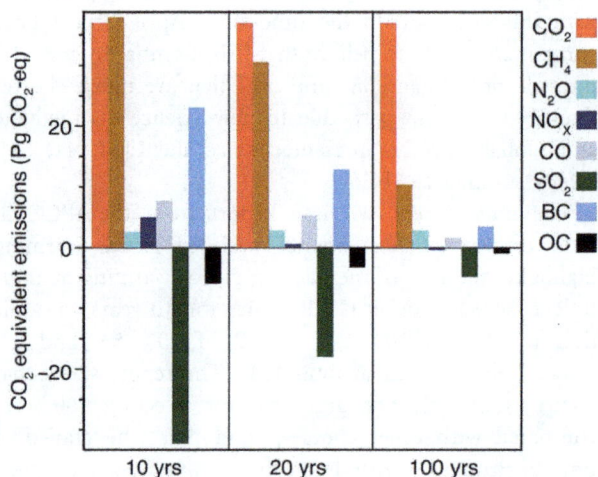

Figure 3. Current global greenhouse gas emissions, as estimated by the IPCC [34], weighted for three different global warming potentials and expressed as carbon dioxide equivalents. At the 10-year time frame, global methane emissions expressed as carbon dioxide equivalents actually exceed the carbon dioxide emissions. Adapted from [34].

methane, unless the emissions of methane lead to tipping points and a fundamental change in the climate system. And that could happen as early as within the next two to three decades.

An increasing body of science is developing rapidly that emphasizes the need to consider methane's influence over the decadal timescale, and the need to reduce methane emissions. Unfortunately, some recent guidance for life cycle assessments specify only the 100-year time frame [47, 48], and the EPA in 2014 still uses the GWP values from the IPCC 1996 assessment and only considers the 100-year time period when assessing methane emissions [49]. In doing so, they underestimate the global warming significance of methane by 1.6-fold compared to more recent values for the 100-year time frame and by four to fivefold compared to the 10- to 20-year time frames [34, 37].

Climate Impacts of Different Natural Gas Uses

In Howarth et al. [8], we compared the greenhouse gas emissions of shale gas and conventional natural gas to those of coal and oil, all normalized to the same amount of heat production (i.e., g C of carbon dioxide equivalents per MJ of energy released in combustion). We also noted that the specific comparisons will depend on how the fuels are used, due to differences in efficiencies of use, and briefly discussed the production of electricity from coal versus shale gas as an example; electric-generating plants on average use heat energy from burning natural gas more efficiently than they do that from coal, and this is important although not usually dominant in comparing the GHGs of these fuels [8, 18–20]. We presented our main conclusions in the context of the heat production (Fig. 1), though, because evaluating the GHGs of the different fossil fuels for all of their major uses was beyond the scope of our original study, and electricity production is not the major use of natural gas. This larger goal of separately evaluating the GHGs of all the major uses of natural gas has not yet been taken on by other research groups either.

In Figure 5 (left-hand panel), I present an updated comparison of the GHGs of natural gas, diesel oil, and coal based on the best available information at this time (April 2014). Values are expressed as g C of carbon dioxide equivalents per MJ of energy released as in our 2011 paper [8] and Figure 1. The methane emissions in Figure 5 are the mean and range of estimates from the recent review by Brandt and colleagues [29] (see Fig. 2), normalized to carbon dioxide equivalents using the 20-year mean GWP value of 86 from the latest IPCC assessment [34]. As noted above, I believe the 20-year GWP is

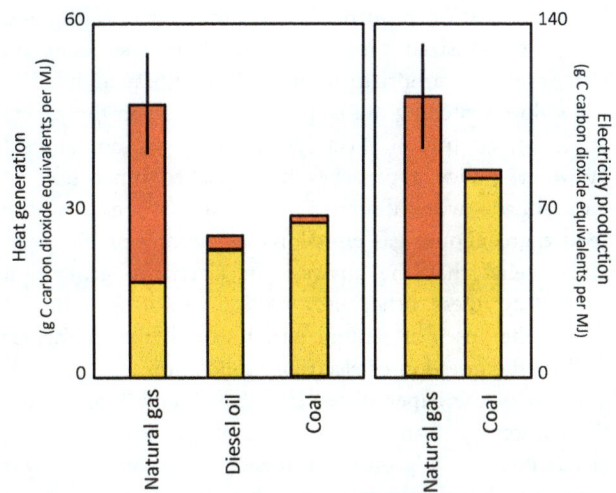

Figure 5. Comparison of the greenhouse gas footprint for using natural gas, diesel oil, and coal for generating primary heat (left) and for using natural gas and coal for generating electricity (right). Direct and indirect carbon dioxide emissions are shown in yellow and are from Howarth et al. [8], while methane emissions shown as g C of carbon dioxide equivalents using the 2013 IPCC 20-year GWP [34] are shown in red. Methane emissions for natural gas are the mean and range for the U.S. national average reported by Brandt and colleagues [29] in their supplemental materials. Methane emissions for diesel oil and for coal are from Howarth et al. [8] For the electricity production, average U.S. efficiencies of 41.8% for gas and 32.8% for coal are assumed [20]. Several studies present data on emissions for electricity production in other units. One can convert from g C of CO_2-equivalents per MJ to g CO_2-equivalents per kWh by multiplying by 13.2. One can convert from g C of CO_2-equivalents per MJ to g C of CO_2-equivalents per kWh by multiplying by 3.6.

an appropriate timescale, given the urgent need to control methane emissions globally. Estimates for coal and diesel oil are from our 2011 paper [8], using data for surface-mined coal since that dominates the U.S. market [20]. The direct and indirect emissions of carbon dioxide are combined and are the same values as in Howarth et al. [8] and Figure 1. Direct carbon dioxide emissions follow the High Heating Value convention [2, 8]. Clearly, using the best available data on rates of methane emission [29], natural gas has a very large GHG per unit of heat generated when considered at this 20-year timescale.

Of the studies listed in Tables 1 and 2 published after our 2011 paper [8], most focused just on the comparison of natural gas and coal to generate electricity, although one also considered the use of natural gas as a long-distance transportation fuel [40]. For context, over the period 2008–2013 in the United States, 31% of natural gas has been used to generate electricity and 0.1% as a transportation fuel [50]. None of the studies listed in Tables 1 and 2, other than Howarth et al. [8], considered the use of natural gas for its primary use: as a source of heat. In the United States over the last 6 years, 32% of natural gas has been used for residential and commercial heating and 28% for industrial process energy [50]. The focus on electricity is appropriate if the only question at hand is "how does switching out coal for natural gas in the generation of electricity affect greenhouse gas emissions?" However, policy approaches have pushed other uses of natural gas – without any scientific support – as a way to reduce greenhouse gas emissions, apparently on the mistaken belief that the analysis for electricity generation applied to these other uses. Before exploring some of these other uses of natural gas, I would like to further explore the question of electricity generation.

Many of the papers listed in Tables 1 and 2 concluded that switching from coal to natural gas for generating electricity has a positive influence on greenhouse gas emissions. Note, though, that for almost all of these papers, the conclusion was driven by a focus on only the 100-year timescale [4, 12–14, 16, 17, 29, 39], on a very low assumed level of methane emission [4, 12–14, 17, 39], or both. The differences in efficiency of use in electric power plants, comparing either current average plants or best possible technologies, are relatively small compared to the influence of the GWP on the calculation [8, 18, 20, 40]. Using a 20-year GWP framework and the methane emission estimates from Howarth et al. [8], the GHG from generating electricity with natural gas is larger than that from coal [8, 18–20]. Alvarez and colleagues [40] concluded that for electricity generation, the GHG of using natural gas was less than for coal for all time frames only if the rate of methane leakage was less than 3.2%. Their analysis used the estimates for the radiative forcing of methane from the IPCC 2007 synthesis [36], and if we correct their estimate for the data in the 2013 IPCC assessment [34], this "break-even point" becomes 2.8%. If we further consider the uncertainty in the radiative forcing of methane of 30% or more [34], this "break-even" value becomes a range of 2.4–3.2%.

In Figure 5 (right-hand panel), I compare the GHGs of natural gas and coal when used to generate electricity, again using the High Heating Value convention [2, 8], the latest IPCC value for the 20-year GWP [34] and the range of methane emission estimates reported by Brandt and colleagues [29]. No distinction is made for less downstream emissions for the pipelines that feed electric power plants, as is assumed in several other studies [12–14, 16], simply because no data exist with which to tease apart downstream emissions specific for electric power generation [51]. This analysis uses the average efficiency for electric power plants currently operating in the United States, 41.8% for gas and 32.8% for coal [20]. The emissions per unit of energy produced as electricity are higher than for the heat generation alone, due to these corrections for efficiency. Although the difference in the foot-

prints for using the two fuels is less for the electricity comparison than for the comparison for heat generation, at this 20-year timescale the GHG of natural gas remains greater than that of coal, even at the low-end methane emission estimate. This conclusion still holds when one compares the fuels using the best available technologies (50.2% efficiency for natural gas and 43.3% for coal [20]); the emissions per unit of electricity generated decrease for both by approximately the same amount.

For the dominant use of natural gas – heating for water, domestic and commercial space, and industrial process energy – the analysis we presented in our 2011 paper [8] and shown in Figure 1 remains the only published study before this new analysis shown in Figure 5 (left-hand panel). The updated version shown here compellingly indicates natural gas is not a climate-friendly fuel for these uses. However, the greenhouse gas consequences may in fact be worse than Figure 5 or Howarth et al. [8] indicate, as I discuss next.

A recent study supported by the American Gas Foundation promoted the in-home use of natural gas over electricity for appliances (domestic hot water, cooking) because of a supposed benefit for greenhouse gas emissions [52]. The report argues that an in-home natural gas appliance will have a higher efficiency in using the fuel (up to 92%) compared to the overall efficiency of producing and using electricity ("only about 40%," according to this study). However, they did not include methane emissions in their analysis, nor did they consider the extremely high efficiencies available for some electrical appliances, such as in-home air-sourced heat pumps for domestic hot water. For a given input of electricity, such heat pumps can produce 2.2-times more heat energy, since they are harvesting and concentrating heat from the local environment [53]. In a comparison of using in-home gas-fired water heaters or in-home high-efficiency electric heat pumps, with the electricity for the heat pumps generated by burning coal, the heat pumps had a lower GHG than did in-home use of gas if the emission rate for methane was greater than 0.7% for a 20-year GWP or 1.3% for a 100-year GWP [51]. Using the mean methane emission estimate from Howarth et al. [8] for conventional natural gas (Fig. 2) and a 20-year GWP, the in-home natural gas heater had a GHG that was twice as large as that of the heat pump [51]. Of course, an in-home heat pump powered by electricity from renewable sources such as wind and solar would have a far smaller GHG yet [54].

What about other uses of natural gas? The "Natural Gas Act," a bill introduced in the United States Congress in 2011 with bipartisan support and the backing of President Obama, would have provided tax subsidies to encourage the replacement of diesel fuel by natural gas for long-distance trucks and buses; the bill did not pass, in part because conservatives opposed it as "market distorting" [55, 56]. In Quebec, industry has claimed that this replacement of diesel by shale gas would reduce greenhouse gas emissions by up to 30% [57]. However, in contrast to a possible advantage in replacing coal with natural gas for electricity generation (if methane emissions can be kept low enough), using natural gas to replace diesel fuel as a long-distance transportation fuel would greatly increase greenhouse emissions [29, 40]. In part, this is because the energy of natural gas is used with less efficiency than diesel in truck engines. Furthermore, although methane emissions from transportation systems have not been well measured, one could imagine significant emissions during refueling operations for buses and trucks, as well as from venting of on-vehicle natural gas tanks to keep gas pressures significantly safe during warm weather. Despite the findings of Alvarez and colleagues published in 2012 [40], the EPA continues to indicate that switching buses from diesel fuel to natural gas reduces greenhouse gas emissions [58].

Concluding Thoughts

By 1950, which is about the time I was born, human activity had contributed enough greenhouse gases to the atmosphere to cause a radiative forcing – the driving factor behind global warming – of 0.57 watts m^{-2} compared to before the industrial revolution [34]. Thirty years later, in 1980 when I taught my first course on the biosphere and global change, this human influence had doubled the anthropogenic radiative forcing, to 1.25 watts m^{-2} [34]. And another 30 years later, the continued release of greenhouse gases by humans has again doubled the forcing, now at 2.29 watts m^{-2} or fourfold greater than just 60 years ago [34]. The temperature of the Earth continues to rise in response at an alarming rate, and the climate scientists tell us we may reach dangerous tipping points in the climate system within just a few decades [34, 41, 42]. Is it too late to begin a serious reduction in greenhouse gas emissions? I sincerely hope not, although surely society has been very slow to respond to this risk. The use of fossil fuels is the major cause of greenhouse gas emissions, and any genuine effort to reduce emissions must begin with fossil fuels.

Is natural gas a bridge fuel? At best, using natural gas rather than coal to generate electricity might result in a very modest reduction in total greenhouse gas emissions, if those emissions can be kept below a range of 2.4–3.2% (based on [40], adjusted for the latest information on radiative forcing of methane [34]). That is a big "if," and one that will require unprecedented investment in natural gas infrastructure and regulatory oversight. For any other

foreseeable use of natural gas (heating, transportation), the GHG is larger than if society chooses other fossil fuels, even with the most stringent possible control on methane emissions, if we view the consequences through the decadal GWP frame. Given the sensitivity of the global climate system to methane [41, 42], why take any risk with continuing to use natural gas at all? The current role of methane in global warming is large, contributing 1.0 watts m^{-2} out of the net total 2.29 watts m^{-2} of radiative forcing [34].

Am I recommending that we continue to use coal and oil, rather than replace these with natural gas? Not at all. Society needs to wean itself from the addiction to fossil fuels as quickly as possible. But to replace some fossil fuels (coal, oil) with another (natural gas) will not suffice as an approach to take on global warming. Rather, we should embrace the technologies of the 21st Century, and convert our energy systems to ones that rely on wind, solar, and water power [59, 60, 61]. In Jacobson et al. [54], we lay out a plan for doing this for the entire state of New York, making the state largely free of fossil fuels by 2030 and completely free by 2050. The plan relies only on technologies that are commercially available at present, and includes modern technologies such as high-efficiency heat pumps for domestic water and space heating. We estimated the cost of the plan over the time frame of implementation as less than the present cost to the residents of New York from death and disease from fossil fuel caused air pollution [54]. Only through such technological conversions can society truly address global change. Natural gas is a bridge to nowhere.

Acknowledgments

Funding was provided by Cornell University, the Park Foundation, and the Wallace Global Fund. I thank Bongghi Hong, Roxanne Marino, Tony Ingraffea, George Woodwell, and two reviewers who have asked to remain anonymous for their valuable comments on earlier drafts of the manuscript.

Conflict of Interest

None declared.

References

1. EIA. 2013. Annual energy outlook 2013 early release. Energy Information Agency, US Department of Energy. Available at http://www.eia.gov/energy_in_brief/article/about_shale_gas.cfm (accessed 27 December 2013).
2. Hayhoe, K., H. S. Kheshgi, A. K. Jain, and D. J. Wuebbles. 2002. Substitution of natural gas for coal: climatic effects of utility sector emissions. Clim. Change 54:107–139.

3. Lelieveld, J., S. Lechtenbohmer, S. S. Assonov, C. A. M. Brenninkmeijer, C. Dinest, M. Fischedick, et al. 2005. Low methane leakage from gas pipelines. Nature 434: 841–842.

4. Jamarillo, P., W. M. Griffin, and H. S. Mathews. 2007. Comparative life-cycle air emissions of coal, domestic natural gas, LNG, and SNG for electricity generation. Environ. Sci. Technol. 41:6290–6296.

5. Harrison, M. R., T. M. Shires, J. K. Wessels, and R. M. Cowgill. 1996. Methane emissions from the natural gas industry. Volume 1: executive summary. EPA-600/R-96-080a. U.S. Environmental Protection Agency, Office of Research and Development, Washington, DC.

6. Personal communication from Roger Fernandez, US EPA. 19 May 2011.

7. EPA. 2010. Greenhouse gas emissions reporting from the petroleum and natural gas industry. Background technical support document. U.S. Environmental Protection Agency, Washington, DC. Available at http://www.epa.gov/climatechange/emissions/downloads10/Subpart-W_TSD.pdf (accessed 24 February 2011).

8. Howarth, R. W., R. Santoro, and A. Ingraffea. 2011. Methane and the greenhouse gas footprint of natural gas from shale formations. Clim. Change Lett. 106:679–690. doi: 10.1007/s10584-011-0061-5

9. U.S. Environmental Protection Agency Office of Inspector General. 2013. EPA needs to improve air emissions data for the oil and natural gas production sector. EPA OIG, Washington, DC.

10. Walsh, B. 2011. People who mattered: Mark Ruffalo, Anthony Ingraffea, Robert Howarth. Time, Person of the Year issue on line, 14 December 2011. Available at http://content.time.com/time/specials/packages/article/0,28804,2101745_2102309_2102323,00.html (accessed 30 December 2011).

11. EPA. 2011. Inventory of U.S. greenhouse gas emissions and sinks: 1990–2009. 14 April 2011. U.S. Environmental Protection Agency, Washington, DC. Available at http://epa.gov/climatechange/emissions/usinventoryreport.html (accessed 25 November 2011).

12. Venkatesh, A., P. Jamarillo, W. M. Griffin, and H. S. Matthews. 2011. Uncertainty in life cycle greenhouse gas emissions from United States natural gas end-uses and its effect on policy. Environ. Sci. Technol. 45:8182–8189.

13. Jiang, M., W. M. Griffin, C. Hendrickson, P. Jaramillo, J. van Briesen, and A. Benkatesh. 2011. Life cycle greenhouse gas emissions of Marcellus shale gas. Environ. Res. Lett. 6:034014. doi: 10.1088/1748-9326/6/3/034014

14. Stephenson, T., J. E. Valle, and X. Riera-Palou. 2011. Modeling the relative GHG emissions of conventional and shale gas production. Environ. Sci. Technol. 45:10757–10764.

15. Hultman, N., D. Rebois, M. Scholten, and C. Ramig. 2011. The greenhouse impact of unconventional gas for electricity generation. Environ. Res. Lett. 6:044008. doi: 10.1088/1748-9326/6/4/044008

16. Burnham, A., J. Han, C. E. Clark, M. Wang, J. B. Dunn, and I. P. Rivera. 2011. Life-cycle greenhouse gas emissions of shale gas, natural gas, coal, and petroleum. Environ. Sci. Technol. 46:619–627.

17. Cathles, L. M., L. Brown, M. Taam, and A. Hunter. 2012. A commentary on "The greenhouse-gas footprint of natural gas in shale formations" by R.W. Howarth, R. Santoro, and Anthony Ingraffea. Clim. Change 113:525–535.

18. Howarth, R. W., R. Santoro, A. Ingraffea. Venting and leakage of methane from shale gas development: reply to Cathles et al. 2012. Clim. Change 113:537–549. doi: 10.1007/s10584-012-0401-0

19. Howarth, R. W., D. Shindell, R. Santoro, A. Ingraffea, N. Phillips, and A. Townsend-Small. 2012. Methane emissions from natural gas systems. Background paper prepared for the National Climate Assessment, Reference # 2011-003, Office of Science & Technology Policy Assessment, Washington, DC. Available at http://www.eeb.cornell.edu/howarth/Howarth%20et%20al.%20–%20National%20Climate%20Assessment.pdf (accessed 1 March 2012).

20. Hughes, D. 2011. Lifecycle greenhouse gas emissions from shale gas compared to coal: an analysis of two conflicting studies. Post Carbon Institute, Santa Rosa, CA. Available at http://www.postcarbon.org/reports/PCI-Hughes-NETL-Cornell-Comparison.pdf (accessed 30 October 2011).

21. Howarth, R. W., and A. Ingraffea. 2011. Should fracking stop? Yes, it is too high risk. Nature 477:271–273.

22. USGS. 2012. Variability of distributions of well-scale estimated ultimate recovery for continuous (unconventional) oil and gas resources in the United States. U.S. Geological Survey, USGS Open-File Report 2012–1118. Available at http://pubs.usgs.gov/of/2012/1118/ (accessed 5 January 2014).

23. Phillips, N. G., R. Ackley, E. R. Crosson, A. Down, L. Hutyra, M. Brondfield, et al. 2013. Mapping urban pipeline leaks: methane leaks across Boston. Environ. Pollut. 173:1–4.

24. Jackson, R. B., A. Down, N. G. Phillips, R. C. Ackley, C. W. Cook, D. L. Plata, et al. 2014. Natural gas pipeline leaks across Washington, DC. Environ. Sci. Technol. 48:2051–2058.

25. Townsend-Small, A., S. C. Tyler, D. E. Pataki, X. Xu, and L. E. Christensen. 2012. Isotopic measurements of atmospheric methane in Los Angeles, California, USA reveal the influence of "fugitive" fossil fuel emissions. J. Geophys. Res. 117:D07308.

26. Pétron, G., G. Frost, B. T. Miller, A. I. Hirsch, S. A. Montzka, A. Karion, et al. 2012. Hydrocarbon emissions characterization in the Colorado Front Range – a pilot

study. J. Geophys. Res. 117:D04304. doi: 10.1029/2011JD016360

27. Karion, A., C. Sweeney, G. Pétron, G. Frost, R. M. Hardesty, J. Kofler, et al. 2013. Methane emissions estimate from airborne measurements over a western United States natural gas field. Geophys. Res. Lett. 40:4393–4397.

28. Caulton, D. R., P. B. Shepson, R. L. Santoro, J. P. Sparks, R. W. Howarth, A. Ingaffea, et al. 2014. Toward a better understanding and quantification of methane emissions from shale gas development. Proc. Natl. Acad. Sci. USA 111:6237–6242.

29. Brandt, A. F., G. A. Heath, E. A. Kort, F. O. O'Sullivan, G. Pétron, S. M. Jordaan, et al. 2014. Methane leaks from North American natural gas systems. Science 343:733–735.

30. Allen, D. T., V. M. Torres, K. Thomas, D. W. Sullivan, M. Harrison, A. Hendler, et al. 2013. Measurements of methane emissions at natural gas production sites in the United States. Proc. Natl. Acad. Sci. USA 110:17768–17773.

31. Shires, T., and M. Lev-On 2012. P. 48 in Characterizing pivotal sources of methane emissions from unconventional natural gas production: summary and analysis of API and ANGA survey responses. American Petroleum Institute, American Natural Gas Alliance, Washington, DC.

32. Miller, S. M., S. C. Wofsy, A. M. Michalak, E. A. Kort, A. E. Andrews, S. C. Biraud, et al. 2013. Anthropogenic emissions of methane in the United States. Proc. Natl. Acad. Sci. USA 110:20018–20022.

33. Romm, J. 2014. By the time natural gas has a net climate benefit, you'll likely be dead and the climate ruined. Climate Progress, 19 February 2014. Available at http://thinkprogress.org/climate/2014/02/19/3296831/natural-gas-climate-benefit/# (accessed 2 March 2014).

34. IPCC. 2013. Climate change 2013: the physical science basis. Intergovernmental Panel on Climate Change. Available at https://www.ipcc.ch/report/ar5/wg1/ (accessed 10 January 2014).

35. IPCC. 1996. IPCC second assessment, climate change, 1995. Intergovernmental Panel on Climate Change. Available at http://www.ipcc.ch/pdf/climate-changes-1995/ipcc-2nd-assessment/2nd-assessment-en.pdf (accessed 22 February 2014).

36. IPCC. 2007. IPCC Fourth Assessment Report (AR4), Working Group 1, the physical science basis. Intergovernmental Panel on Climate Change. Available at http://www.ipcc.ch/publications_and_data/ar4/wg1/en/contents.html (accessed 22 February 2014).

37. Shindell, D. T., G. Faluvegi, D. M. Koch, G. A. Schmidt, N. Unger, and S. E. Bauer. 2009. Improved attribution of climate forcing to emissions. Science 326:716–718.

38. Wigley, T. M. L. 2011. Coal to gas: the influence of methane leakage. Clim. Change Lett. 108:601–608.

39. Skone, T. J., J. Littlefield, and J. Marriott. 2011. Life cycle greenhouse gas inventory of natural gas extraction, delivery and electricity production. Final report 24 October 2011 (DOE/NETL-2011/1522). U.S. Department of Energy, National Energy Technology Laboratory, Pittsburgh, PA.

40. Alvarez, R. A., S. W. Pacala, J. J. Winebrake, W. L. Chameides, and S. P. Hamburg. 2012. Greater focus needed on methane leakage from natural gas infrastructure. Proc. Natl. Acad. Sci. USA 109:6435–6440. doi: 10.1073/pnas.1202407109

41. Shindell, D., J. C. I. Kuylenstierna, E. Vignati, R. van Dingenen, M. Amann, Z. Klimont, et al. 2012. Simultaneously mitigating near-term climate change and improving human health and food security. Science 335:183–189.

42. UNEP/WMO. 2011. Integrated assessment of black carbon and tropospheric ozone: summary for decision makers. United Nations Environment Programme and the World Meteorological Organization, Nairobi, Kenya.

43. Hansen, J., M. Sato, P. Kharecha, G. Russell, D. W. Lea, and M. Siddall. 2007. Climate change and trace gases. Philos. Trans. R. Soc. A 365:1925–1954.

44. Hansen, J., and M. Sato. 2004. Greenhouse gas growth rates. Proc. Natl. Acad. Sci. USA 101:16109–16114.

45. Schaefer, K., T. Zhang, L. Bruhwiler, and A. Barrett. 2011. Amount and timing of permafrost carbon release in response to climate warming. Tellus 63:165–180. doi: 10.1111/j.1600-0889.2011.00527.x

46. Zimov, S. A., E. A. G. Schuur, and F. S. Chapin. 2006. Permafrost and the global carbon budget. Science 312:1612–1613.

47. BSI. 2011. Specification for the assessment of the life cycle greenhouse gas emissions of goods and services. British Standards Institute, Lond.

48. WRI/WBCSD. 2012. Product life cycle accounting and reporting standard. World Resources Institute, Washington, DC.

49. EPA. 2014. Overview of greenhouse gases. US Environmental Protection Agency. Available at http://epa.gov/climatechange/ghgemissions/gases/ch4.html (accessed 17 February 2014).

50. EIA. 2014. Natural gas consumption by end use. Energy Information Agency, US Department of Energy. Available at http://www.eia.gov/dnav/ng/ng_cons_sum_dcu_nus_a.htm (accessed 3 March 2014).

51. Hong, B., and R. W. Howarth. In review. Assessing an acceptable level of methane emissions from using natural gas: domestic hot water example.

52. IHS CERA. 2014. Fueling the future with natural gas: bringing it home. Executive Summary. January 2014. Available at www.fuelingthefuture.org/assets/content/AGF-Fueling-the-Future-Study.pdf (accessed 2 March 2014).

53. American Council for and Energy-Efficient Economy. 2014. Water heating. Available at http://www.aceee.org/ consumer/water-heating (accessed 3 February 2014).

54. Jacobson, M. Z., R. W. Howarth, M. A. Delucchi, S. R. Scobies, J. M. Barth, M. J. Dvorak, et al. 2013. Examining the feasibility of converting New York State's all-purpose energy infrastructure to one using wind, water, and sunlight. Energy Policy 57:585–601.

55. Weis, D. J., and S. Boss. 2011. Conservatives power big oil, stall cleaner natural gas vehicles. Center for American Progress, 6 June 2011. Available at http://www. americanprogress.org/issues/2011/06/nat_gas_statements. html (accessed 2 March 2014).

56. Dolan, E. 2013. What stands in the way of natural gas replacing gasoline in the US? OilPrice.com, 8 January 2013. Available at http://oilprice.com/Energy/Natural-Gas/ What-Stands-in-the-Way-of-Natural-Gas-Replacing-Gasoline-in-the-US.html (accessed 2 March 2014).

57. Beaudine, M. 2010. In depth: shale gas exploration in Quebec. The Gazette, 15 November 2010.

58. EPA. 2014. Sources of greenhouse gas emissions. US Environmental Protection Agency. Available at http://www. epa.gov/climatechange/ghgemissions/sources/ transportation.html (accessed 21 February 2014).

59. Jacobson, M. Z. 2009. Review of solutions to global warming, air pollution, and energy security. Energy Environ. Sci. 2:148–173.

60. Jacobson, M. A., and M. A. Delucchi. 2009. A path to sustainable energy by 2030. Scientific American, November 2009.

61. Jacobson, M. A., and M. A. Delucchi. 2011. Providing all global energy with wind, water, and solar power, Part I: technologies, energy resources, quantities and areas of infrastructure, and materials. Energy Policy 39: 1154–1169.

9

Models and experiments for energy consumption and quality of green tea drying

Nickson Langat[1], Thomas Thoruwa[2], John Wanyoko[3], Jeremiah Kiplagat[1], Brian Plourde[4] & John Abraham[4]

[1]Kenyatta University, P.O. Box 43844-00100, Nairobi, Kenya
[2]Pwani University, P.O. Box 195-80108, Kilifi, Kenya
[3]Tea Research Foundation of Kenya, P.O. Box 820-20200, Kericho, Kenya
[4]University of St. Thomas, St. Paul, Minnesota, 55105-1079

Keywords

Energy efficiency, fluidized bed, food processing, heat transfer, tea drying, tea processing.

Abstract

An experimental apparatus has been developed to evaluate the drying process of green tea leaves. Tea drying is an energy-intensive process which results in the removal of leaf moisture; it is essential to the quality of the final product. In order to more efficiently use process energy, a prototype drying system has been built and tested. The prototype incorporates a rotating perforated drum which helps speed the drying process. Experiments were carried out with multiple temperatures, airflow rates, and drum rotation rates; a subset of those results is shown here. In particular, the impact of airflow rate on the process was studied. It was found that as the airflow increased, the drying rate increased, as expected. However, the efficiency of energy use, which was quantified by the Specific Energy Consumption rate, varied considerably with flow. While higher flows led to faster drying, it resulted in a lower energy efficiency. Also, a two parameter predictive model was developed that was able to accurately match the moisture removal rates for a very wide range of flows. This predictive model, which is based on thermal-fluid fundamentals, can be used to extrapolate the presented results to cases which were not considered.

Introduction

Tea is the most commonly consumed nonwater beverage in the world. It is consumed throughout virtually in all regions and has experienced recent significant growth rates in its consumption [1]. Approximately two-third of the world's tea is categorized as black tea with the remainder mainly comprising green teas with small amounts of others such as oolong, jasmine, and Pu-erh teas constituting the balance.

The volumes of tea produced in various parts of the world are large. For instance, worldwide ~4.1 million metric tons (4.1×10^9 kg) are produced with nearly 650,000 tons (6.5×10^8 kg) produced in the continent of Africa. With respect to green teas, the corresponding numbers are ~1 million metric tons (1×10^9 kg) globally.

The production of tea involves multiple energy-intensive processes. The large volumes of material and significant energy costs motivate studies into the improvements of the processing and the present study. The processing of green tea, in particular, requires four stages: 1, steaming; 2, shaping; 3, drying; and 4, postprocessing. Briefly, the steaming process is employed to prevent aeration and to halt oxidation. The duration of the steaming process impacts the quality of the resulting product. While steaming processes vary, they often last for 30–150 sec.

Next, shaping of the leaves is carried out through various machine-driven processes or by human labor. The rolling output dictates the shape of the tea leaves and impacts the release of flavors during steeping.

Third, tea leaves are dried to reduce the moisture content within the cellular structure. A variety of drying methodologies are available including spray drying, rotary drying, fluid-bed drying, infrared heating, microwave heating and others. The most common approach uses fluidized bed systems, which leads to a uniform drying of leaves to a low moisture content. Fluidized bed drying (FBD) is characterized by a high rate of moisture and heat transfer with a high degree of control. Despite these characteristics, FBD can lead to poor quality products if defluidization of the tea leaves occurs, an issue that is particularly prevalent with high moisture-content teas such as green tea [2]. In order to alleviate this problem, some researchers have experimented with agitation of the leave bed during the processing [3].

Despite the general widespread use of FBD, in many regions, particularly in developing regions, it has not yet been successfully employed. Consequently, there is a need for improvement to the FBD method with a focus on improving the quality of the product and reducing energy utilization during the processing.

In particular, the goal is to design an improved drying system which is agitated (rather than static). It is hoped that a bed, agitated through rotary motion, will speed the rate of drying, reduce fluidization velocities, without sacrificing efficiency. In particular, measurements of the efficiency will be given based on a thermodynamic analysis of the heat transferred during the drying process per unit of mass transfer. More details will be given later.

Another motivation for the device which is presented here is that the drying method will preserve the tea aroma compared to conventional drying methods.

The goal of this study is the implementation of FBD for green tea drying specifically and the quantification of energy use during the drying stage. Furthermore, a quantitative model will be developed that allows the prediction of the performance for other situations not studied in this project.

For further background, interested readers are directed to [4–11]. That collection of papers deals with some of the fundamental heat transfer and fluid flow concepts which govern cases such as this.

FBD Experiments

Great care was taken to ensure that the results from the FBD experiments were of the highest quality. First, tea leaves were collected from plants in a consistent manner. Six high-yielding tea clones were identified and at least 1 kg of fresh leaves was collected daily from each clone. Two leaves and a bud were removed from each plant, the standard harvesting method [12]. Harvesting was performed in the morning and evening hours to reduce moisture loss from the leaves.

The leaves were spread on perforated trays that promoted air circulation. In [3], three drying methods were then employed in replicate (three times). Those methods were microwave drying, conventional FBD, and drying with a newly invented prototype FBD technique. Only results for the prototype will be conveyed in this manuscript. The prototype FBD incorporated a rotating perforated drum to hold the leaves and promote a more uniform drying. The rotary method also was found to reduce energy utilization [2].

A simple schematic has been prepared in Figure 1 to show the essential components of the system. In the figure, a shaft is shown which connects the circular drum to a driving motor. Hot air enters the system from an upstream heating section. The drum diameter and length are 279 and 660 mm, respectively. The hole size for the mesh which contains the tea leaves was 1 mm. This size was large enough to permit airflow, yet small enough to avoid spillage of tea-leaf components.

The drying drum was conditioned by six 1 kW heating elements which were collectively enclosed in a cylindrical casing. One effect of the casing was that the air temperature emerged at a uniform, well-mixed temperature. The heating system was surrounded by an asbestos-cement material whose thermal conductivity was 2.07 W/m°C. A final enclosing layer of concrete was used to contain the asbestos. Air motion was caused by 1 kW centrifugal forward-curved fans which were able to deliver up to 10 m³/sec of airflow.

Moisture content measurements were taken with a Mettler-Toledo-HR83 device with a resolution of 0.01%. Air speed was measured in the ducting system at locations away from bends and obstructions. Measurements made at multiple locations in the cross section displayed

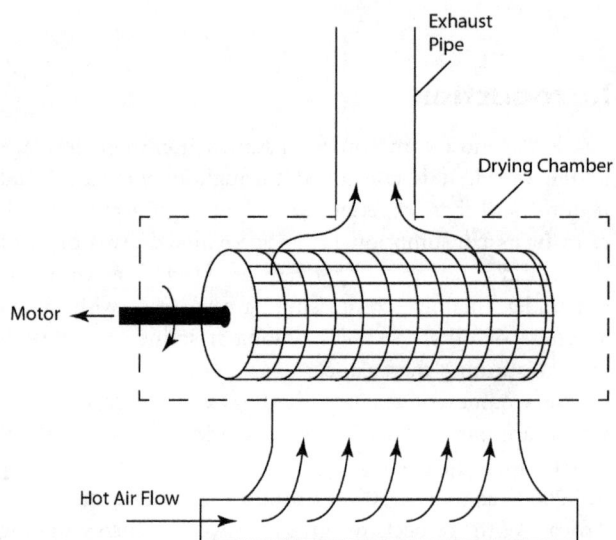

Figure 1. Schematic diagram showing rotating tea drying perforated drum and airflow patterns.

a high degree of velocity uniformity. The measurements were taken with a digital hot-wire anemometer. With the motor control of the rotating drum, it was possible to control the rotational speed of the prototype dryer. Results for various rotating speeds will be presented.

For all experiments, the local humidity of the inlet air was very low compared to that of the air exiting the drying chamber. The initial leaf moisture content varied among the individual experiments but it was ~80% by mass for all cases. The electrical power provided to the heaters were set to result in air temperatures of 80°C. Ten different flowrates were examined that varied from 41 to 167 L/sec. Leaf moisture content was measured at 600, 1200, 1800, and 2400 sec. Results from the experiments are compared with a mathematical model that will be presented in the next section. The rotation was achieved through the use of induction motors. The speed of the motor was controlled by power input through an inverter and was varied from 0 to 21.3 rad/sec (0–203 rpm). It was found that 12.56 rad/sec (120 rpm) was required for proper agitation of the leaves and maximal drying rate. All of the results which are presented here correspond to 120 rpm.

Predictive Model

It is useful to develop a model that is capable of predicting the time-wise reduction of leaf moisture. To develop this model, it is necessary to describe the fundamental heat and mass transfer processes. During drying processes, there are typically three stages of transfer. Initially, the material to be dried (tea leaves) are heated from a starting temperature to an elevated value. During this phase, mass transfer occurs but it increases as the heating process proceeds. When a more or less constant temperature is reached, the mass transfer occurs at a relatively steady rate. The actual rate of transport is dictated by processes at the interface between the fluid and the solid object. During this phase, resistance to mass transport within the material (diffusion) is significantly lower than surface transport resistances. As the amount of water within the object further decreases, the internal resistance grows and can exceed surface transport resistances. The focus here is on the second phase which is characterized by constant drying. This phase is the longest duration and most moisture transport occurs here.

The first law of thermodynamics requires that energy transfer from the leaves is matched by energy loss from the preheated air. A control volume, defined so that it encompasses the airstream but not the leaves will incorporate heat transfer processes between the air stream and leaves as a boundary condition. A simple schematic of the control volume is shown in Figure 2. There, a dashed line

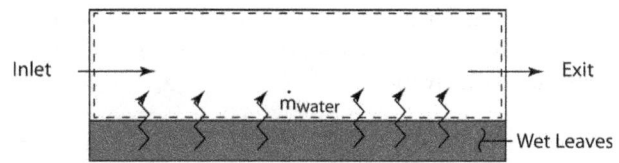

Figure 2. Simplified diagram showing the thermodynamic control volume.

signifies the control–volume boundary. Flows across the boundary occur at the inlet, exit, and the leaf–volume interface.

With the control volume identified, it is possible to express the first law of thermodynamics as

$$q = \sum (\dot{m}h)_{\text{exit}} - \sum (\dot{m}h)_{\text{inlet}} \tag{1}$$

Here, the symbol q represents the rate of heat transfer across the boundary (excluding latent energy transfer). The symbols $\dot{m}$ represent mass flowrates and the h terms are specific enthalpies. While there will be some heat loss to the environment and also some heat transfer between the leaves and the airstream, both of these constituents are expected to be small. Heat loss to the environment would be that heat which travels through the insulation wall by conduction. A simple one-dimensional heat conduction analysis shows that this term is orders of magnitude below the rates of heat exchange between the fluid and the leaves. Consequently, it was neglected. With these assumptions, equation (1) can be rewritten as

$$(\dot{m}h)_{\text{exit}} = \dot{m}_{\text{water}}h_{\text{water}} + (\dot{m}h)_{\text{inlet}} \tag{2}$$

Equation (2) can further be reduced by recognizing that the mass flowrates are a combination of the air and water vapor components in the two airstreams, so that

$$\dot{m} = \dot{m}_a + \dot{m}_v, \tag{3}$$

where the subscripts a and v represent dry air and vapor. Combination of equations (2) and (3), along with cancellation gives

$$(1 + \omega_{\text{exit}})h_{\text{exit}} = (\omega_{\text{exit}} - \omega_{\text{inlet}})h_{\text{water}} + (1 + \omega_{\text{inlet}})h_{\text{inlet}} \tag{4}$$

The symbols ω are the specific humidity levels at the respective locations. In equation (4), all terms aside from h_{exit} and ω_{exit} were known prior to the experiments.

Proper formulation of the energy balance is essential for the energy optimization of the system. A primary goal of the device was to maximize the rate of moisture removal for a quantity of input energy. To create a performance metric, the specific energy consumption rate (SEC) was introduced [13] which is the rate of heat transfer to the air prior to its introduction into the dryer for each unit mass of moisture loss. This metric can be expressed mathematically as

$$SEC = \frac{\dot{m}_{air}(h_{inlet} - h_{ambient})}{\dot{m}_{water}} \qquad (5)$$

It certainly is true that other performance measures can be utilized to quantify the efficiency of a tea-leaf drying system. The SEC was chosen as the most appropriate measure for the present study.

The movement of moisture from a solid object to a fluid stream is governed by internal diffusive processes and by external convective processes. At different times during the drying, one or the other of these two processes dominates. When drying is convective limited, it means that there are more or less constant surface moisture levels. On the other hand, when moisture transport is diffusively dominated, significant moisture gradients are developed within the object and the surface becomes drier than the interior. In the first of these situations, the rate of mass transfer is approximately constant. In the second of these, the rate of mass transfer decreases with time.

Consequently, during the drying process, there are three primary stages, each governed by different phenomena. The first stage is typically a short initial transient wherein the temperature of the drying object is increased by the warm drying fluid. Next, there is a constant drying rate period where the moisture transfer is dominated by convection. Finally, the last stage where the drying rate decreases until it ultimately becomes zero. These stages and their behaviors have been described in prior literature [14]. Often, the largest loss of mass occurs in the second (constant-rate) stage. This observation has been reported in the literature and in experiments carried out in support of this research project.

One of the goals of the present manuscript is to provide the ability to simply model the moisture content of tea leaves during the constant rate drying stage. During this portion of the drying process, the rate of moisture transfer was steady and the expression in equation (5) can be rewritten as

$$SEC = \frac{\dot{m}_{air}(h_{inlet} - h_{ambient})t}{M_{water}}, \qquad (6)$$

where M_{water} is the total water transfer across the control volume boundary in some duration t.

While the aforementioned expressions allow a comparison of performance, they do not provide a predictive means of estimating the moisture content in the leaves for various flowrates. In order to create a predictive method, it is necessary to express the water mass transfer as a constitutive equation. That results in

$$\frac{dm_{leaves}}{dt} = \dot{m}_{water} = h_{mass}A_{surf}(\Delta\omega)_{B.L.} \qquad (7)$$

Here, $\Delta\omega_{B.L.}$ represents the difference in specific humidity across the mass-transfer boundary layer separating the tea leaves from the air stream. The term A_{surf} is the surface area of the leaves and h_{mass} is the mass transfer coefficient. From equation (7), the requirements for a steady rate of moisture loss is evident. The mass transfer coefficient (h_{mass}), the surface area of the leves (A_{surf}) and the difference in specific humidity between the surfaces of the leaves and the airstream $\Delta\omega_{B.L.}$ must all be constant in time. Inasmuch as h_{mass} depended on the specifics of the fluid flow (similar to the convective coefficient between a surface and a fluid), it is common to relate the transfer coefficient to the flowrate [15] such as

$$Sh = C \cdot Re^n Sc^{1/3} \qquad (8)$$

Here Sh, Re, and Sc are the Sherwood, Reynolds, and Schmidt numbers, respectively. The symbols C and n are experimentally determined constants from external convective experiments carried over the past nine decades [15]. Since by definition, the Sherwood number is related to the mass transfer coefficient through

$$Sh = \frac{h_{mass}L_c}{D}, \qquad (9)$$

where L_c and D are a characteristic length of the leaf and the mass diffusivity of water vapor in air. The length was measured with a 300-mm ruler; it was taken to be the distance from the bud to the leaf tip. When it is recognized that the Reynolds number scales with flowrate, equations (7)–(9) can be combined and simplified to

$$moisture_{final}(t) = moisture_{initial} - C' \cdot Q^n \cdot t, \qquad (10)$$

for the portion of the drying process where the rate of drying is constant. To test the validity of the model, ten separate experiments were performed for flowrates that varied from 41 to 167 L/sec and with drum rotation rates of 120 rpm. In equation (10), a number of constant multipliers have been combined into a new symbol C'. The symbol n is identical to that of equation (8). Values of $C' = 0.0016$ and $n = 0.8$ are used to provide the fit to that data. It should be noted that the value of n used here is in good agreement with the values reported in [15] for other external-flow convective situations. This agreement lends credibility to the present results.

Here, moisture content is a mass fraction of liquid in the leaves (measured on a wet-basis), Q is volumetric flowrate in m^3/sec, and time is in seconds. With only these two parameters, the results in Figure 3A–J were obtained. The specific flowrates for each comparison are

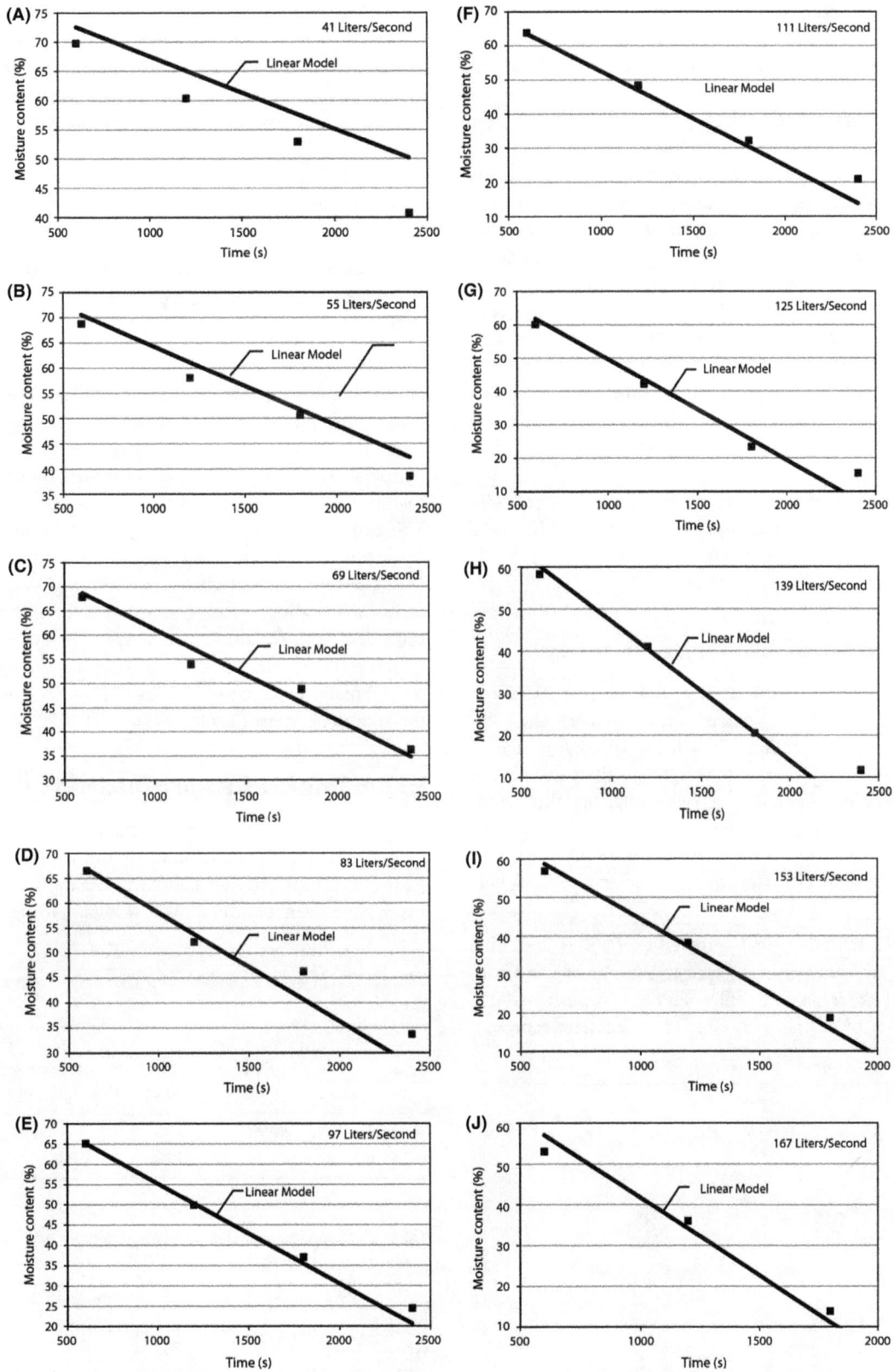

Figure 3. Comparisons of linear two-parameter model with experiments for ten different flowrates. Each flowrate are listed as annotations within the images.

listed in the image. For the results shown in the image, initial moisture contents were 80%. Also, in the experiments, the inlet air temperature was set to 80°C for all cases.

It is seen that there is very good agreement between this simple model and the experimental results. This finding suggests that the dependency of the mass transfer on flowrate is adequately found through the power relationship. Close inspection of the images reveals that for higher flowrates, there was a slowing of the evaporation process when the moisture content drops to values below the 10–15% range. This finding is expected as the constant drying regime gives way to a diffusion limited drying process. In recognition of this, the linear model should not be used to predict moisture contents below 10–15%. At those levels, internal diffusion dominates the process and the rate of mass transfer decreases. Since final green tea moisture levels are often below 10–15%, it is necessary for practitioners to make an account for this. Nevertheless, the methodology employed here is useful for rotating drum drying systems and can be extended to other systems as well.

Comparison of Energy Efficiency

Next, a comparison of the energy efficiency for the above reported cases will be made. In particular, the SEC expressions from equation (6) will be found for each of the ten cases in Figure 3. The results are then normalized to the energy use for the lowest flowrate. So, the definition of the normalized SEC is

$$(\text{Normalized SEC})_{\text{Flow rate } i} = \frac{\text{SEC}_{\text{Flow rate } i}}{\text{SEC}_{\text{lowest flow}}}. \quad (11)$$

The results are shown in Figure 4. There it is seen that flowrate significantly impacts energy use. All of the results of Figure 4 correspond to inlet air temperatures of 80°C and 2400 sec of drying. However, the results are represen-

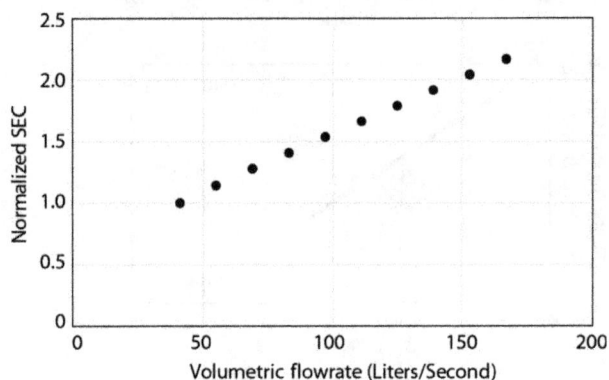

Figure 4. Impact of flowrate on energy usage.

tative of other inlet temperatures and other durations. It is clearly seen that if the desire is to dry with minimal energy requirements, lower flowrates should be employed. On the other hand, if the goal is to obtain very low moisture rates for a specified drying duration, higher flowrates should be used. To quantify the results, flowrates of 170 L/sec will more than double the energy requirement for each evaporated unit of moisture.

Comparison of the new rotating dryer to a conventional drying system

In order to justify the effort of the new system, it must provide superior performance compared to the simpler alternative (a static system). The comparison will be made using a number of metrics: drying time, evaporation rate, and SEC. Experiments were carried out at multiple air temperature settings to elucidate sensitivity of the results to those parameters. The following results were obtained. The experiments were carried out so that similar overall mass transfer was achieved with the two drying methods. It is seen that in all cases, the rotating system outperforms the static device. The rotating drying time is 2.5 times less than for the static system. The evaporation rate of the static system is ~55% of that for the rotating system. Finally, in all cases, the use of energy was lower for the rotating system (Tables 1–3).

Experimental Reproducibility

To provide some assessment of the quality for the experiments and the reproducibility of the results set forth in Figure 3, values of the standard deviation from various cases have been obtained. The results, shown in Table 4,

Table 1. Comparison of static and rotating dryer with 80°C air temperature.

Dryer type	Drying time (sec)	Evap. rate (g/m² per sec)	SEC (kJ/kg)	Ratio of drying time	Ratio of evap. rate	Ratio of SEC
Static	2280	0.41	2760	2.5	0.54	1.92
Rotating	900	0.76	1440			

Table 2. Comparison of static and rotating dryer with 100°C air temperature

Dryer type	Drying time (sec)	Evap. rate (g/m² per sec)	SEC (kJ/kg)	Ratio of drying time	Ratio of evap. rate	Ratio of SEC
Static	1800	0.52	2970	2.5	0.55	2.33
Rotating	720	0.95	1270			

Table 3. Comparison of static and rotating dryer with 120°C air temperature.

Dryer type	Drying time (sec)	Evap. rate (g/m² per sec)	SEC (kJ/kg)	Ratio of drying time	Ratio of evap. rate	Ratio of SEC
Static	1200	0.78	1650	2.5	0.55	1.43
Rotating	480	1.4	1150			

Table 4. Comparison of static and rotating dryer with 120°C air temperature.

	Standard deviation of moisture content			
Airflow (L/sec)	600 sec	1200 sec	1800 sec	2400 sec
41	1.084	0.665	0.451	0.501
55	1.178	0.266	0.465	1.135
69	0.445	0.148	0.769	0.582
83	0.827	0.580	0.873	1.122
97	1.555	0.235	0.251	0.290
111	1.490	0.274	0.665	0.601
125	1.717	4.289	0.484	0.684
139	0.421	0.238	0.467	0.499
153	0.948	0.548	2.113	1.406
167	0.841	0.750	0.636	0.393

reveal that the standard deviation for the moisture content is very small compared to the magnitude of the measurements. The quantification of the uncertainty lends credibility to the conclusions which are derived from the experiments.

Concluding Remarks

Here, a set of experiments have been performed to assess the performance of a prototype rotating tea drying system. The system allows variation of the rotating rate of the drum which holds the tea, incoming air temperature, and air flowrate. It was found that for the system under consideration, 120 rpm was required to properly agitate the tea leaves. Experiments focused on elucidating the impact of airflow on drying performance were completed.

It was found that higher airflow rates led to faster drying of the tea leaves. While this result was expected, it was also found that higher airflows led to significantly more energy use (lower energy efficiency) than lower airflows. In fact, the thermal performance varied linearly with airflow. A secondary benefit which is expected but was not quantified as part of this research is that the rotating drying method will better preserve the aroma of green tea compared to static drying beds.

The experiments spanned the three main drying regimes (initial drying where tea leaves are brought to an elevated and steady temperature, main drying during which surface mass transfer is the main resistance, and mass loss in steady rate over time, and final drying where tea-leaf moisture decreases substantially so that internal diffusion resistances are dominate). The most important stage is the main drying phase. A simple two-parameter model was developed to model moisture loss in this region. Despite the wide range of air flowrates, the two-parameter model was sufficient for meaningful predictions. In fact, it can be used to predict moisture loss for air flowrates that were not considered here. While the two parameters are expected to be dictated by the specific operation of the dryer, it is believed that other drying systems can similarly develop two-parameter models that can be used for similar purpose.

Conflict of Interest

None declared.

References

1. International Tea Limited, 2008. Tea trade statistics and research, 1 Carlton House Terrace, London.
2. Langat, N. 2014. Development of an improved fluid bed dryer system for green tea drying in the industry, PhD Thesis, Kenyatta University.
3. Sadeghi, M., and M. H. Khoshtaghaza. 2012. Vibration effect on particle bed aerodynamic behavior and thermal performance of black tea in fluidized bed dryers. J. Agric. Sci. Technol. 14:781–788.
4. Udell, K. S. 1985. Heat transfer in porous media considering phase change and capillarity – the heat pipe effect. Int. J. Heat Mass Transf. 28:485–495.
5. Vafai, K. 1984. Convective flow and heat transfer in variable-porosity media. J. Fluid Mech. 147: 233–259.
6. Alazmi, B., and K. Vafai. 1999. Analysis of variants within the porous media transport models. J. Heat Transfer 122:303–326.
7. Vafai, K., R. L. Alkire, and C. L. Tien. 1985. An experimental investigation of heat transfer variable in porous media. J. Heat Transfer 107:642–647.
8. Mickley, H. S., and C. A. Trilling. 1949. Heat transfer characteristics of fluidized beds. Ind. Eng. Chem. 41:1135–1147.
9. Vafai, K., and R. Thiyagaraja. 1987. Analysis of flow and heat transfer at the interface region of a porous medium. Int. J. Heat Mass Transf. 30:1391–1405.
10. Amiri, A., and K. Vafai. 1998. Transient analysis of incompressible flow through a packed bed. Int. J. Heat Mass Transf. 41:4259–4279.
11. Minkowycz, W. J., A. Haji-Sheikh, and K. Vafai. 1999. On departure from local thermal equilibrium in porous media

due to a rapidly changing heat source: the Sparrow number. Int. J. Heat Mass Transf. 42:3373–3385.

12. Harold, N., and P. D. Graham. 1992. Green tea composition, consumption, and polyphenol chemistry. J. Prev. Med. Hyg. 3:334–350.

13. Strumillo, C., and T. Kudra. 1986. Drying: principles, applications, and design. Gordon and Breach Scientific Publishers, New York, NY.

14. Kawai, S. 1993. Granulation and drying of powdery or liquid materials by fluidized-bed technology. Drying Technol. 11:719–731.

15. Sparrow, E. M., J. P. Abraham, and J. C. K. Tong. 2004. Archival correlations for average heat transfer coefficients for non-circular and circular cylinders and for spheres in cross flow. Int. J. Heat Mass Transf. 47:5285–5296.

10

Nuclear fusion as a massive, clean, and inexhaustible energy source for the second half of the century: brief history, status, and perspective

Joaquin Sánchez

Laboratorio Nacional de Fusion, CIEMAT, 28040 Madrid, Spain

Keywords

Fusion, Energy, Plasma

Abstract

Fusion energy, based on the use of broadly available inexhaustible resources as lithium and deuterium and with minimal impact to the environment, aims at a change in the energy supply paradigm: instead of its current dependence on natural resources and environmental impact, energy would become a technology-dependent resource with unlimited adaptive availability and whose unit cost should decrease as technology progresses. This article intends to give a picture of where fusion research stands today and the perspectives: the achievements, the difficulties, the current status, marked by the construction of the ITER experiment which will demonstrate the scientific feasibility of fusion power, and the perspectives toward the first demonstration power plant, DEMO, which, according to the European Roadmap, could start the construction shortly after the full power experiments in ITER (<2030) and be in full operation, generating net electricity into the grid, by 2050.

Introduction

It is widely recognized that the energy supply is one of the largest challenges that mankind will be facing during this century. Population growth and increasing per capita consumption of goods and services in the emerging countries will lead to a likely twofold energy demand in a couple of decades, despite the efforts toward efficiency and energy savings in the developed countries. In addition, new demands will appear derived from the need for massive water supply and food production or large-scale recycling of basic materials.

In this scenario, we will need to count on massive sources of energy, environmentally friendly, and based on abundant primary resources. Nuclear fusion intends to be one of these sources, its main objective being to transform the energy paradigm: from today's dependence on natural resources and environmental impact into a technology-dependent resource, in the same way as we see today Internet access, mobile communications, or computer power: a resource whose availability can grow easily with demand and whose cost per unit decreases as technology progresses.

Nuclear fission, the basic process in today's nuclear power plants, consists on breaking a large nucleus into medium size ones, nuclear fusion is based on the opposite reaction: the union of two small nuclei in order to generate a larger one, but still small. In both cases, the mass of the reaction products is slightly smaller than that of the original nuclei and this lost mass is converted into energy according to Einstein's equation $E = mc^2$. The nuclear forces involved in the process are much larger than the electromagnetic forces which are the basis of standard combustion of fuel and so is the capability of energy production per unit mass of fuel: one gram of fusion fuel, equivalent to about 7 tons of oil [1] would be enough to provide the full energy consumption of an average person during more than 1 year.

Fusion is the reaction which powers the sun and all the stars. In the center of a star, the incredible high gravitational forces generate the conditions for the fusion of hydrogen nuclei into deuterium as the first of a chain of reactions in which deuterium will fuse with remaining hydrogen into helium (He^3) and later the helium with itself (to generate He^4). On Earth, where we cannot count on such strong gravitational forces, we will need to look for more accessible reactions, though still very hard to achieve. The fusion reaction with larger cross section under reasonably achievable conditions is the fusion of deuterium and tritium, two isotopes of hydrogen. The reaction produces as a result a helium nucleus and a neutron and releases 17.6 MeV (mega electron volt) of energy, (91,000 kWh per gram of fuel). Of this energy, 4/5 is carried by the neutron and the remaining 1/5 by the helium nucleus.

A fusion power plant would be essentially a thermal plant. The energy released by the fusion reaction is absorbed by a coolant and extracted to the heat exchangers and to the electricity-producing turbines. The fusion fuel would be composed of two species: deuterium and tritium. Deuterium exists in natural water in a fraction of 33 mg/L. On the other hand, tritium, another, heavier, hydrogen isotope, is unstable and does not exist in nature. It is usually a secondary product of fission power plants. Fusion reactors would generate in situ the tritium they would consume by means of neutron bombardment of lithium, another chemical element. Lithium is also very abundant in nature and, given the fact that the required quantities are very small in comparison with the amount of energy obtained, it could be extracted at affordable cost from salts solved in seawater. The estimated reserves of lithium in seawater would be sufficient to satisfy the world's energy needs during many million years and it is expected that in the future technologies for mastering the deuterium–deuterium reaction would become available, thus extending the availability of fusion fuel beyond the expected life of the solar system.

The main exhaust product resulting from the reaction is helium, the very same gas we use to fill balloons for children. This element is harmless for people and the environment, it does not contribute to the greenhouse effect and, in fact it does not even accumulate in the atmosphere: due to its low weight it escapes to the space. In addition, the quantities produced would be very small: if all the energy in the world would come from fusion, the amount of helium produced worldwide would be in the order of several thousand tons a year, to be compared with the ten billion ton CO_2 per year released currently.

Safety is a major concern on every industrial facility and in nuclear plants in particular. One of the advantages of a nuclear fusion plant would be its intrinsic safety: fusion plants will be safe not just because they will be carefully designed and operated, they will be safe because the physical properties of the process make impossible an uncontrolled fusion reaction. As we will discuss later, the very high temperatures required in the reactor, in the order of several hundred million degrees, are impossible to sustain in case any malfunction arises, for example, an air leak into the reactor would immediately bring down the temperature and extinguish the fusion reaction. Another element to be taken into account is that, whereas in a fission reactor the amount of fuel inside the reactor could sustain the reaction for months (and therefore it might be more difficult to manage if control is lost), in a fusion plant the fuel contained inside the reactor would last for only a few seconds if the supply from outside is interrupted. As an example: the cooling system, whose failure was the cause of the problems at the Fukushima reactor, was not even an important safety component in the large experiment ITER because there was no significant residual heat when the operation stops.

The main safety concern in a fusion plant would be the existence of tritium, which is radioactive and, even if it is not a long-term pollutant (its half life is 12.3 years), it is dangerous if inhaled or ingested as tritiated water. Fortunately, tritium is only used as a transition element and the main supply comes as lithium, but still the storage of several kilograms of tritium is difficult to avoid and, as it would happen with any other dangerous substance in an industrial facility, safety measures are required to prevent any release to the environment. The current designs would guarantee that in the worst case accident, there would be no need to evacuate people staying outside the facility fence.

The main drawback of fusion as a potential source of energy is the difficulty to generate and sustain the reaction. In order to achieve the reaction, the two colliding nuclei must get close enough to allow the short range nuclear forces to act, this can only be achieved if their energy is high enough to overcome the electrostatic repulsion of the two positively charged nuclei. Accelerating deuterium or tritium ions to these energies, 15–20 keV, is not particularly difficult, the difficulty arises when we try to get energy gain from the process: launching an ion beam against a target at the required energy would produce just a very small fraction of fusion reactions because in most cases the long range coulombian repulsion will deviate the ions, which will miss the target, and also many of them will release their energy in collisions with the electrons. The only way to achieve an efficient process is to be able to confine the accelerated ions in a closed space, in such a way that, after having gained the required energy, they have many opportunities to collide before their energy is lost. The main problem is the availability of a suitable recipient: a gas where the average particle energy is 15 keV has a temperature of 170 million degree.

Since 1950s, scientists have been trying to find this kind of recipient and have developed two main families of experiments: inertial confinement, based on a fast heating of the fuel so that it enters the fusion reaction before it has time to expand, and magnetic confinement, based on the fact that at such high temperatures the gas, in state of "plasma", is composed of charged particles, which can be confined by magnetic fields.

Inertial fusion uses an ion beam or a laser, the preferred option nowadays, as the means of delivering a big amount of energy in a very short time to the deuterium–tritium (DT) target. The target is illuminated with spherical symmetry and this produces a pressure wave which converges toward its center. At a given moment, a fusion "spark" should be generated in the center and the heat generated by these initial fusion reactions would propagate back the fuel burn toward the rest of the target. The most advanced inertial fusion experiment is currently the National Ignition Facility (NIF) located in the Lawrence Livermore National Laboratory, (Livermore, CA). Recently, experiments have been reported where the initial spark of fusion has been found [2]. Its extension to the whole target has not yet been achieved and the energy production is still a small fraction of that delivered by the lasers, however it is a promising result. A similar experiment, the "Laser Mégajoule" is under construction in France, essentially with military purposes: inertial fusion experiments can be used to validate the models which are the basis for the computer simulation of thermonuclear explosions.

In parallel, the largest worldwide effort toward fusion energy has been and is being devoted to the so called "magnetic confinement". The DT fuel at such high energies is on "plasma" state, a gas where ions and electrons move separately, and can, therefore, be confined by a magnetic field, which essentially allows particles to move freely along the field lines but forces them to move in small circles when trying to go in perpendicular direction. The next step is to construct a configuration where field lines close on themselves, so, ideally, particles would stay indefinitely moving along those closed trajectories. These configurations have been implemented since the 1960's along two main families of toroidally shaped devices: "stellarators" and "tokamaks" which differ in the way they generate the necessary complementary field which is required to avoid particle drift in a toroidal geometry. The third approach, a linear configuration with "magnetic mirrors" at both ends, turned to be much less effective and has been less developed.

Fusion Devices: The "Tokamak" and ITER

The "tokamak" – word derived from the Russian expression for "toroidal chamber with magnetic coils" – was first developed by I. Tamm and A. Sakharov (who later received the Nobel peace prize) in the early 1960's and was rapidly adopted by researchers around the world. Thirty years later (1991), the Joint European Torus (JET) – a tokamak experiment owned by the European Commission and located in Culham, near Oxford (UK) – carried out the first D-T experiment toward controlled fusion, providing a substantial amount of energy from the fusion reactions [3], few years later (1996–97), JET and the TFTR tokamak (Princeton, NJ) reached fusion power levels in the order of 10–15 MW, with a ratio of fusion power to heating power of 60% [4, 5].

Despite the criticism to fusion researchers to be "always 40 years away" from the goal of fusion power, the reality is that the efficiency of tokamak experiments, measured as the "triple product" of ion temperature, ion density, end energy confinement time ($T_i.n_i.\tau_E$), was growing at comparable pace to that of microprocessors between 1960 and 2000 and will hopefully recover when the large ITER experiment will start (see Fig. 1). However, it is necessary to realize the magnitude of the challenge: the magnetic field confines very well a single particle, but, as the many particles collide, there is diffusion across field lines and both particle and energy flow slowly away. In order to minimize these losses we have two essential tools: one is to increase the magnetic field, but this has a limit for the superconductor coils which generate it, so it is difficult to envisage a device with an average field above 6–7 Tesla; the second tool is to increase the machine size. "Wind tunnel" comparisons with tokamaks of similar geometry and increasing size have shown that in order to achieve "ignition" conditions, a situation where the energy generated in the fusion reactor can compensate the losses and maintain the required high temperatures which sustain the reaction, we need a hot plasma volume in the order of 1000 m^3 for a standard magnetic field value of 5–6 T. This means very large, complex, and expensive devices, with development times in the range of 10–20 years.

After the success of JET and TFTR, the next step will be ITER, a joint experiment of seven parties which represents more than half the world population: China, India, Japan, Korea, Russia, the United States, and Europe, which acts as a single party and is represented by the European Commission. ITER (from the latin word "iter", the way), with nearly 1000 m^3 of hot plasma, will aim at demonstrating the scientific feasibility of fusion as an energy source. The specific objective is to obtain energy gain $Q = 10$, which means that ITER will generate ~500 MW of fusion power with 50 MW of external power being injected to heat the plasma. The gain $Q = 10$ would be sustained during 400 sec periods; as a second objective, a less demanding value of $Q = 5$ would be sustained for periods of 1500 sec (see details in Table 1) In

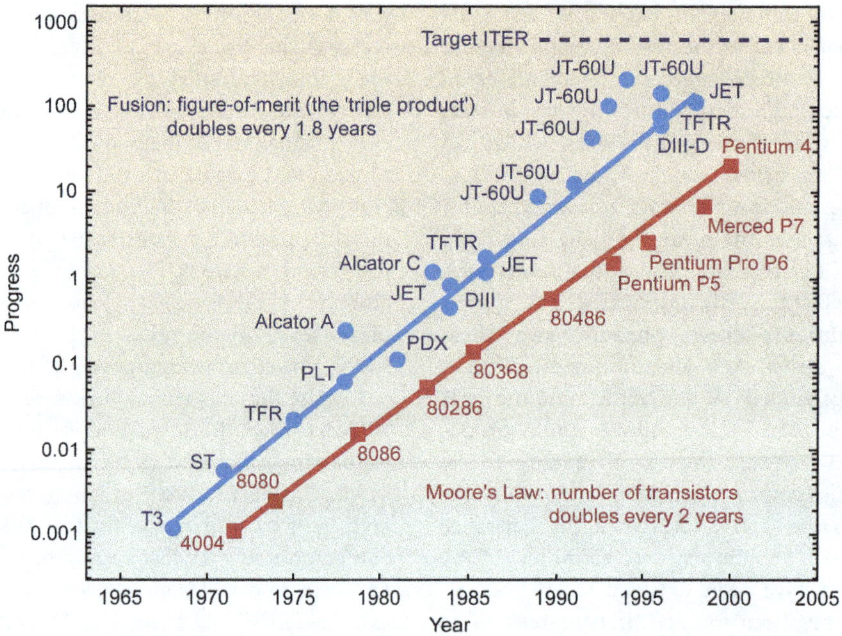

Figure 1. Progress of the fusion triple product $T_i.n_i.\tau_E$.

Table 1. Main ITER parameters.

Fusion power	500 MW
Fusion power gain (Q)	>10 (for 400 sec inductively driven burn)
	>5 (1500 sec)
Plasma major radius (R)	6.2 m
Plasma minor radius (a)	2.0 m
Plasma vertical elongation	1.70/1.85
Plasma current (Ip)	15 MA
Toroidal Field at 6.2 m radius	5.3 T
Installed auxiliary heating	73 MW
Plasma volume	830 m^3
Plasma surface area	680 m^2
Plasma cross-section area	22 m^2

addition, ITER will carry out a number of experiments to test the technology developments necessary for a power plant, in particular, the "breeding blanket" modules which will test the technology for tritium generation from lithium (Fig. 2).

ITER is a large extrapolation in volume (10 times larger than JET) and also in technology. In addition to the use superconducting coils, cooled at 1.4 K while located less than a meter away from the million degree hot plasma, the largest challenges come from the goal to operate long pulses at full fusion power: all the internal elements need active cooling as well as neutron-resistant functional materials, particularly insulators.

The challenge in science and technology is formidable but it is not the only one: ITER is also to some extent a social experiment. With magnetic fusion research being a declassified activity both by the eastern and western countries during the cold war, the undertaking of a large joint experiment was one of the agreements between presidents Reagan and Gorbachev on their summit of November 1985 at Geneva. The European Union and Japan joined the project immediately and the design evolved slowly during the following years in a process which included the temporary withdrawal of the USA and the decision in 1998 to redesign the device in order to make it more affordable and ended with the delivery of the final design report in 2001. In 2004, the USA returned to the project and China and Korea, as well as later India, expressed their interest to join, which was welcome in order to share the multibillion costs of the project.

The main drive for the interest of the parties was fusion energy as a long-term goal, but, in the shorter term, their interest was also focused in the important technology developments around ITER. This led to an organization based on a moderate size central team, located at the project site in Cadarache (France) and in charge of about 15% of the total ITER budget, and seven, smaller but still strong, teams, named "domestic agencies", located at the different parties' headquarters and in charge of delivering in-kind components to the central team for about 85% of the budget. Europe as the host party would provide nearly half of the total budget and the remaining part would be covered by the other six parties.

Figure 2. Artist view of the ITER device.

The organization based on in-kind contributions allowed for an a priori distribution of the participation and could accommodate the wish of the parties to participate in the technologies of their interest, in addition, this was the way to allow for the emerging countries to have lower costs by developing components with their own workforce. On the other hand, the system, based on a central team which prescribes the design but does not have responsibility on the cost of construction of these components and the domestic agencies which have to procure and pay for the components, is prone to produce internal discussions and delays in the decisions.

The ITER agreement was signed 21 years after the idea was launched, in November 2006 and the first estimate of the construction period was 9 years. Soon it became evident that between the report delivered in 2001 and the necessary constructive design there was much more distance than originally estimated. The report had concentrated in the main machine parameters, the related physics and the design of the critical high-technology components, but ITER was a very complex industrial plant, subject to a nuclear license, not as a nuclear power plant but as a nuclear facility, and with a very demanding integration process into the buildings' design. The consequence of having concentrated on the critical components, necessary to guarantee the feasibility of the project, but having overlooked the more conventional parts of the facility was an underestimation of cost, which essentially doubled after an in depth revision, and construction time.

Although the cost has been kept within reasonable bounds after the 2010 revision, suffering moderate increases but remaining within the limits of the originally foreseen contingency, the schedule seems difficult to control. Subsequent revisions of the baseline schedule have led to an estimate for the "first plasma", which would mark the end of the construction period, to happen in 2022–23. This delay is the accumulation of several causes: lack of a finalized design, lack of manpower at the central team imposed by the budget restrictions, delays in Japanese components after the 2011 earthquake damaged some key facilities, additional licensing requirements derived from the post–Fukushima revision of all the nuclear procedures, etc., but a significant part of the delay comes from the extremely complex organization of the project and the distribution of roles and responsibilities. A typical example is when a component design performed by the central team is felt as an over specification which rises cost by the domestic agency in charge of the construction: the domestic engineers will come back with redesigns aiming to lower the cost while the central ones will be just worried by confidence in the functional role of the component, thus entering on a loop with no clear outcome. Many of those organizational problems have been highlighted by the management assessment report commissioned recently by the ITER Council.

In the meantime, the good news is that most of the high-tech components like the superconducting coils or the vacuum vessel, have undergone the final designs, with the corresponding design reviews, and the related construction contracts have been awarded to industry, which is so far progressing without known major difficulties. The major technical problem happened with the central solenoid superconducting cable, which in 2012 was showing degradation with operation time in the samples tested. Fortunately, further R&D by the Japanese team in charge of this cable provided, on time for the coils construction at the USA, a new design which was successfully tested without showing any degradation.

Other elements of confidence have been provided from the physics side by the research carried in supporting experiments around the world. As an example, one of the elements of concern with the original design of ITER, which used a carbon inner wall, was the problem of tritium accumulation in the form of hydrocarbons deposited in remote parts of the device, which could lead to the requirement to stop operation after every few experiments and undertake a complex tritium removal procedure. On the other hand, the use of carbon, due to its good behavior at high temperature and low atomic number, was capital for an efficient operation from the physics point of view and no clear alternative was at sight. Fortunately, tests of plasma operation with a full tungsten wall carried out in the recent years in the German device ASDEX-U and with a tungsten divertor and beryllium wall (the same combination of ITER) in JET have demonstrated reliable efficient operation without the tritium retention problem. Now ITER has changed its design and the lower part of the inner wall, the so called "divertor", where the interaction with the plasma concentrates, will use tungsten as plasma-facing material.

In parallel, developments in the control of the periodic busts of power to the wall (the so called Edge Localised Modes, ELMs), the progress in the understanding of energy and particle confinement and its extrapolation to ITER size or the achievement of reliable operation at the high plasma densities projected for ITER reinforce our confidence in the operational success of the experiment.

Still some of the original concerns remain, for instance the need to avoid and mitigate the so called "disruptions", rapid losses of confinement which could lead to damage of internal components, but progress is steady in all those fronts.

This situation, with organizational delays in one side but smooth progress in the most critical components, and physics projections in the other side, makes us to be relatively optimistic toward the actual success of the project and encourages us to work in order to find the right organizational frame to avoid further delays.

With ITER starting operation in 2023, the critical high-gain results with $Q = 10$ will come shortly before 2030. One of the answers we expect to get from these experiments is the efficiency of the plasma heating by the high-energy He ions generated at the fusion reaction, also called "alpha particles". This is crucial for the future of the tokamak as a fusion reactor because we need to use these alpha particles to maintain the high temperatures which sustain the reaction. In the fusion reaction, neutrons carry 80% of the released energy, they cannot be retained by the magnetic field, and therefore they cannot contribute to sustain the reaction (in the power plant their energy will be extracted by the coolant and used to drive the turbines). Alphas will carry only 20% of the fusion power but they are charged particles which can be retained in the plasma by the magnetic field and contribute to sustain the plasma temperature. The problem is that, whereas the plasma particles have an average energy of 15–30 keV, the alphas are born with an energy of 3.5 MeV, hundred times higher, and they would escape quickly unless the energy transfer mechanism by means of collisions is efficient enough. Preliminary experiments in JET [6] as well as theoretical predictions show that, very likely, the alphas will indeed heat the plasma efficiently, but the ultimate test will be performed in ITER. With $Q = 10$, the power generated by fusion will be ten times the heating power injected externally, then the alphas, which carry 20% of this power will provide twice the externally injected heating, leading to a clear effect that will serve as a concluding test of the alpha heating.

Toward the Demonstration Reactor

ITER will demonstrate the scientific feasibility of fusion as energy source and will also test key technologies for the reactor but ITER will not yet be a real power plant. The main differences between ITER and a demonstration power plant, the so called "DEMO", from "demonstration", would be: tritium self-sufficiency, full plant energy efficiency, use of low-activation neutron-resistant materials, and reliable continuous operation. In the following pages, we will address the status and perspective of the related developments.

Tritium Self-Sufficiency

As explained in the introduction, fusion plants would need to generate in situ the tritium they will consume by bombarding lithium with the fusion-generated neutrons, through the reaction: $n+Li^6 \rightarrow T+He^4$. This function will be performed by the so called "breeding blanket" which will surround the plasma. The breeding blanket has also two main additional functions: to extract the power of the neutrons conveying it to the steam generators and the turbines and to shield the sensitive components, in particular the superconducting coils, from the neutron flux which would heat and damage them. This makes of the breeding blanket a very demanding nuclear component.

ITER will not be equipped with a full breeding blanket, and therefore it will not be self-sufficient in tritium. The current plan is to purchase the tritium that it will consume from an external source, essentially the Canadian nuclear fission programe, but it will have a number of smaller blanket modules which will be used to test different tritium-generation technologies.

The ITER blanket modules will test different options for the three main elements in a breeding blanket. First, there is the choice of breeding material, the main options being molten eutectic lithium–lead, with 90% Li^6 enrichment or lithium salt pebbles (Li_4SiO_4 or Li_2TiO_3) with 30–60% Li^6 enrichment. Secondly, we need a neutron multiplier, usually beryllium or the lithium–lead itself, because a fraction of the neutrons generated in the fusion reaction will fail to hit the breeding material. Fortunately, the neutron energy required for the breeding reaction is relatively low and using a neutron multiplier each single 14 MeV fusion neutron can generate several secondary neutrons able to produce tritium. The third element is the coolant, which must extract the energy deposited and generated in the blanket (the breeding reaction is exothermic), here the options are: water cooling, helium cooling, or dual cooling by helium and lithium–lead. [7].

The integral test of breeding blanket modules in ITER will be a crucial experiment in order to validate the different technologies. The strategic value of those designs is such that the breeding blanket program is not part of the ITER agreement, which foresees that all the knowledge generated in the project will be shared among the seven parties, but a separate activity whose results would be private intellectual property. It has to be coordinated due to the host role of ITER but the information obtained in the experiments will be sole property of the related party and in principle would not be shared.

The European Roadmap toward fusion electricity [8] includes a breeding blanket technology programe parallel to the preparation of the validation tests of the ITER blanket modules. The four technologies selected are the two which Europe will test in ITER, lithium–lead, and ceramic pebbles both cooled by helium, plus two additional options. The water cooled lithium–lead, as a shorter term option, has the advantage of avoiding the use of helium, which might become scarce if thousands of fusion plants need to use it, and the high cooling capacity of water; on the other hand, water generates corrosion problems as well as safety issues and its temperature operation window (280–325°C) is hardly compatible with the low-activation neutron-resistant materials which we have at hand today. The dual coolant, a longer term option uses a faster circulation of the liquid LiPb to use it as high-temperature coolant. The use of insulating inserts and an additional helium cooling system allow for the structural material to remain at lower temperature than the main coolant, which has a much higher temperature and leads to a higher plant efficiency.

High Gain for Plant Efficiency and the Energy Extraction Problem

The $Q = 10$ power gain of the ITER plasma will not be enough for having a real energy gain in the full balance of the plant, which must take into account all the energy consumption of the coils, cryogenic systems, and other auxiliary systems as well as the wall plug efficiency of the plasma heating systems and the efficiency of the thermal cycle. Overall, an efficient power station would require Q in the order of 50, which means a device with either a more efficient physics, a higher magnetic field (difficult to achieve due to the limitations in the superconductors) or a larger size.

Typical European designs of a demonstration fusion power plant [9] consider the total fusion power in the range of 2000 MW thermal and have a linear size, 1.5 times of ITER. A device with this size and power is a significant challenge, in particular, on what concerns the extraction of the power.

As explained before, the neutrons carry 80% of the power, they escape the plasma isotropically and cross the wall of the tokamak and are absorbed volumetrically in the coolant and blanket structures. The thermal power is very high but this broadly distributed load can be tolerated by the materials. The neutrons can generate a number of other issues in the materials but the thermal load is not a serious problem.

On the other hand, the remaining 20% of the power, carried by the alpha particles, together with externally injected power, also carried by charged particles, flows slowly toward the wall. The magnetic field can delay this flow but once the steady state is reached there is a continuous flux of energy toward the inner wall of the device.

All this power is conveyed to a small fraction of the wall, the so called "divertor". This is necessary to avoid the penetration of sputtered wall particles into the hot plasma that would quench the high temperature but it generates a serious problem: all this power is deposited in a narrow, several cm, ribbon along the torus, leading to thermal loads in excess of 20 MW/m², twice higher than the current engineering limit.

The possible solutions to what the experts see nowadays as the main challenge toward the success of fusion energy, operate from the two sides of the problem: cooling the plasma edge by emission of radiation in the visible and ultraviolet range, which distributes the load over a larger surface, and designing "divertor" geometries and materials which can handle the power.

Cooling of the plasma edge can be achieved by injecting gases like nitrogen, krypton, and argon [10], the goal is that a large fraction of the power is radiated at the edge while preserving the good core confinement. In addition, geometries which expand the interaction area in order to decrease the power density are being developed, like the "super X" [11] or the "snowflake" [12] divertors. The final tool to overcome this challenge is the choice of materials, the basic reference material is tungsten but liquid metal alternatives, using lithium, gallium, or tin, which offer "self-repairing" walls, are also being considered.

Neutron Resistant Materials

The power carried by the neutrons is not a big problem as explained above, however, the high fluence of energetic 14 MeV neutrons generates a different set of problems in the material. The problems will not be present in ITER, at least for the structural materials, due to the relatively low accumulated neutron fluence, but will be very severe for DEMO and for the commercial fusion plants.

Firstly, each neutron impact will give rise to a cascade of collisions which will displace many atoms from their positions. This is measured on "displacements per atom" or dpa's, one dpa meaning that, on average, every atom within the material has been displaced once. The structural material of the blanket and first wall in a fusion reactor will suffer an excess of 100 dpa's during the component lifetime, in addition, the 14 MeV neutrons, distinctly to the neutrons in a standard fission reactor, will generate transmutation reactions in the material which will produce helium and hydrogen and create blisters as well as material swelling. All these phenomena can degrade significantly the mechanical properties of the material, but there is one more adverse effect: the irradiated material becomes radioactive and will have to be treated as radioactive waste.

The materials which adapt best to the 14 MeV neutron bombardment are: vanadium alloys, titanium alloys, silicon carbide, a long-term promise but still difficult to use as structural material and, the current reference material which has achieved the highest technological maturity, the RAFM (Reduced Activation Ferritic-Martensitic) steels. As iron is relatively resilient to neutron bombardment and suffers little activation, RAFM steels, like the Japanese F82H or the European EUROFER, are based on the suppression of problematic impurities (Ni, Cu, Al, Si, Co, etc.) and the substitution of problematic alloying components (Mo, Nb) by other elements which play the same chemical role in the alloy but have a more benign nuclear behavior (Ta, W). RAFM steels would suffer less activation than the standard ones although they would still be an activated material after decommissioned from the fusion reactor. The current studies foresee that the components could be recycled after ~100 years under custody as medium-low level radioactive waste, as opposed to ~100,000 years for standard steel components under equivalent conditions. The possibility to further reduce this period depends on the level of impurity suppression technically, and economically, achievable. A fast activation

decay is also observed for vanadium alloys [13] but vanadium currently lacks industrial development and has some negative effects, like corrosion, Tritium permeation, and narrower operating temperature.

One of the problems in the development of materials for fusion reactors is the absence of intense sources of 14 MeV neutrons which could allow us to test the behavior of the material in similar conditions to those in the fusion plant [14]. EUROFER has shown good performance under irradiation in fission reactors, which essentially reproduce the dpa's effect but there is little knowledge about the effect of He and H accumulation.

One possibility is to use theoretical modeling of the irradiation effects. Activation is relatively easy to determine, as it essentially depends on the concentration of the different elements and neutron propagation calculations are possible. However, the structural changes are nearly impossible to compute starting from first principles: we are in a problem where the number of particles is in the order of Avogadro's number and the changes must be tracked in picoseconds scale for periods of many seconds (which are the characteristic times of the changes in the mechanical properties). The modeling is performed using a multiscale approach, but the approximations are such that the experimental tests of every scale model as well as an overall test of the complete modeling are necessary.

A family of 14 MeV neutron sources under consideration is based on the use of reduced size fusion reactors with modest Q but with substantial DT reaction rate sustained by external injected power and equipped with a full breeding blanket in order to self generate the tritium. The so called CTF's (Component Test Facilities) belong to this family and there are several proposals under study both in the China and the USA [15–17].

The second family of sources is based on accelerator-driven neutron generation. For example, the reference proposal, IFMIF (International Fusion Materials Irradiation Facility) considers two 40 MeV deuteron beams of 125 mA each which hit a liquid lithium target producing a neutron spectrum very similar to that of a fusion reactor. IFMIF, a 1500 M€ project, would produce 20–50 dpa/year in a reduced volume of 0.5 L and smaller rates in the wider adjacent space. It is considered as the ultimate tool to qualify materials for the fusion power plants. Currently, Europe and Japan are carrying validation developments for IFMIF components and a complete accelerator with all the basic elements will be tested in Rokkasho (Japan) in 2017. The possibility to use this prototype accelerator in an early reduced version of IFMIF is currently gaining momentum. This source could be available by the early 2020's in order to qualify components at 20 dpa for an earlier phase of DEMO.

In the meantime, the fusion materials programe is strongly involved in the development of new materials and the consolidation of the reference ones, for example, one of the current limitations of EUROFER type steels is their reduced operation temperature window (350–550°C) which might be expanded by using ODS (Oxide Dispersion Strengthened) versions with yttrium oxide. Limited irradiation tests are also carried in fission reactors (use of boron doped material or the inclusion of some amount of ^{56}Fe can simulate the He generation by 14 Mev neutrons) or using multiple ion beams to produce simultaneous dpa's and He/H implantation, at a very fast rate but in very reduced sample volumes. Those experiments can complement the theoretical models as well.

Maintenance Issues and Plant Systems

RAMI (Reliability, Availability, Maintainability, and Inspectability) will be a key issue in a complex facility as a fusion power plant if we want it to operate under economically sustainable conditions. In particular, given the fact that the structure will become activated soon after the start of operation, most maintenance operations will have to be done by remote control manipulation. This means that all components inside the vacuum vessel and many of the components inside the cryostat, even for ITER, will have to be designed compatible with Remote Handling (RH) operations: size and weight of the components, assembly method, assembly sequence, interfaces with the RH tools…etc. Today, devices like JET have shown the feasibility of complex RH operations like the full substitution of the divertor or the first wall (see Fig. 3), however, the replacement times need to be significantly shortened in a commercial reactor and this would imply to evolve from today's man-driven operation to automatic operation for many of the actions.

A lot of technology development would also be required in plant systems: tritium extraction, isotope separation systems, He, and liquid metal heat exchangers, as well as advanced thermal cycles are among the systems which are currently being developed as part of the fusion technology programes worldwide.

Steady State Reactors: The Stellarator

The "tokamak" concept, on which ITER and most fusion devices are based, is a very clever design with optimal confinement properties. In this configuration, the confined plasma contributes itself to the construction of the confining magnetic configuration, this is achieved by

Figure 3. JET remote handling system.

inducing a strong electric current in the highly conductive hot plasma. With this contribution from the plasma, some complex additional magnets that otherwise would be necessary are spared. The current also contributes to heating the plasma by Joule effect. This solution offers some advantages and disadvantages as compared with the other family of devices, the "stellarator" which assumes no help from the plasma and configures the complete magnetic field by means of additional 3-dimensionally shaped magnets.

The advantages of the tokamak: comparative simplicity and very good confinement properties, makes this configuration the best option for a fusion ignition prototype like ITER or even a first DEMO device, however, the tokamak has also some limitations derived from the strong coupling of the plasma and its confinement. First of all, the plasma current (up to 15 MA on ITER) is usually induced with a transformer effect, which is impossible to sustain in steady state. Today's tokamaks are pulsed devices and this might have implications on the management of the supply to the electric network and the components fatigue when used as a power plant. Some progress has been achieved in the development of noninductive current drive systems, but there is still a long way ahead in the path toward the complete steady state. The second problem, derived from the plasma–confinement coupling, is the existence of scenarios where plasma and confinement drop suddenly together in a very fast positive feedback process (milliseconds) which ends in a tremendous thermal release to the wall and the quench of the >10 MA plasma current. In these events, called "disruptions", very strong electromagnetic forces are generated and jets of fast electrons can achieve multi MeV energy, becoming a potential threat for the integrity of the internal components.

On the other hand, Stellarators are inherently steady state devices which could operate under stationary conditions for months and stop only for maintenance purposes. As the confinement is decoupled from the plasma, stellarators are also free of disruptions.

Stellarators are, in fact, older than tokamaks, first devices were developed by Spitzer in the early 1950's, but the simplicity and good results of the tokamak soon relegated them to a secondary role. In the 1980's, new design tools and constructive techniques, together with the introduction of radiofrequency plasma heating systems which could substitute the traditional Joule heating based on plasma current, allowed for a relaunching of the stellarators and the results from devices like the German W7AS and the Japanese LHD, a superconductor-based device, have shown the strong potential of this configuration, overcoming the main limitations in confinement that hindered the progress with earlier devices. In 2015, the large superconducting stellarator W7X (Fig. 4), currently under construction in Greifswald (Germany), will start operation. The results of this experiment might strongly reinforce the potential of the stellarator as a long-term option for the commercial reactor units, on which the engineering complexity will play a secondary role compared with the simplicity and smoothness of the operation.

Fission– Fusion Hybrid Systems

The 14 MeV DT fusion neutrons can be used to irradiate uranium 238 or thorium 232 and generate fissile material,

which could be used either in a pure fission reactor, in this case the fusion system would be a way to produce fission fuel, or in the fusion reactor blanket playing the role of energy amplification. The same DT neutrons could be used to just irradiate and "burn" the nuclear radioactive waste accumulated during the complete history of fission energy generation.

Those three applications have intermittently gained and lost attention since the idea was conceived in the 1950's and later relaunched by H. Bethe in 1979 [18]. In principle, a fusion gain $Q = 5$, complemented with a $10\times$ amplification from the fission blanket would suffice for having an efficient power plant, which means that from the fusion side, a device like ITER, and even a bit smaller, could do the job. Those who support the idea see as the main advantage the simplification of the fusion core and a faster process toward energy generation. For those opposing, the hybrids just bring together all the problems of fusion, in particular complexity, and fission: less waste but still significant, proliferation, handling of highly active material. A very interesting discussion, which includes the opinion of a "skeptics group", can be found in ref. [19]. Currently, there is no effort on hybrids in the European Roadmap, which focus on "pure fusion", but there are active groups in China and the USA and significant activity and interest have been reported by the Russian programe [20].

The Roadmap to Fusion Electricity

The parallel effort of ITER and the technology programes should converge in the construction of a DEMO power reactor. The concept of DEMO varies in the different

Figure 4. Assembly process of the W7X stellarator.

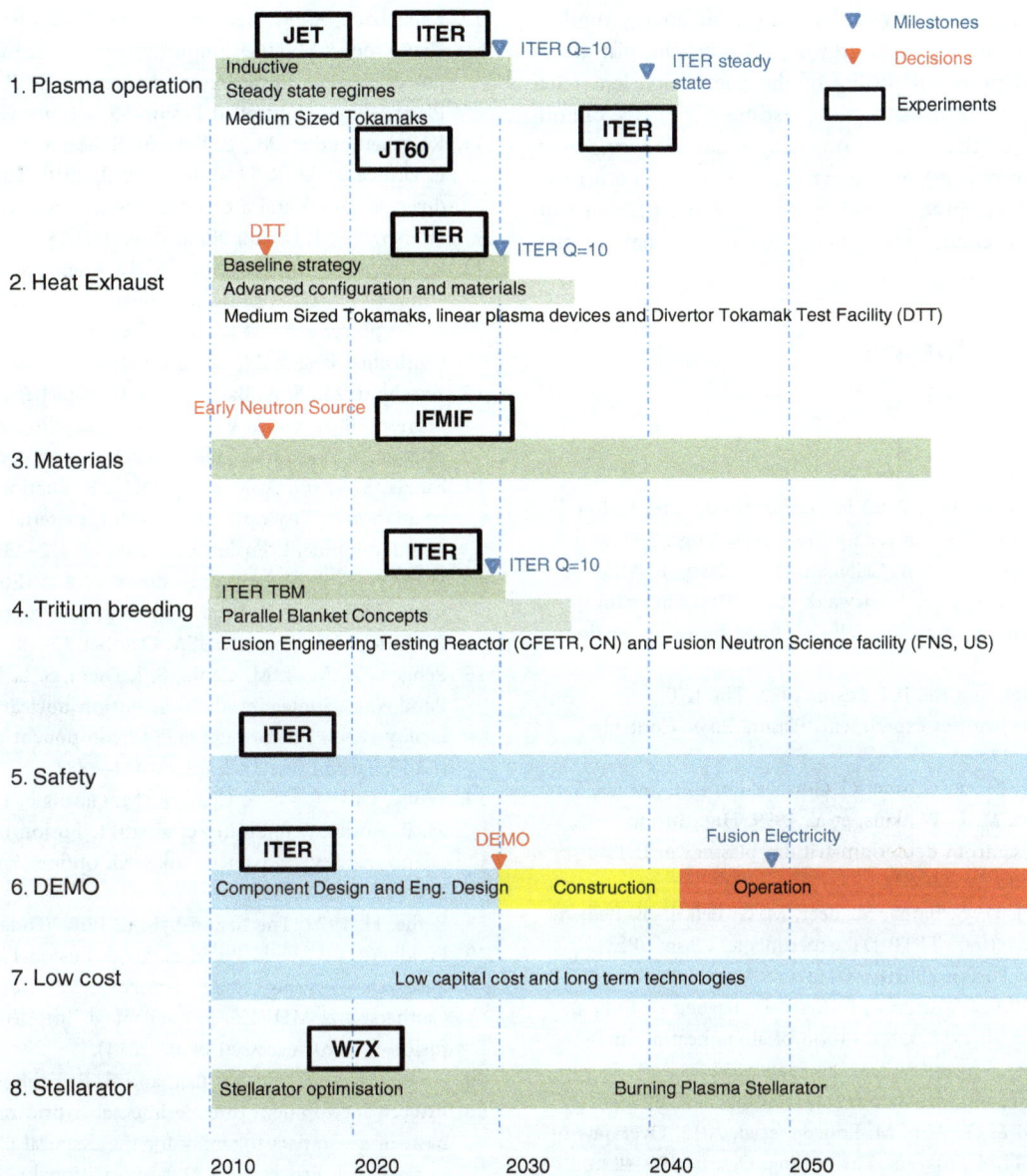

Figure 5. European Fusion roadmap [8].

world programes and it is not even clear whether DEMO would be a single worldwide collaborative experiment like ITER or several competing developments running in parallel in different countries, looking for a leading position in a phase where the economic profit of fusion might be at sight.

The European DEMO concept [21] sees the device as the last experimental facility before industry takes the lead in the construction of commercial fusion plants. As described above, it should be self-sufficient in tritium, use advanced low-activation materials and provide in the order of 500 MW of net electricity to the grid during operational periods of several weeks.

With the ITER high Q experiments foreseen for the late 2020's and the results of the 20 dpa materials irradiation available by the same dates, the DEMO construction could start by the mid 2030's and should be able to start net electricity generation before 2050 (Fig. 5). By that time, we expect that the materials irradiation facility IF-MIF would have been built and provided the necessary data for full qualification of low-activation structural materials under >100 dpa's. Those data might also be complemented with results from the current projects for CTF. From this point, we will enter the situation where private investors and industry will engage in the construction of the first commercial plants. When will this happen is difficult to predict, it will depend on the energy market situation and the overall energy supply scenario, but given the size and potential profits of the energy market, (the full cost of ITER construction, estimated 12–15 billion euro, is about the cost of one single day of worldwide energy consumption) we expect that this might be a relatively fast development, leading to a significant share of fusion in the energy mix during the second half of the century.

Conflict of Interest

None declared.

References

1. Freidberg, J. P. 2007. P. 12 *in* Plasma physics and fusion energy. Cambridge University Press, New York, NY.
2. Hurricane, O. A., D. A. Callahan, D. T. Casey, P. M. Celliers, C. Cerjan, E. L. Dewald, et al. 2014. Fuel gain exceeding unity in an inertially confined fusion implosion. Nature 506:343–347.
3. Rebut, P.-H, and the JET Team. 1992. The JET preliminary tritium experiment. Plasma Phys. Control. Fusion 34: 1749.
4. Keilhacker, M., A. Gibson, C. Gormezano, P. J. Lomas, P. R. Thomas, M. L. Watkins, et al. 1999. High fusion performance from deuterium-tritium plasmas in JET. Nucl. Fusion 39:209–234.
5. Strachan, J. D., S. Batha, M. Beer, M. G. Bell, R. E. Bell, A. Belov, et al. 1997. TFTR DT experiments. Plasma Phys. Controlled Fusion 39:B103–B114.
6. Thomas, P. R., P. Andrew, B. Balet, D. Bartlett, J. Bull, B. De Esch, et al. 1998. Observation of alpha heating in JET DT plasmas. Phys. Rev. Lett. 80:5548–5551.
7. Giancarli, L. M., M. Abdou, D. J. Campbell, V. A. Chuyanov, M. Y. Ahn, M. Enoeda, et al. 2012. Overview of the ITER TBM program. Fusion Eng. Des. 87:395–402.
8. Romanelli, F., P. Barabaschi, D. Borba, G. Federici, L. Horton, R. Neu, et al. 2012. A roadmap to the realisation of fusion energy. EFDA document available at http://www.efda.org/wpcms/wp-content/uploads/2013/01/JG12.356-web.pdf
9. Maisonnier, D., I. Cook, P. Sardain, L. Boccaccini, L. Di Pace, L. Giancarli, et al. 2006. DEMO and fusion power plant conceptual studies in Europe. Fusion Eng. Des. 81:1123–1130.
10. Kallenbach, A., M. Bernert, R. Dux, L. Casali, T. Eich, L. Giannone, et al. 2013. Impurity seeding for tokamak power exhaust: from present devices via ITER to DEMO. Plasma Phys. Controlled Fusion 55, art. no. 124041
11. Kotschenreuther, M., P. Valanju, S. Mahajan, L. J. Zheng, L. D. Pearlstein, R. H. Bulmer, et al. 2010. The super X divertor (SXD) and a compact fusion neutron source (CFNS). Nucl. Fusion 50, art. no. 035003
12. Ryutov, D. D., R. H. Cohen, T. D. Rognlien, and M. V. Umansky. 2012. A snowflake divertor: a possible solution to the power exhaust problem for tokamaks. Plasma Phys. Controlled Fusion 54, art. no. 124050
13. Zucchetti, M., S. A. Bartenev, A. Ciampichetti, and R. Forrest. 2007. A zero-waste option: recycling and clearance of activated vanadium alloys. Nucl. Fusion 47:S477–S479.
14. Zinkle, S. J., and A. Möslang. 2013. Evaluation of irradiation facility options for fusion materials research and development. Fusion Eng. Des. 88:472–482.
15. Wan, Y. 2012. Mission & readiness of a facility to bridge from ITER to DEMO. 1st IAEA-DEMO Program workshop, Los Angeles, USA, October 15–18.
16. Peng, Y. K. M., J. M. Canik, S. J. Diem, S. L. Milora, J. M. Park, A. C. Sontag, et al. 2011. Fusion nuclear science facility (FNSF) before upgrade to component test facility (CTF). Fusion Sci. Technol. 60:441–448.
17. Wong, C. P. C., V. S. Chan, A. M. Garofalo, J. A. Leuer, M. E. Sawan, J. P. Smith, et al. 2011. Fusion nuclear science facility – advanced tokamak option. Fusion Sci. Technol. 60:449–453.
18. Bethe, H. 1979. The fusion hybrid. Phys. Today 32:44–51.
19. Freidberg, J.P., Fink, Ph, et al. 2009. Fusion-Fission Research Workshop (2009) September 30–October 2, Gaithersburg, MD (USA). Available at http://web.mit.edu/fusion-fission/ (accessed 04 07 2014).
20. Azizov, E. A., G. G. Gladush, and E. P. Velikhov. 2013. Project development of experimental hybrid reactor on the basis of a compact tokamak for the disposal of spent nuclear fuel. Proceedings 21st International Conference on Nuclear Engineering, July 29–August 2, Chengdou, China.
21. Federici, G., R. Kemp, D. Ward, C. Bachmann T. Franke, S. Gonzales, C. Lowry, M. Gadomska, J. Harman, B. Meszaros, C. Morlock, F. Romanelli, and R Wenninger. 2014. Overview of EU DEMO design and R&D activities, Fusion Engineering and Design 89:882–889.

11

Pilot-scale processing with alkaline pulping and enzymatic saccharification for bioethanol production from rice straw

Motoi Sekine[1], Yukiharu Ogawa[1], Nobuhiro Matsuoka[1] & Yoshiya Izumi[2]

[1]Chiba University, Matsudo, Chiba, Japan
[2]Biomaterial in Tokyo Co., Ltd., Kashiwa, Chiba, Japan

Keywords

Alkaline pulping, bioethanol, enzymatic saccharification, pilot scale, rice straw

Abstract

We examined pilot-scale bioethanol production from rice straw using sodium carbonate pulping as the alkaline pulping method and enzymatic saccharification. The yield of prewashed rice straw after the crushing and prewashing stage decreased with an increase in the input in rice straw. The pulp yield after alkaline cooking was 66–68% at kappa number ranging from 32 to 36, which was comparatively higher than the laboratory-scale study. The yield of enzymatic saccharized glucose was decreased with the increase in washed pulp and its saccharification rate was approximately 20%. We successfully produced approximately 100 liters of 95% ethanol from 1000 Bone-Dry kg (BDkg) rice straw. The results of our pilot-scale study indicated that the relationship between resource input and product yield for each operation exhibited exponential or logarithmic curves, rather than linear decreases or increases, which could suggest a high-cost structure for bioethanol production when the resource input is larger. However, we established an optimum quantity of resource input, approximately 2000–3000 BDkg in our pilot-scale study, for higher efficiencies.

Introduction

In recent decades, harvested material residues, including rice straw, have been the focus of bioethanol production. However, the use of rice straw faces several challenges before ethanol generation can be accomplished cost effectively [1]. One of the largest problems is the requirement for large quantities of energy for ethanol conversion. Therefore, highly efficient conversion techniques, which reduce energy requirements, need development [2]. Typically, enzymatic saccharification, which isolates glucose and xylose from polysaccharides such as a cellulose and hemicellulose, is necessary to produce ethanol from botanical biomass resources. In general, the cellulose and/or hemicellulose contained in rice straw exist as lignocellulose, which is difficult to

degrade by enzymes [3]. Therefore, a suitable method to enhance the chemical reactivity between enzyme molecules and lignocellulose has to be applied during a processing phase for bioethanol production from rice straw.

A pulping method, where lignin in the form lignocellulose is partially removed and/or denatured, is used as pretreatment processing in paper manufacturing. Alkaline pulping, which can be subdivided into kraft pulping, soda pulping, and sodium carbonate pulping, is one pulping method for paper manufacturing from wooden chips. Alkaline pulping was also used as a paper manufacturing method from rice straw almost 100 years ago in Japan [4], though it is not used nowadays, it has still remained in some countries for manufacturing specialized papers from rice straw. Kraft pulping, in which

NaOH, Na$_2$S, Na$_2$CO$_3$ are mainly employed as pulping chemicals, is the most advantageous alkaline pulping method, because a kraft pulp made by the kraft pulping method has better characteristics for paper manufacturing. However, kraft pulping generates a bad smell caused by sulfur compounds during the pulping operations. This problem is particular to kraft pulping and can be prevented using large preventive devices, which have high equipment costs and could be disadvantageous for large-scale bioethanol production. Sodium carbonate pulping, which is hard to employ as a wood pulping method because of its weak alkalinity, can be used for rice straw pulping for bioethanol production [5]. The alkaline chemicals used in sodium carbonate pulping can be recycled as black liquor, which is produced during the alkaline cooking operation and used as a resource for heat generation by burning it [6], thus, it can be regarded as a low-emission method and particularly good for bioethanol production. Pretreatment processing followed by sodium carbonate pulping is, therefore, a comparatively useful and efficient method for bioethanol production processing from rice straw.

In general, research about bioethanol production from harvested material residues has examined enzymatic saccharification and fermentation processes in the laboratory. However, few studies have reported on bioethanol production at larger processing scales with pretreatment processes, which is vital to elucidate cost effectiveness, because manufacturing costs of plant production are proportional to the 0.6th power at the production scale by the six tenths rule [7]. Commercial chemical plants are developed with a scale up approach, that is, from laboratory and pilot to commercial scales. Laboratory-scale studies are used to estimate basic examinations, and develop process design. Pilot-scale studies are applied to confirm the reproducibility of laboratory-scale examinations and yield versus input for larger resource amounts, and economic efficiency. Few reports, however, provide estimates of the scaled up pretreatment processing efficiency of alkaline pulping rice straw for bioethanol production.

In this study, an actual pilot-scale bioethanol production process with sodium carbonate pulping as an alkaline pulping method and enzymatic saccharification operation was performed and examined to investigate the yield efficiency of processed materials versus resource input for each operation.

Materials and Methods

Materials

Rice straw (*Oryza sativa* L., cv. Koshihikari) cultivated in Kashiwa, Chiba, Japan in 2009 was used for this study.

Harvested plants were threshed, and the straw was placed on the field and subsequently field dried in the sun for 3 months, facilitated by the climate conditions around our test site and to reduce operating costs. The dried straw was baled and tied with sisal-twine (30 × 40 × 80 cm size bale). The average weight of a single bale was approximately 12.5 kg, and the average bale moisture content before examination was approximately 16.2%. Cellulase (Accellerase DUET, Nagase ChemteX Co., Osaka, Japan) was used as the hydrolysis enzyme for enzymatic saccharification, and bakery yeast (Oriental yeast, Oriental Yeast Co., Tokyo, Japan) was employed for fermentation procedures.

Chemical analysis method

Total sugar was measured using phenol–sulfuric acid method [8]. Monosaccharide was measured using TAPPI Standards T12 os-75 and T249 cm-09 [9]. Acid-insoluble lignin was measured using LAP-003 [10]. Ash, acid-soluble ash, and acid-insoluble ash were measured using high-temperature firing at 575°C [11].

Bioethanol production processing

A flowchart of the bioethanol production process applied to this study is depicted in Figure 1. Baled off rice straw was pretreated for the saccharification and fermentation operations. The straw was shredded, then crushed, and prewashed. The prewashed straw was cooked with alkaline chemicals, and the cooked pulp was subsequently washed. After pretreatment, the slurry state washed pulp was treated to enzymatic saccharification, which was subsequently treated to an alcoholic fermentation, distilled, and dehydrated to obtain a 95% ethanol solution. The bioethanol production process employed in this study was developed and constructed in Kashiwanoha, Chiba, Japan. Specifications of each device are provided in Table 1, and photographs of the main devices and operations are shown in Figure 2.

Crushing and prewashing

Baled off rice straw was shredded and crushed using a masher (E-Gimmick, Taizen Co., Shizuoka, Japan), fitted with a rotary blade knife. The crushed straw was prewashed with running 80°C hot water. The rice straw to masher input was 500 kg/h. The prewashed straw (Fig. 2A) was drained by self-weight consolidation, and packed into a flexible 1 m^3 volume container bag. The bag filled with prewashed straw was weighed using a crane scale, and the straw moisture content was measured to calculate the dry weight of input prewashed rice straw

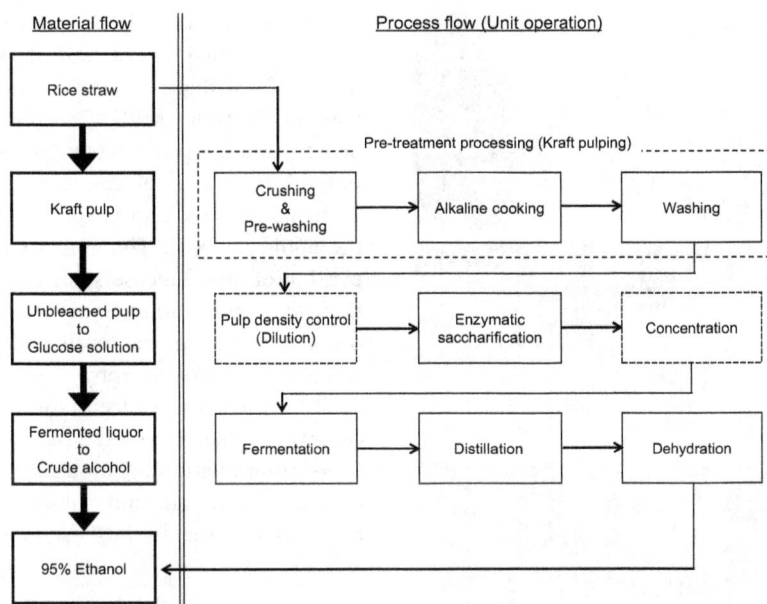

Figure 1. A flowchart of bioethanol production processing applied to this study.

Table 1. Specifications of integrated devices into bioethanol production processing.

Operation	Appliance	Performance (×unit)	Manufacturer
Crushing and prewashing	Musher	500 kg/h × 1	Taizen Co., Ltd.
Alkaline cooking	Digester (pressure vessel)	1.8 m³ × 1	Kato Seikan Co.
Pulp washing	Drum-type washer	460 kg/h × 1	Taizen Co., Ltd.
(Pulp-density control)	Dilution tank	5.0 m³ × 2	Masuko Sangyo Co., Ltd.
Enzymatic saccharification	Saccharification tank (with agitator)	3000 L × 4	KK. Takasugi Seisakusho
(Concentration)	Evaporator	200 kg/h × 1	Okawara MFG Co., Ltd.
Fermentation	Fermenter (with agitator)	1000 L × 1	KK. Takasugi Seisakusho
Distillation and dehydration (for confirmation test)	Thermo vapor distillation	20 L × 1	KK. Takasugi Seisakusho
Heating (utility)	Once-through boiler	2000 kg/h × 1	Miura Co., Ltd.

for weight calculation of alkaline chemicals to alkaline cooking.

Alkaline cooking

Less than 200 BDkg of prewashed straw was added to the glove rotary digester (inner volume 1.8 m³), with a maximum input of 200 BDkg per batch digester. Na_2CO_3 and a NaOH were also added to the digester, and the concentrations, respectively, diluted to 3.45% (0.5 mol/L) and 0.07% (0.026 mol/L) with water. The ratio of alkaline liquor to dry weight of prewashed rice straw was 5:1. The digester was subsequently closed, and heated to 165°C by steam injection with 1.0 rpm of rotation for approximately 1 h. After heating, the digester was air-cooled for 1 h while maintaining its rotation. The rice straw pulp, so-called cooked pulp, and the cooking spent liquor, so-called black liquor, were the products of this procedure.

The cooked pulp was separated from the black liquor using a drainer with self-compression (Fig. 2B). This batch process was carried out repeatedly for larger amounts of prewashed straw than 200 BDkg, and the total mass for each cooking operation was divided by the mass of prewashed straw input to calculate cooked pulp yield. A pulp kappa number was calculated, which had a correlation with residual lignin amount in pulp; TAPPI Standard T236m-60 (same to JIS P 8211 and ISO 302:2004) was applied for this calculation.

Pulp washing

Cooked pulp separated from the black liquor was moved to a 5 m³ dilution tank, and diluted with tap water to form pulp slurry. The quantity of added water was approximately 10 times the cooked pulp weight. The pulp slurry was moved into a drum-type washer (Taizen Co.,

Figure 2. Photographs of the main devices and operations. (A) Discharging of prewashed straw from masher, (B) digester and drainer, (C) drum-type washer, (D) inside of the drum-type washer during washing and separating of washed pulp with rotation, (E) saccharification jar, (F) saccharified liquid.

Fig. 2C) using a tube pump for viscous fluid, and the drum was rotated. The washer input rate was 460 kg/h. The washed pulp was separated from pulp slurry during drum rotation (Fig. 2D), collected into a drainer, and drained by self-weight consolidation. The black liquor and the washed pulp drainage were tested by the TAPPI Test Method T625ts-64 for estimation of residual alkali [12].

Enzymatic saccharification

Washed and dehydrated pulp was transferred into a saccharification jar (Fig. 2E), diluted with water up to 20 times of the dry weight of dehydrated pulp, and adjusted to pH 6.0 using a diluted H_2SO_4 solution. The jar was heated to 50°C with indirect heating of continuous steam, then 101 g of hydrolysis enzyme per 1 kg of dried pulp was added to the jar, and stirred for 24 h to accelerate the enzymatic reaction. A Brix meter (PAL-BX/RI, Atago, Tokyo, Japan) was used to measure a glucose concentration in the saccharified mixture.

Fermentation and postfermentation

The saccharified mixture was released from the jar bottom stop valve, and separated into saccharified liquid containing glucose (Fig. 2F) and residues using a drainer. The saccharified liquid was moved to a concentrating device (centrifugal-flow thin-film vacuum evaporator, CEP-5S, Okawara MFG. Co., Shizuoka, Japan), and condensed to an approximately 20% glucose solution. A volume of 1000 liters of condensed liquid were moved into fermentation jars, and bakery yeast was added to ferment the liquid at 28°C. The amount of added yeast was 20 g per 1 L of 20% glucose solution. After 24 h of fermentation, the fermentation liquor was sampled, and the ethanol concentration was evaluated by J. TAPPI No.42-84 using gas chromatography [13]. The liquor was distilled until 90% ethanol concentration was achieved. The concentrated ethanol solution was dehydrated for 10 h by PSA-column with a 3A molecular sieve up to 95% of ethanol concentration, and subsequently evaluated for adequacy as a biofuel by JASO standard (JIS K 2190:2011).

Results and Discussion

The relationships between rice straw input and prewashed rice straw yield after crushing and prewashing operations are shown in Figure 3. The relationship with the moisture content of prewashed rice straw is also shown in this figure. The yield exponentially decreased with increases in input. In contrast, moisture content showed a logarithmic increase. The approximate exponential and logarithmic curves exhibited high correlation values ($R^2 = 0.98$, prewashed rice straw yield to rice straw input; $R^2 = 0.86$, moisture content of prewashed rice straw to rice straw input, respectively); therefore, yield and moisture content should become constant at larger inputs. In this case, yield tended to show constant values (96–97%) over 3000 BDkg input. The moisture content should also show

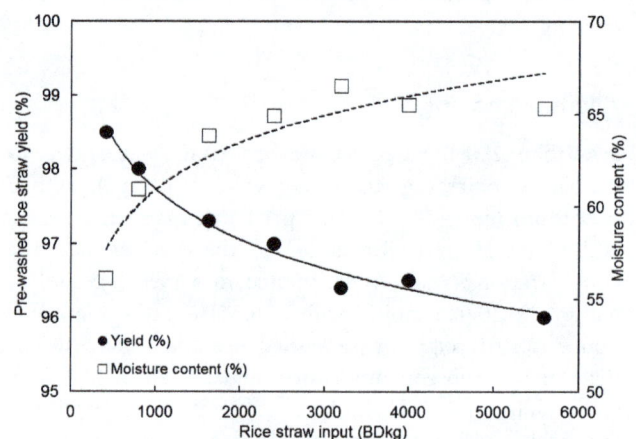

Figure 3. Relationships between rice straw input and prewashed straw yield and its moisture content at the crushing and prewashing operation.

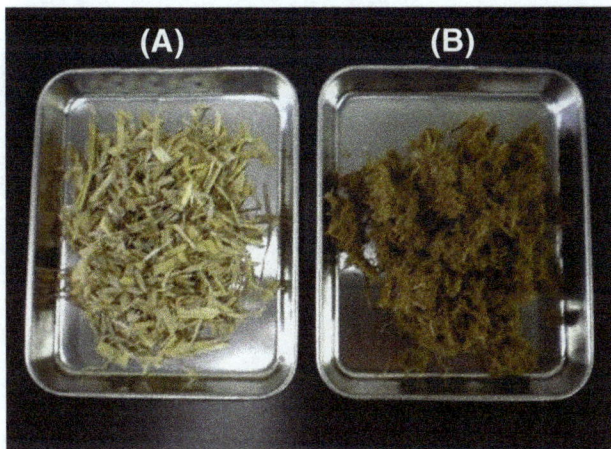

Figure 4. Images of shredded (A) and prewashed (B) rice straw samples.

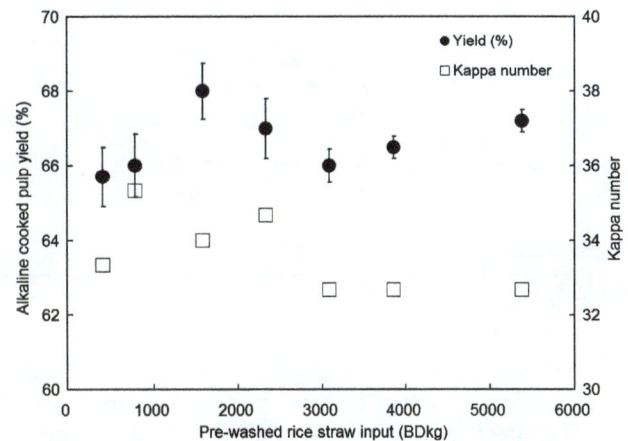

Figure 5. Relationships between prewashed straw input and alkaline cooked pulp yield and kappa number at the alkaline cooking operation. Error bars show standard deviation (*n* = 3–30).

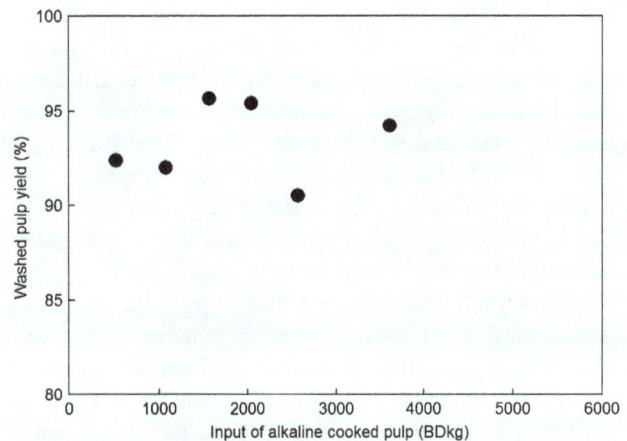

Figure 6. Relationships between input on alkaline cooked pulp and washed pulp yield at the pulp washing operation.

constant values over 3000 BDkg input, which was approximately 65%. Shredded (A) and prewashed (B) rice straw sample images are shown in Figure 4. The prewashed sample was darker in color, due to washing in hot water, and high moisture content. The prewashed sample also exhibited a wadding-like state compared with the shredded sample, which had low handling properties for processing. Overall, the straw samples were fragmented and pulverized by the cutter mill or atomizer in the laboratory-scale experiments, which is analogous to the coarse-crushing operation, thus results in the wadding-like state (Fig. 4B) cannot be produced and observed. Furthermore, the yield from crushing and prewashing operations cannot be estimated from laboratory-scale experiments [14]. The pilot-scale processing study provided a value for rice straw resource losses, and results indicated that prewashed rice straw exhibits the possibility of low handling properties. In addition, the prewashed straw wadding-like state is associated with its moisture content.

The relationships between prewashed rice straw input and alkaline cooked pulp yield following the alkaline cooking operation are shown in Figure 5. The relationship with cooked pulp Kappa number is also provided. The yield shown in this pilot-scale study was 66–68% at kappa number ranging from 32 to 36. This was comparatively higher than the typical laboratory-scale study [5], which reported the yield was 60–62% at same kappa number range. A continuous digester has usually been employed to cook large amounts of resources during the alkaline cooking operation. The digester employed in this pilot-scale study was a comparatively smaller batch type for total cost reduction. However, results for yield and kappa number in each batch seemed to be stabilized.

Relationships between input of alkaline cooked pulp and washed pulp yield resulting from the pulp washing opera-

tion are indicated in Figure 6. The washed pulp yielded between 90% and 95% throughout the input range (0–4000 BDkg). The fiber length range of pretreated rice straw pulp in this study was approximately 0.3–0.6 mm, which was less than the fiber of hardwood bleached kraft pulp (LBKP) (0.6–0.7 mm) for ordinary paper resources [15]. A mesh size of drainage screen incorporated into the pulp washing machine was designed for LBKP; therefore, the washed pulp would not yield 100% and device modification should be optimized for the rice straw attributes. Straw sample compositions for each operation and yields for the rice straw resource at each operation in the pretreatment processing are provided in Table 2. Lignin content in alkaline cooked pulp was less than in prewashed straw, and washed pulp lignin content was less than in cooked pulp. These results indicated lignin content decreased during the

Table 2. Straw sample composition and yield for each operation in the pretreatment processing.

Sample	Composition (%)					Yield at each operation (%) (vs. rice straw, weight base)
	Cellulose	Hemicellulose	Lignin	Ash	Others	
Rice straw (bale)	29.1 ± 1.0	21.3 ± 1.0	13.4 ± 3.0	24.8 ± 3.0	11.4 ± 8.0	100.0
Prewashed rice straw	27.0 ± 1.0	16.5 ± 1.5	15.3 ± 4.0	23.4 ± 2.0	17.8 ± 8.5	95.0
Alkaline cooked pulp	51.1 ± 0.5	17.1 ± 1.0	5.05 ± 0.5	23.9 ± 0.5	2.9 ± 2.5	63.8
Washed pulp	51.5 ± 0.5	17.1 ± 0.5	1.7 ± 2.0	25.5 ± 1.0	4.2 ± 4.0	60.1

Mean ± standard deviation (n = 3).

alkaline cooking and washing operations; in contrast, cellulose and hemicellulose contents increased relative to the other contents following cooking operations. Ash content, showed no change throughout the pretreatment processing, and remained at 23% to 25%. These results indicated that the ash, similar to silica contained in rice straw, would dissolve at the same ratio as the rice straw dissolved during pretreatment processing. The remaining ash content might affect processing efficiency.

The relationships between washed pulp input and the saccharification rate during enzymatic saccharification are shown in Figure 7. The saccharification rate exhibited an exponential decrease with increased input (R^2 = 0.9761). The saccharified mixture exhibited high stirring resistance by the unreacted pulp fibers, therefore the actual stirring rate for the mixture in the saccharification jar was 20–30 rpm using a direct drive-type stirrer. This result also suggested that larger-scale processing would need a stronger agitator for the enzymatic saccharification operation. Relationships between rice straw input and converted 95% ethanol are shown in Figure 8, which were produced by multiple distillations and dehydrations after the fermentation process. The resulting 95% ethanol showed a logarithmic increase consistent with an increased straw quantity (R^2 = 0.9313). The pilot-scale processing employed in this study produced approximately 100 L of 95% ethanol per 1000 BDkg of rice straw when the rice straw quantity exceeded 2000 BDkg. This result showed a comparatively lower bioethanol production efficiency than laboratory-scale bioethanol processing [16].

Conclusions

In general, construction costs, operating costs, and energy consumption of bioethanol production processing increases when the system scale and devices become larger. However, larger systems might result in increased efficiency by implementing larger biomass resource inputs, and subsequently reduced costs. In this pilot-scale study, the relationship between input and yield for each operation indicated exponential or logarithmic curves

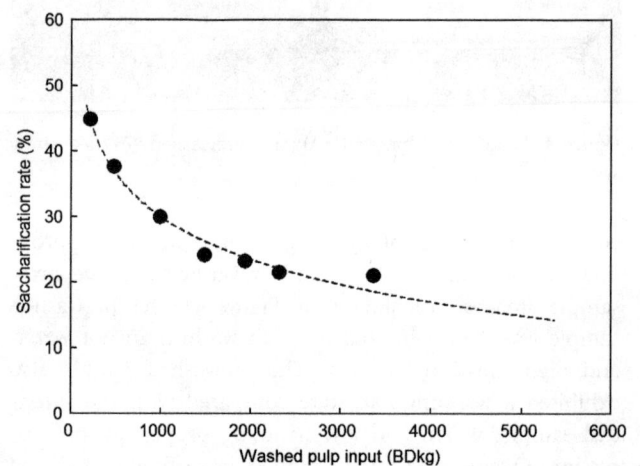

Figure 7. Relationships between washed pulp input and saccharification rate during enzymatic saccharification.

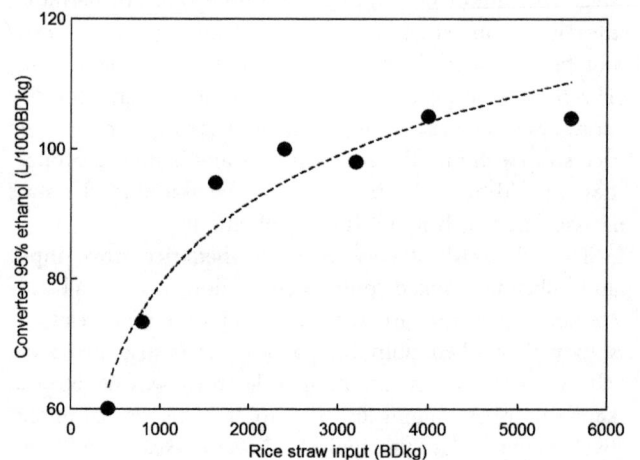

Figure 8. Relationships between rice straw input and converted 95% ethanol in the pilot-scale bioethanol production processing.

rather than linear decreases or increases, suggesting a high-cost structure for bioethanol production when the resource inputs are larger. However, there is an optimum quantity of resource input, approximately 2000–3000 BDkg in our pilot-scale study, for higher efficiencies at each operation. This should be confirmed by larger quantities of resource input with optimum modifications for each device in future studies.

Acknowledgment

The soft cellulose-based bioethanol project of the Ministry of Agriculture, Forestry and Fisheries, Japan provided partial support for this study.

Conflict of Interest

None declared.

References

1. Binod, P., R. Sindhu, R. R. Singhania, S. Vikram, L. Devi, S. Nagalakshmi, et al. 2010. Bioethanol production from rice straw: an overview. Bioresour. Technol. 101:4767–4774.
2. Koizumi, T. 2013. The Japanese biofuel program-developments and perspectives. J. Clean. Prod. 40:57–61.
3. Zhao, X., L. Zhang, and D. Liu. 2012. Biomass recalcitrance. Part I: the chemical compositions and physical structures affecting the enzymatic hydrolysis of lignocellulose. Biofuels Bioprod. Biorefin. 6:465–482.
4. Sugiya, K. 1999. Dyeing property of non-wood pulp. Jpn. Tappi J. 53:64–69.
5. Yang, L., J. Cao, Y. Jin, H. Chang, H. Jameel, R. Phillips, et al. 2012. Effects of sodium carbonate pretreatment on the chemical compositions and enzymatic saccharification of rice straw. Bioresour. Technol. 124:283–291.
6. Andersson, E., and S. Harvey. 2007. Comparison of pulp-mill-integrated hydrogen production from gasified black liquor with stand-alone production from gasified biomass. Energy 32:399–405.
7. Humbird, D., R. Davis, L. Tao, C. Kinchin, D. Hsu, A. Aden, et al. 2011. Process design and economics for biochemical conversion of lignocellulosic biomass to ethanol. Technical Report NREL/TP-5100-47764, Contract No. DE-AC36-08GO28308.
8. Hodge, J. E., and B. T. Hofreiter. 1962. Determination of reducing sugars and carbohydrates. Pp. 380–394 in R. L. Whistler and M. L. Wolfrom, eds. Methods in carbohydrate chemistry. Vol. 1. Academic Press, New York.
9. TAPPI. 2002. Carbohydrate composition of extractive-free wood and wood pulp by gas–liquid chromatography, T 249 cm-09.
10. Templeton, D., and T. Ehrman. 1995. Determination of acid-insoluble lignin in biomass (LAP-003), standard biomass analytical methods. Technical report NREL/MRI.
11. Sluiter, A., B. Hames, R. Ruiz, C. Scarlata, J. Sluiter, and D. Templeton. 2008. Determination of ash in biomass. Technical Report NREL/TP-510-42622, Contact No. DE-AC36-99-GO10337.
12. Parker, J. L., R. P. Hensel, and C. L. Wagoner. 1970. Measurement of total solids in black liquors. TAPPI 53:874–877.
13. J TAPPI No.42-84. 1984. Method for determining the carbohydrate composition of wood pulp and extractive-free wood by gas chromatography. Kamipa Gikyo-shi 38:361–364.
14. Small, W. M. 1969. Scaleup problems in reactor design. Chem. Eng. Prog. 65:81–82.
15. Guo, S., H. Zhan, C. Zhang, S. Fu, A. Heijnesson-Hulten, J. Basta, et al. 2009. Pulp and fiber characterization of wheat straw and eucalyptus pulps: a comparison. Bioresources 4:1006–1016.
16. Brodin, F. W., Y. Sonavane, and H. Theliander. 2013. Preparation of absorbent foam based on softwood kraft pulp: advancing from gram to kilogram scale. Bioresources 8:2099–2117.

12

In situ synchrotron-based X-ray powder diffraction and micro-Raman study of biomass and residue model compounds at hydrothermal conditions

Morgan L. Thomas[1,2], Ian S. Butler[3] & Janusz A. Kozinski[1,2]

[1]Lassonde School of Engineering, York University, 4700 Keele Street, Toronto, Ontario, Canada M3J 1P3
[2]Department of Chemical and Biological Engineering, University of Saskatchewan, 57 Campus Drive, Saskatoon, Saskatchewan, Canada S7N 5A9
[3]Department of Chemistry, McGill University, 801 Sherbrooke Street West, Montreal, Quebec, Canada H3A 2K6

Keywords
Biomass, diamond-anvil cell, hydrothermal, Raman, supercritical water, X-ray diffraction

Abstract

The applications of synchrotron X-ray powder diffraction (XRPD) and laser micro-Raman techniques in an examination of the dissolution, transformation, and gasification of pure cellulose and models for biomass residue under hydrothermal conditions in a diamond-anvil cell are reported. The results contribute to the measurement of in situ time-resolved profiles of biomass reactions, catalyst stability, and residue formation that occur in aqueous fluids at near- and supercritical conditions.

Introduction

The study of hydrothermal systems for combustion and related processes has been an active research area for many years, with there even being some limited examples of industrial application [1–3]. The results presented here represent an advancement of our efforts to develop the hydrothermal diamond-anvil cell (HDAC) as a valuable experimental tool for in situ measurements under hydrothermal or supercritical water (SCW) conditions. Our particular interest is focused on the use of SCW as a reaction medium for biomass conversion and combustion [4–10]. The potential for hydrothermal processing as a unique and efficient method for biomass conversion has previously been demonstrated for various other experimental setups and biomass varieties [11–14]. Synchrotron radiation offers great potential for fundamental and applied study of hydrothermal processes. We report here on our successful efforts to perform in situ synchrotron X-ray powder diffraction (XRPD) and laser micro-Raman spectrometry analyses of cellulose and biomass residue under hydrothermal conditions using the HDAC.

In this communication, we will describe a new approach and discuss our results, which agree with established mechanisms for cellulose and biomass evolution, and allow us to propose, for the first time, a more detailed hypothesis for biomass transformation within a SCW-dominated environment in terms of particulate behavior. We first discuss the physicochemical data of cellulose and biomass materials. Next, we explain why microstructural and crystallographic evidence suggests that cellulosic biomass loses crystallinity with an increase in temperature. We then discuss a plausible mechanism which we believe satisfactorily accounts for both the chemical and structural data. In conclusion, we suggest some practical implications of the proposed approach.

Experimental

All of the work described here was performed using a Bassett-type HDAC (Foxwood Instruments, Ithaca, NY, USA) [15]. The HDAC permits in situ measurements at high temperature and pressure, and has been described in detail elsewhere [16, 17]. All chemicals were purchased from Sigma-Aldrich (Burlington, ON, Canada) and were used without further purification. A stainless-steel gasket, 250 μm in thickness, with sample chamber of 250 μm o.d., was used in all the experiments. The sample chamber was loaded with distilled water and analyte under a constant flow of nitrogen, which was maintained during the heating cycle of the cell.

Raman microscopy setup and data acquisition/processing

Raman spectra were recorded on a Renishaw inVia Raman microscope system using either 514.5- or 785-nm laser

excitation. Calibration of the Raman spectrometer was performed prior to each set of measurements with a standard Si wafer. The proprietary software Renishaw WiRE™ (version 3.2, Wotton-under-Edge, Gloucestershire, UK), was used for data acquisition and processing. Focusing was achieved with a super long-working distance 20× objective (Olympus SLMPLN20X; Olympus, Waltham, MA, USA).

XRPD setup

XRPD was performed using the hard X-ray micro analysis (HXMA, 06ID-1) beamline at the Canadian Light Source (Saskatoon, Saskatchewan), Canada's national synchrotron radiation facility. The cell was mounted on a motorized swivel stage (SA05B-RL; Kohzu Precision, Kawasaki, Japan), allowing for angular rotation of the cell, which was itself mounted on a second stage to provide vertical and horizontal motion. The cell was fixed in position on the stage using an insulating U-shaped mount, machined from Grade A lava stone (Maryland Lava, Bel Air, MD, USA), to prevent heat dissipation from the cell to sensitive components. A 250 μm pinhole was used to prevent interferences from beam divergence and scattering. The layout of the associated equipment in the experimental hutch has been described in detail earlier [18]. Shown in Figure 1B and C is an image and sketch of the setup used in our experiments.

XRPD and Raman probe data acquisition and processing

X-ray diffraction images were recorded on a MAR345 image plate detector (Marresearch, Norderstedt, Germany) at the HXMA beamline. Exposure times were 60 s

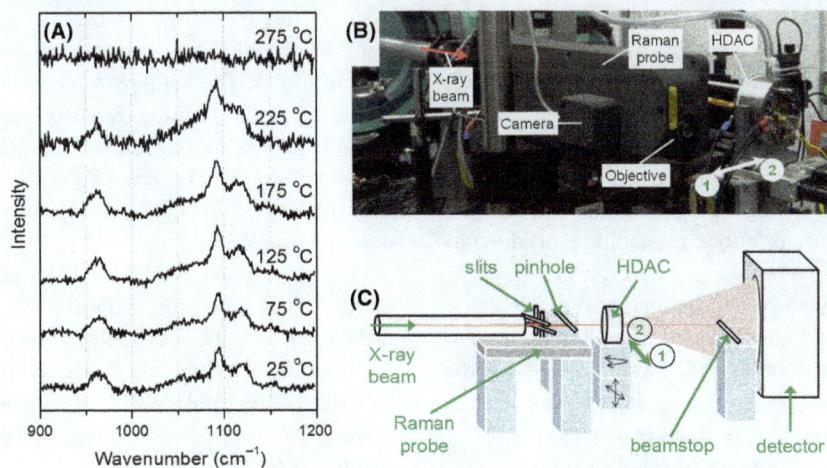

Figure 1. (A) Raman spectra (785 nm laser line) for cellulose with increasing temperature showing the loss of crystallinity at temperatures >225°C, (B) image and (C) sketch (not to scale) of synchrotron setup for XRPD measurements in the HDAC at the HXMA beamline. The cell was shuttled between position 1 (Raman/camera) and position 2 (X-ray) for these measurements. XRPD, X-ray powder diffraction; HDAC, hydrothermal diamond-anvil cell; hard X-ray micro analysis.

unless otherwise noted, with $\lambda = 0.509175$ Å. Calibration was performed using a NIST lanthanum hexaboride (LaB$_6$) standard. The raw image files were processed and integrated with Fit2D (V12.077) software [19]. Baseline correction (spline interpolation) and final processing of the resultant powder patterns were performed with Full-Prof Suite (1.10) software [20].

Raman spectroscopic measurements at the beamline were conducted using a Renishaw RM 2000 Raman microscope system (Renishaw, Wotton-under-Edge, Gloucestershire, UK) coupled to a Renishaw RP20 fiber optic probehead with 514.5-nm laser excitation. The probe head also incorporated a Philips ToUcam PRO II (1.2 Megapixel) CCD camera, which was used to obtain images of the sample in the HDAC. A super long-working distance 20× objective (Olympus SLMPLN20X) was mounted on the probehead. The probe was positioned parallel to the X-ray beam path and the HDAC was moved between the two positions as shown in Figure 1B and C to measure images (position 1) and X-ray patterns (position 2).

Discussion

We have previously reported on the decomposition of glucose, cellulose, and lignocellulose under hydrothermal conditions [4, 21, 22]. Our approach in the present work has been to extend the study of high temperature high-pressure aqueous systems with the use of more advanced analytical techniques. In the initial stage, we measured the micro-Raman spectra of microgranular cellulose in the HDAC. As this organic material is a rather weak Raman scatterer, ideally a prolonged measurement must be performed (e.g., 30 min) to achieve a reasonable signal-to-noise ratio. In this work, the approach taken to avoid long data acquisition times was to select small spectral windows in an attempt to identify chemical changes as they occurred. Having identified two key regions of interest (namely 300–500 and 900–1200 cm^{-1}, see Fig. S1), we proceeded with an investigation of the changes in the spectra with increasing temperature in the HDAC, as shown in Figure 1A. In the 900–1200 cm^{-1} region, we observed a clear loss of cellulose crystallinity or dissolution of cellulose at temperatures greater than 225°C.

Our initial synchrotron experiments, using the setup shown in Figure 1B and C, employed these Raman data as a basis for experimental validation. Further details of the experimental methods are provided in the Supporting Information. Figure 2 shows an analogous experiment for measurement of X-ray diffraction of cellulose with increasing temperature in the HDAC. Heating rates for both the Raman and diffraction measurements were ~10°C/min. Again, we observed a clear loss of signal at temperatures greater than 225°C, corresponding to a loss of crystallinity.

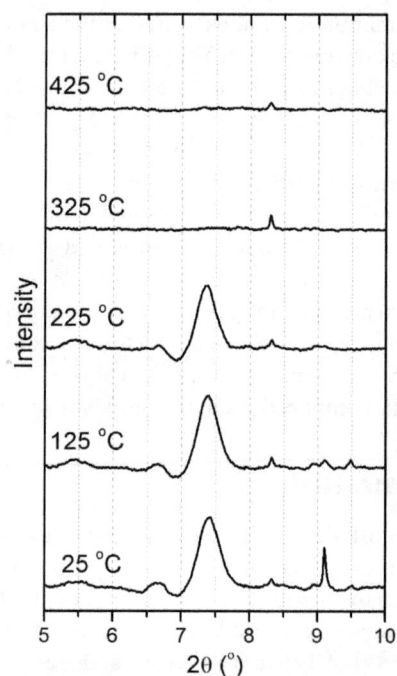

Figure 2. XRPD patterns for cellulose with increasing temperature showing the loss of crystallinity at temperatures >225°C. XRPD, X-ray powder diffraction.

Notably, the initial diffraction pattern (and indeed the initial Raman spectrum) was not recovered upon cooling of the sample, implying that the change is irreversible. This is in agreement with the mechanisms we proposed earlier for cellulose conversion under hydrothermal conditions, with an initial step of cellulose dissolution and depolymerization (hydrolysis), followed by reaction of the component sugars [7–10, 23–25]. The significance of this measurement relates to defining a means to identify the point of loss of crystallinity, presumably immediately preceding dissolution of the cellulosic component of biomass, and possible use in refining existing kinetic models for this transformation. We note that the identification and quantification of the products of the cellulose decomposition, widely reported in the literature, is outside of the scope of this manuscript.

Our subsequent work has dealt with the measurement of XRPD data for other important solid materials that are present in biomass residue, measurement of pressure, and the ongoing development of the Raman system for simultaneous (same spot) XRPD and micro-Raman measurements. A brief discussion on the limitation of the combined technique, in particular with regard to the reliable measurement of pressure under reaction conditions is provided in the Supporting Information.

We employed several different calcium-based compounds as models for the residue that may be produced from the hydrothermal decomposition reactions of lignocellulosic materials. These model compounds, viz., calcium oxalate, calcium hydroxyapatite, and calcium carbonate, were selected based on earlier reports of their transformation under hydrothermal or high-pressure conditions, the prevalence of calcium-containing deposits from SCW biomass conversion [5] and the high proportion of calcium in biomass, particularly in certain woody species [26–28]. No data are actually reported for calcium carbonate as, unfortunately, this substance proved to be too weak a Raman scatterer to produce reliable data.

Calcium oxalate is a species known to be present in a wide variety of types of biomass and plays a key role in structural and biochemical processes. In some cases, it is present in crystalline form in plant tissues. Figure 3 shows the visible changes occurring in the cell during the heating experiment. With a suitable choice of heating rate (~10°C/min), it was possible to collect the images at position 1 (see Fig. 1B and C regarding the location of position 1 and 2), move the cell while heating to position 2, perform the XRPD measurement and finally return the cell to position 1 while reading the detector image plate. Thus, the data collection can be performed in a relatively short time period (i.e., a total heating time of 45–60 min).

The XRPD data measurements performed for calcium oxalate under hydrothermal conditions show a dramatic change in the powder pattern at a temperature close to the critical temperature of pure water ($T_c = 374°C$), as shown in Figure 4. Similar to the earlier observations with cellulose, the crystalline component of the sample appears to react and is not present in the pattern obtained after cooling. Moreover, there was an increased amount of gaseous products present in the sample volume after cooling (see Fig. 3). We suggest that calcium oxalate under these conditions reacts irreversibly via a decarboxylation or decarbonylation pathway, producing CO_2 or CO, respectively.

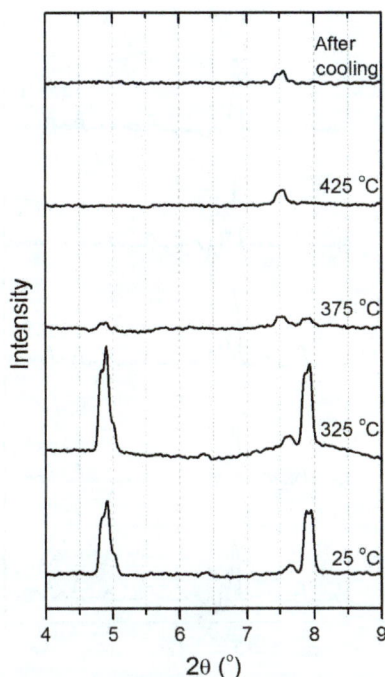

Figure 4. XRPD patterns for calcium oxalate with increasing temperature showing the change in structure at around 375°C (patterns between 25 and 325°C omitted for clarity). XRPD, X-ray powder diffraction.

Calcium hydroxyapatite and related materials are known to undergo structural transformations at hydrothermal conditions and at high pressures [29, 30]. In our XRPD study of this material, however, we observed no significant changes with increasing temperature, as shown in Figure 5. It should be emphasized that our study was principally concerned with the understanding of rapid processes for application in chemical reactors, whereas the structural changes reported in the literature are most likely due to less rapid transformations. The stability of

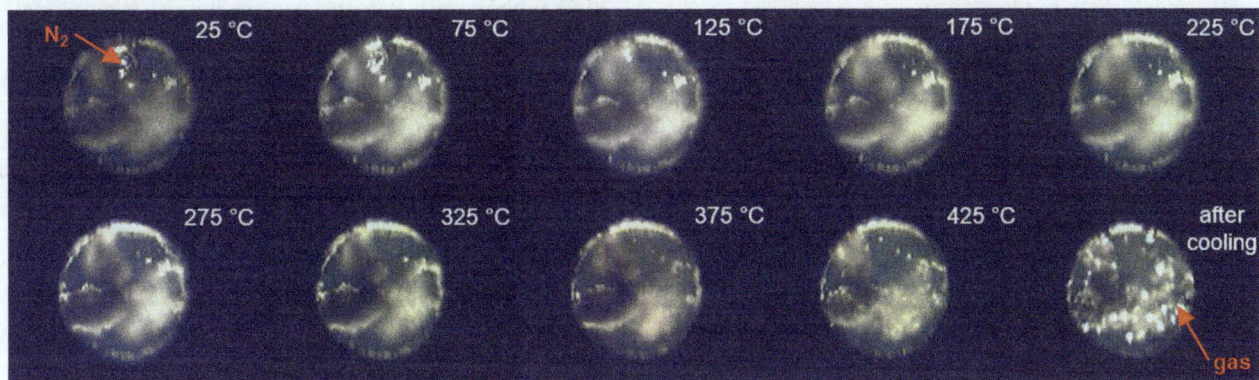

Figure 3. Images of calcium oxalate in HDAC with increasing temperature showing qualitatively the change in the material during heating, and the N_2 bubble dissolution used for pressure determination ($T_h = 150°C$, $\rho = 917.01$ kg/m^3). HDAC, hydrothermal diamond-anvil cell.

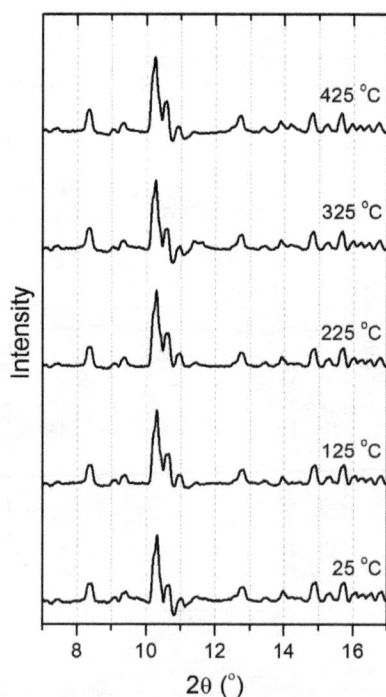

Figure 5. XRPD patterns for hydroxyapatite with increasing temperature (T_h = 225°C, ρ = 833.75 kg/m³) showing little change over the studied range. XRPD, X-ray powder diffraction.

this material under the tested conditions provides an indication of the types of material that may be present in the residue following prolonged hydrothermal treatment.

Conclusions

SCW remains a promising alternative to conventional chemical and materials processing technologies. We have demonstrated in this work some key analyses to aid in our understanding of the fundamental processes that occur during the hydrothermal transformations of biomass models. Our results concur with established mechanisms for cellulose reactivity and provide new insights into the chemical pathways involved in the production of residues in these experiments. We have also demonstrated the usefulness of synchrotron science for biomass conversion research. Specifically, the use of XRPD as an in situ measurement tool is invaluable and, more generally, synchrotron science offers a wealth of opportunities for fundamental and applied research for hydrothermal processes.

The data we have presented here will be greatly enhanced by the proposed simultaneous (i.e., same spot) XRPD/micro-Raman facility at the Canadian Light Source. Such a setup, demonstrating the advantages of simultaneous measurement, has already been developed at

several synchrotron facilities [31, 32], and the addition of this technique will be advantageous in developing the HDAC experimental capabilities. In this work, we have primarily employed both Raman spectroscopic measurements and XRPD to determine the crystallinity of relevant moieties. Although the current work has not fully highlighted the advantages, in future work, the complementary nature of these two techniques will be fully employed, that is, to measure the functional group transformations via Raman spectroscopy and the consequent crystalline sample changes using XRPD.

Our observations suggest possible explanations for a number of important phenomena. For example, the loss of crystallinity of the cellulose during dissolution of the cellulose prior to reaction of component sugars and the processes involved in the formation of calcium-containing residues. However, the general scheme we have proposed for biomass rearrangements in hydrothermal environment is as yet incomplete. It is clear from our analyses that many questions remain to be answered before the picture of biomass structure evolution can be described in detail. What stimulates loss of crystallinity of cellulose other than thermochemical interaction? What role is played by catalysts and volatilization of selected species? Are hydrolysis and dissolution alone responsible for biomass conversion, or do they interact with other processes (e.g., pyrolysis)? The results of this study also provide the possibility for measurement of in situ time-resolved profiles of biomass reaction, catalyst stability, and residue formation in hydrothermal biomass processing.

Acknowledgments

We thank Jason Maley, Sophie Brunet, and Ramaswami Sammynaiken (Saskatchewan Structural Sciences Centre, University of Saskatchewan), Kelly Akers (ProSpect Scientific Inc.), and Richard Bormett (Renishaw Inc.) for their kind support with Raman instrumentation. We are grateful for the assistance of our colleagues Gaëlle Dupouy (York University) and Chang-Yong Kim (beamline scientist, Canadian Light Source) in developing our beamline experiments. We also thank John Tse (University of Saskatchewan) and William Bassett (Cornell University) for their helpful suggestions related to diamond-anvil cells and measurement. We are grateful to the assistance given by technical staff at the College of Engineering, University of Saskatchewan. The research was supported by Discovery, Strategic and Research Tools and Instrumentation grants to I. S. B. and J. A. K. from the Natural Sciences and Engineering Research Council of Canada and as part of the Agricultural Biorefineries Innovation Network (Canada).

Conflict of Interest

None declared.

References

1. Hodes, M., P. A. Marrone, G. T. Hong, K. A. Smith, and J. W. Tester. 2004. Salt precipitation and scale control in supercritical water oxidation – Part A: fundamentals and research. J. Supercrit. Fluids 29:265–288.

2. Marrone, P. A., S. D. Cantwell, and D. W. Dalton. 2005. SCWO system designs for waste treatment: Application to chemical weapons destruction. Ind. Eng. Chem. Res. 44:9030–9039.

3. Bermejo, M. D., and M. J. Cocero. 2006. Supercritical water oxidation: a technical review. AIChE J. 52:3933–3951.

4. Fang, Z., T. Minowa, C. Fang, R. L. Smith, H. Inomata, and J. A. Kozinski. 2008. Catalytic hydrothermal gasification of cellulose and glucose. Int. J. Hydrogen Energy 33:981–990.

5. Bocanegra, P. E., C. Reverte, C. Aymonier, A. Loppinet-Serani, M. M. Barsan, I. S. Butler, et al. 2010. Gasification study of winery waste using a hydrothermal diamond anvil cell. J. Supercrit. Fluids 53:72–81.

6. Calahoo, C., M. M. Barsan, M. L. Thomas, J. A. Kozinski, and I. S. Butler. 2011. Hydrothermal Raman microscopy studies of manganese carbonyls. Vib. Spectrosc. 57:2–7.

7. Sobhy, A., I. S. Butler, and J. A. Kozinski. 2007. Selected profiles of high-pressure methanol-air flames in supercritical water. Proc. Combust. Inst. 31:3369–3376.

8. Fang, Z., H. Assaaoudi, A. Sobhy, M. M. Barsan, I. S. Butler, R. I. L. Guthrie, et al. 2008. Use of methanol and oxygen in promoting the destruction of deca-chlorobiphenyl in supercritical water. Fuel 87:353–358.

9. Sobhy, A., R. I. L. Guthrie, I. S. Butler, and J. A. Kozinski. 2009. Naphthalene combustion in supercritical water flames. Proc. Combust. Inst. 32:3231–3238.

10. Xu, S. K., I. Butler, I. Gokalp, and J. A. Kozinski. 2011. Evolution of naphthalene and its intermediates during oxidation in subcritical/supercritical water. Proc. Combust. Inst. 33:3185–3194.
   Matsumura, Y., T. Minowa, B. Potic, S. R. A. Kersten, W. Prins, van Swaaij W. P. M., et al. 2005. Biomass gasification in near- and super-critical water: Status and prospects. Biomass Bioenergy 29:269–292.

12. Kruse, A., P. Bernolle, N. Dahmen, E. Dinjus, and P. Maniam. 2010. Hydrothermal gasification of biomass: consecutive reactions to long-living intermediates. Energy Environ. Sci. 3:136–143.

13. Letellier, S., F. Marias, P. Cezac, and J. P. Serin. 2010. Gasification of aqueous biomass in supercritical water: a thermodynamic equilibrium analysis. J. Supercrit. Fluids 51:353–361.

14. Hammerschmidt, A., N. Boukis, E. Hauer, U. Galla, E. Dinjus, B. Hitzmann, et al. 2011. Catalytic conversion of waste biomass by hydrothermal treatment. Fuel 90:555–562.

15. Bassett, W. A., A. H. Shen, M. Bucknum, and I. M. Chou. 1993. A new diamond-anvil cell for hydrothermal studies to 2.5 GPa and from 190°C to 1200°C. Rev. Sci. Instrum. 64:2340–2345.

16. Smith, R. L., and Z. Fang. 2009. Techniques, applications and future prospects of diamond anvil cells for studying supercritical water systems. J. Supercrit. Fluids 47:431–446.

17. Syassen, K. 2008. Ruby under pressure. High Pressure Res. 28:75–126.

18. Desgreniers, S., and C.-Y. Kim. 2011. La matière condensée sous conditions extrêmes: un programme de recherche au Centre canadien de rayonnement synchrotron. Physics in Canada / La Physique au Canada 67:3–6.

19. Hammersley, A. P., S. O. Svensson, M. Hanfland, A. N. Fitch, and D. Häusermann. 1996. Two-Dimensional Detector Software: From Real Detector to Idealised Image or Two-Theta Scan. High Pressure Res. 14:235–248.

20. Rodríguez-Carvajal, J. 1993. Recent advances in magnetic structure determination by neutron powder diffraction. Physica B. 192:55–69.

21. Hashaikeh, R., Z. Fang, I. S. Butler, and J. A. Kozinski. 2005. Sequential hydrothermal gasification of biomass to hydrogen. Proc. Combust. Inst. 30:2231–2237.

22. Hashaikeh, R., Z. Fang, I. S. Butler, J. Hawari, and J. A. Kozinski. 2007. Hydrothermal dissolution of willow in hot compressed water as a model for biomass conversion. Fuel 86:1614–1622.

23. Fang, Z., and J. A. Kozinski. 2000. Phase behavior and combustion of hydrocarbon-contaminated sludge in supercritical water at pressures up to 822 MPa and temperatures up to 535 degrees C. Proc. Combust. Inst. 28:2717–2725.

24. Fang, Z., and J. A. Kozinski. 2001. Phase changes of benzo (a)pyrene in supercritical water combustion. Combust. Flame 124:255–267.

25. Fang, Z., T. Minowa, R. L. Smith, T. Ogi, and J. A. Kozinski. 2004. Liquefaction and gasification of cellulose with Na2CO3 and Ni in subcritical water at 350 degrees C. Ind. Eng. Chem. Res. 43:2454–2463.

26. Meerts, P. 2002. Mineral nutrient concentrations in sapwood and heartwood: a literature review. Ann. For. Sci. 59:713–722.

27. Misra, M. K., K. W. Ragland, and A. J. Baker. 1993. Wood ash composition as a function of furnace temperature. Biomass Bioenergy 4:103–116.

28. Hudgins, J. W., T. Krekling, and V. R. Franceschi. 2003. Distribution of calcium oxalate crystals in the secondary phloem of conifers: a constitutive defense mechanism? New Phytol. 159:677–690.

29. Xu, J., D. F. R. Gilson, I. S. Butler, and I. Stangel. 1996. Effect of high external pressures on the vibrational spectra

of biomedical materials: calcium hydroxyapatite and calcium fluoroapatite. J. Biomed. Mater. Res. 30:239–244.

30. Parthiban, S. P., K. Elayaraja, E. K. Girija, Y. Yokogawa, R. Kesavamoorthy, M. Palanichamy, et al. 2009. Preparation of thermally stable nanocrystalline hydroxyapatite by hydrothermal method. J. Mater. Sci. Mater. Med. 20:S77–S83.

31. Davies, R. J., M. Burghammer, and C. Riekel. 2005. Simultaneous microRaman and synchrotron radiation microdiffraction: tools for materials characterization. Appl. Phys. Lett. 87:264105.

32. Boccaleri, E., F. Carniato, G. Croce, D. Viterbo, W. van Beek, H. Emerich, et al. 2007. In situ simultaneous Raman/high-resolution X-ray powder diffraction study of transformations occurring in materials at non-ambient conditions. J. Appl. Crystallogr. 40:684–693.

Supporting Information

Additional Supporting Information may be found in the online version of this article:

Figure S1. Raman spectrum (785 nm laser line) for microgranular cellulose in HDAC. The highlighted region was studied further at high temperature and pressure and shown in Figure 1A.

Figure S2. Ruby fluorescence with increasing temperature (514.5 nm excitation) for ruby and water in HDAC showing shift and increasing broadening.

Figure S3. Raman spectra (-OH stretch region) for pure water in HDAC with increasing temperature showing shift and increasing background.

13

Why the molten salt fast reactor (MSFR) is the "best" Gen IV reactor

Darryl D. Siemer

Nuclear Energy Department, Idaho State University, 921 S. 8th Ave Mail Stop 8060 Pocatello, ID 83209-8060, USA

Keywords

Breeder reactors, GEN IV reactor options, molten salt fast reactor, nuclear renaissance, thorium

Abstract

A simultaneously "nuclear", permanent, and in-time solution to mankind's energy-related problems would require the relatively *rapid* manufacture of 10,000–30,000 genuinely sustainable, full-sized (~1 GW$_e$) reactors. This "nuclear renaissance" would have to be implemented with breeder reactors because today's commercial nuclear fuel cycle is unsustainable and based upon a fuel (^{235}U) that is intrinsically expensive and politically problematic. The purpose of this paper is to point out why a simple/cheap "minimal reprocessing" implementation of the European Union's (EU's) molten salt fast reactor (MSFR) concept represents the most promising way to implement that technical fix:

• It would be relatively simple/cheap to both build and operate,
• Its fuel cycle is genuinely sustainable (no fuel shortages "forever"),
• Radwaste management would also be relatively simple and cheap,
• Operation would neither generate nor require huge amounts of transuranic (TRU) elements,
• The consequences of accidents (fuel spills, etc.) would be relatively benign,
• When steady state is achieved, the world would no longer need its uranium enrichment facilities.

Its primary drawback is that it would require virtually everyone currently involved with managing, researching, implementing, regulating, or "helping" the USA's nuclear power industry to embrace a massive paradigm shift.

Introduction

A simultaneously "nuclear", permanent, and in-time solution to mankind's energy-related problems[1] would require the relatively *rapid* manufacture of 10,000–30,000[2] genuinely sustainable, full-sized (~1 GW$_e$), not "modular" (small) reactors. The reasons why such a "nuclear renaissance" would have to be implemented with breeder-type reactors[3] were identified by Alvin Weinberg and H. E. Goeller four decades ago [1]; that is, because the fissile consumed by today's "converter" reactors (^{235}U) comprises only ~0.2% of the world's potential nuclear fuel supply[4] and is both expensive and politically problematic to obtain,[5] it is too costly to represent a truly sustainable energy (fuel) source for everyone. Two years ago, a book written by the managers of Argonne National Laboratory's (ANLs) "Integral Fast Reactor" (IFR) program reiterated that message and described how the USA had gone

about developing/promoting a sustainable nuclear fuel cycle based upon the conversion of ^{238}U to ^{239}Pu [2]. Unfortunately, that/their program was canceled two decades ago and none of the US Department of Energy Office of Nuclear Energy's (USDOE-NE's) current research priorities address the development of anything capable of addressing the world's long-term energy needs.

The rationale[6] for why the US-initiated Generation IV International Forum (GIF) program included molten salt reactors (MSRs) among the six (now seven) advanced reactor concepts[7] identified for cooperative development included: (1) they *can*[8] be operated in ways that would generate very little long-lived TRU waste; (2) their inventories of fissile materials per unit of energy *can* be lower than those of other reactors; (3) the dispersible inventory [source term] of radionuclides within them *can* be less than that of any other reactor; (4) both fuel and operating costs *could* be very low compared to solid-fuel reactors; and (5) there are large economics-of-scale with the potential to build very large reactors with extremely low per-megawatt capital costs [3]. Unfortunately, despite the fact that virtually all of the research revealing these characteristics had been performed at the USA's Oak Ridge National Laboratory (ORNL)[9] several decades earlier, the USA decided to cede GEN–IV MSR development work to its European collaborators in order to attempt another revival of ANL's Liquid Metal-cooled Fast Breeder Reactor (LMFBR, a.k.a., IFR) program. Its latest version would comprise a liquid metal-cooled, metal-fueled, fast-spectrum "burner" reactor (not a breeder – no ^{238}U-containing blanket around the core) fed with TRU "waste" (mostly ^{239}Pu) extracted from spent commercial light water reactor (LWR) fuel via a modified version of the pyrochemical fuel reprocessing/waste treatment system originally developed for the IFR. The resulting "sodium fast reactors" (SFRs) were to have become the USA's chief contribution to the Global Nuclear Energy Partnership (GNEP).

While critical reviews [4] soon led to a drastic downsizing of the USA's GNEP/Advanced Fuel Cycle Initiative NE research and development (R&D) programs [5], its GIF collaborators/partners continued to support/fund theirs, the most promising of which has turned out to be the MSFR concept jointly developed/studied by research teams in seven different counties +EURATOM+Russia. The remainder of this paper will seek to support my contention that it represents the "best" (most practical/cheap/clean/simple/safe) way to realize Weinberg/Geoller's technical fix for the consequences of a burgeoning human population's addiction to fossil fuels. That judgment is based upon a comparison of sustainability, cost, and waste management-related considerations.

Evolution of the Molten Salt Breeder Reactor

The first research performed to devise a molten salt "power" (electricity-generating) reactor capable of generating as much fissile (^{233}U) as it consumes is summed up in chapters 11–17 of "Fluid Fueled Reactors"[6]. Figure 1[10] depicts ORNL's "reference reactor" at the time that book was written (1958). It consists of a roughly spherical Hastelloy N "core" (or "reactor") tank through which the fissile-containing fuel salt stream (primarily ^{233}UF$_4$ in a low-melting solvent comprised of a 2:1 molewise mix of ^{7}LiF and BeF$_2$) is pumped, temporarily experiences criticality which generates heat energy which is then transferred to a second nonradioactive molten salt stream via an external heat exchanger. The core tank is situated within a larger "blanket" tank containing a similarly molten blanket salt containing thorium (as ThF$_4$)

that absorbs neutrons leaking though the core tank's wall (a.k.a., "barrier") and is thereby transmuted to ^{233}Pa. The 27-day half-life ^{233}Pa then decays to generate fresh fissile (^{233}U) that is transferred back to the fuel salt stream. In principle, it can operate indefinitely with no fuel other than makeup thorium *if* it is close coupled to a chemical reprocessing system capable of simultaneously transferring the fissile generated in its blanket salt to the core and preventing excessive fission product (FP) neutron poison[11] build up without excessive loss of thorium and ^{7}Li. However, ORNL's calculations indicated that this configuration could not achieve break-even fissile regeneration (isobreeding) with a core tank large enough to generate a worthwhile amount of power (>100 MW$_e$) unless thorium was also present in the fuel-side salt, which, in turn, would raise seemingly insuperable "reprocessing issues" due to its chemical similarities to rare earth element (REE) FP neutron poisons. This, plus the fact that Herbert McPherson, formerly National Carbon's foremost graphite expert,[12] had assumed management of ORNL's molten salt research program [7], shifted emphasis to the graphite moderated[13] two-salt system depicted in Figures 2, 3.

In principle, a full-sized (~1 GW$_e$) reactor featuring that concept's complex interlaced graphite pipe core configuration could achieve a CR ≥ 1 utilizing a relatively easy-to-reprocess (meaning thorium-free) fuel salt. Unfortunately, like many "paper reactors",[14] it would have been virtually impossible to either build or operate due to graphite's physical characteristics (e.g., highly anisotropic, unweldable, modest strength, poor ductility, etc.) and its propensity to first shrink and then swell upon fast neutron irradiation.

Those considerations plus an especially promising breakthrough in reprocessing technology [8] shifted ORNL's attention to the more simply configured, also graphite-moderated, one-salt reactor by the end of the 1960's (Fig. 4).[15] This is the "classic MSBR" being investigated when the AEC decided to axe Weinberg's "chemist's reactor" program. ORNL operated an ~8 MW$_t$ Molten Salt Reactor Experiment (MSRE) pilot plant from 1965 to 1969 to test theories/materials, demonstrate fissile recycle (both ^{233}U and ^{235}U), and determine generic MSR operational characteristics [7]. Due to fiscal constraints, it represented only the MSBR's central core (no surrounding "undermoderated"[16] Th-containing blanket zone) and therefore could not "breed".

The primary weakness of the one-fluid MSBR concept (Fig. 5) is that achieving both the degree and kind[17] of reprocessing necessary to achieve break-even fissile regeneration proved to be extremely difficult/expensive – and still would be if attempted today. Another drawback is the fact that the ~300 tons of radiologically contaminated,

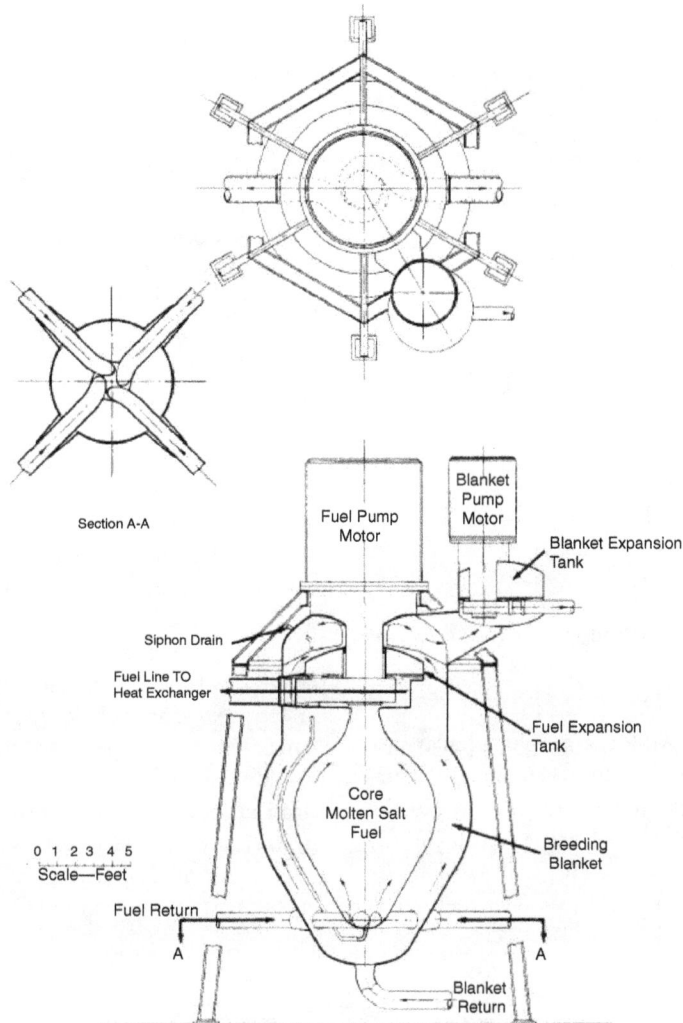

Figure 1. Mid-1950's ORNL unmoderated (no graphite) 2-fluid reactor (from chapter 17 of Ref. 6).

neutron-damaged graphite within its core would have to be replaced every 3–4 years, which, in the absence of a suitable repository, raises significant waste issues.

Consistent with GIF's sustainability and safety[18] goals [9], ORNL's classic MSBR concept gradually evolved in the direction of unmoderated (no graphite) fast-spectrum systems [10] and eventually (in 2008), the two-fluid (blanket equipped) MSFR (Fig. 6) became the EURATOM Consortium's (EVOL's) "reference" MSR.[19] While most of the previous studies of fast-spectrum MSRs had assumed a chloride[20]-based solvent salt in order to facilitate Pu breeding [11], fertile-free TRU burning, or U-supported TRU burning [12], EVOL's leadership assumed a somewhat less fast fluoride salt-based system in order to capitalize upon ORNL's extensive materials science and fuel salt chemistry experience [13].

Finally, because it soon became apparent that both CR and reactor durability[21] would be improved by sur-

rounding its entire core region with blanket salt rather than just along its sides, recent EVOL papers [14] often describe optimized MSFRs that look much like ORNL's original sphere-within-sphere concept – compare Figures 7 and 1.[22]

Everything that the rest of this paper has to say about what could be achieved with the EVOL's reference MSFR would also apply to such an optimized system.

Reasons Why the MSFR is the "Best" Gen IV Reactor

This section only addresses features relevant to its use as a genuinely sustainable thorium-fueled reactor operated to produce cheap, clean, electricity. References [15–17] give more detailed descriptions of both the reactor itself and its characteristics if utilized as a waste treatment technology for LWR-derived TRU.

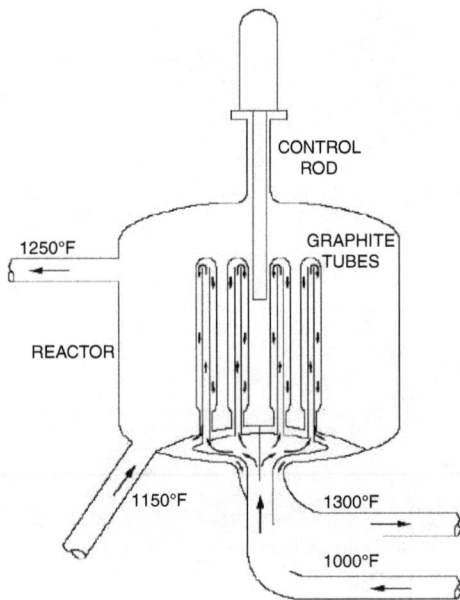

Figure 2. Mid-1960's ORNL graphite-moderated two-salt reactor, from ORNL 4528.

The core of the reference MSFR is a right circular cylinder with diameter equal to height (each 2.17 m), surrounded axially by 1-m thick steel reflectors, and radially by a 50-cm thick fertile blanket salt layer, a boron carbide layer, and a steel neutron reflector. It is filled with the fuel salt with no core internal structures. The fuel circu-

lates out of the core through 16 external loops each of which includes a pump and heat exchanger. A geometrically safe overflow tank accommodates salt expansion/ contraction due to temperature changes. A salt draining system connected to the bottom of the core allows core dumping to passively cooled criticality-safe tanks to facilitate maintenance or respond to emergencies. This system includes freeze valves that would melt as soon as electric power is lost or the salt seriously overheats. The entire primary circuit, including the gas reprocessing unit, would be contained within a secondary containment vessel. Figure 5 is a schematic of the primary loop's layout and Table 1 summarizes its core parameters.

Its primary fuel salt circuit is connected to two others which serve to keep the salt sufficiently free of the FP "ash" that would otherwise scavenge/absorb too many neutrons. The first of these would be a gas sparging system like that utilized by ORNL's MSRE pilot plant [7]. It would add a small stream of helium bubbles to the salt upstream of the heat exchanger (HX) pump(s) to strip-out both gaseous and nonsoluble FP.[23] A 30-second extraction time is generally assumed for both, although, in fact, extraction of nongaseous FPs will take somewhat longer (a noble metal extraction time of several days would have a limited impact on CR [19]). The second clean up circuit is the "reprocessing" system that removes intrinsically soluble ("salt seeking") FPs (e.g., REE, alkaline earths, alkalis, etc.) from the fuel salt either continuously or batch-wise on a daily basis.

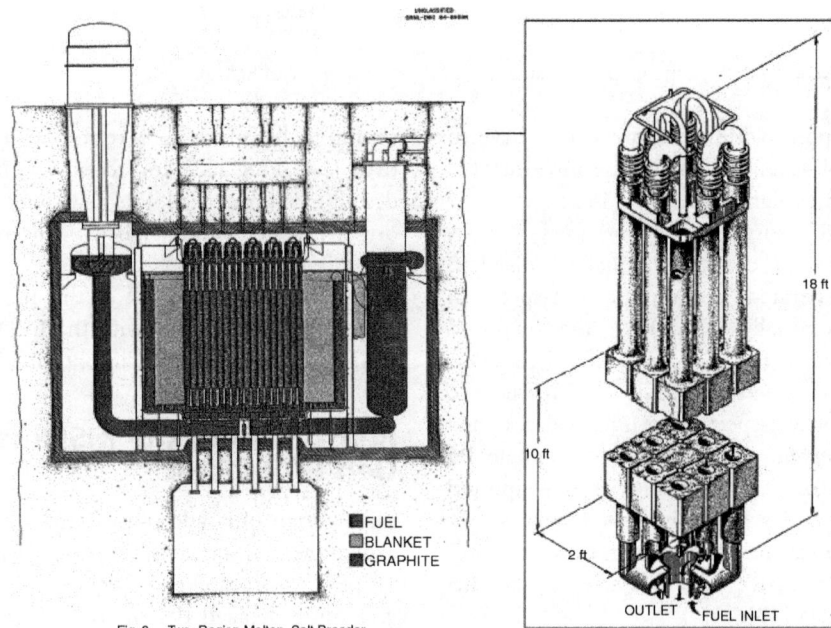

Fig. 2. Two–Region Molten–Salt Breeder.

Figure 3. Detailed depiction of ORNL's two-salt MSBR.

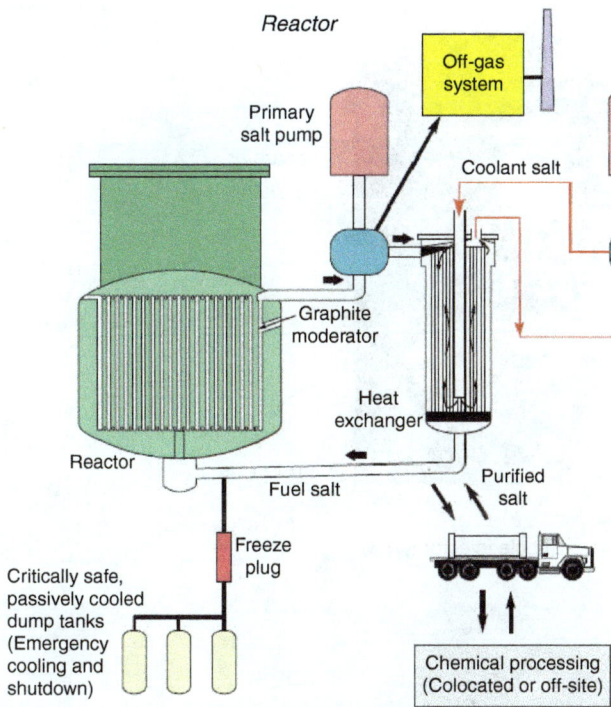

Figure 4. Current depiction of ORNL's graphite-moderated one-salt MSBR (from Ref. 3).

Low reprocessing requirement

This paper's chief technical contribution to the state of the art[24] is that it points out that a sustainable (CR ≡ 1) MSFR fuel cycle would require so little salt-seeker removal (see Figs. 8, 9) that doing such reprocessing is unnecessary – everything but the uranium[25] (primarily ^{233}U) in the fuel salt requiring that treatment could simply be discarded. The reasons for this include: (1) the

Table 1. MSFR core parameters [18].

Thermal/electric power	3000 MWt/1500 MWe
Core inlet/outlet temperatures	923/1023 K
Fuel salt volume	18 m³
Fraction of salt inside the core	50%
Number of loops for heat exchange	16
Flow rate	4.5 m³/sec
Salt velocity in pipes assuming 0.3 m diameter	~4 m/sec
Blanket thickness	50 cm
Blanket salt volume	7.3 m³
Boron carbide layer thickness	20 cm

value of the nominally costly ^{7}Li and Th in it would be less than the cost of separating/recovering them; (2) discarding TRU rather than recovering/recycling it would minimize the generation of both Pu and minor actinides (MA); (3) the total amount of radwaste generated would be very low; and (4) the resulting much simplified fuel cleanup process would be intrinsically safer to operate and significantly reduce the cost of producing electricity.

Consequently, satisfying the reprocessing requirements of a MSFR isobreeder would be much simpler, cheaper, and safer than those of either ORNL's classic MSBR or any of the other *potentially*[26] sustainable GEN IV.reactors (SFR/IFR, LFR & GFR). It would also obviate one of the primary drivers for the development of the too-simple "deep burn"-type MSRs that might eventually pose more problematic/costly radioactive waste management issues than does today's nuclear fuel cycle.[27]

Table 2 compares the MSFR to both of ORNL's graphite-moderated Th-breeders (Figs. 2–5). Note the differences engendered by the shift to a faster neutron spectrum. A point not addressed by Table 2 is that a moderated, one-salt MSBR's ^{233}Pa isolation/hold system

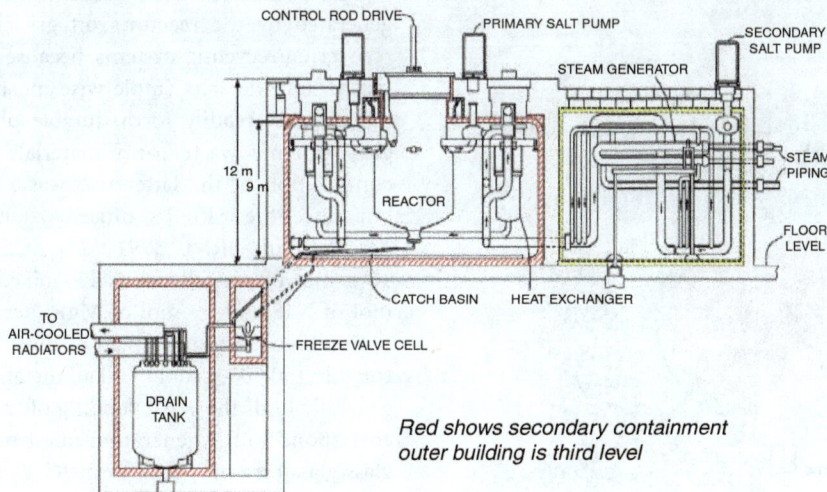

Red shows secondary containment outer building is third level

Figure 5. ORNL's depiction of its one-salt MSBR (early 1970's).

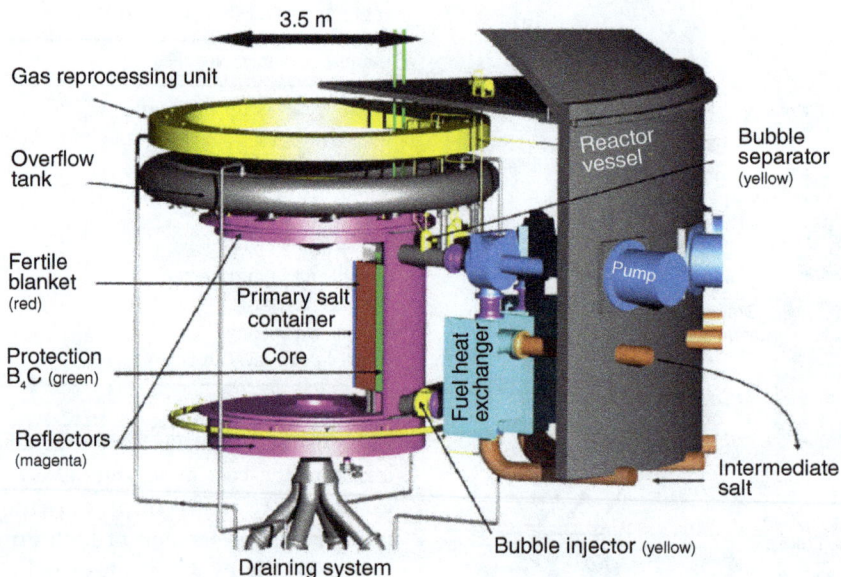

Figure 6. Reference MSFR (circa 2009–2013, from Ref. 16).

creates an almost realistic "proliferation issue".[28] Another is that graphite would likely enhance the corrosion rate of the reactor's metallic components [20].

Waste management-related advantages

Wastes generated by this particular MSFR implementation scenario would consist of: (1) everything in the 6 L[29] / day of reprocessed fuel salt except uranium; (2) waste generated by the reactor's off gas cleanup and uranium recycling systems; and (3) an occasional "worn out" reactor core and/or blanket salt tank.

The first of these waste streams is simple to characterize because at steady state, the reactor's salt cleanup systems

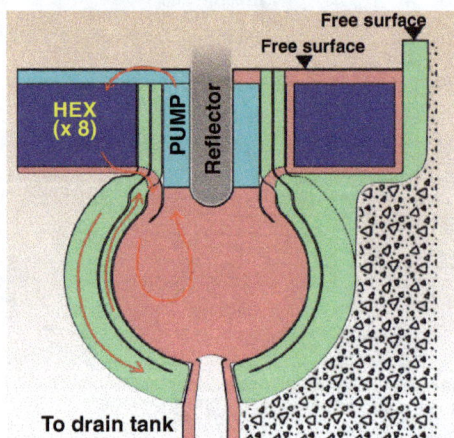

Figure 7. "Optimized" MSFR circa 2013 (from Ref. [14]).

must remove FP as fast as it is generated. Since 2 g mol of FP are generated by the fissioning of each mole of fissile, and 3 GW$_e$ worth of thermal energy at 200 mev/fission requires the consumption of ~3.13 kg (13.4 g mol) of ^{233}U per day, this corresponds to ~26.8 g mol total FP/day. Since roughly 15% of the FP will consist of inert gases and another 15% will comprise noble metal scum, about 70% or about 20 g mol (about 2 kg) of the FP will accompany the ~1.6 kg of uranium, ~13 kg of thorium, ~1.5 kg of ^{7}Li, and ~7 kg of fluoride in the ~6 L of fuel salt being reprocessed every day. Everything but the uranium in it (about 2 kg) would be fed to a small on-site glass melter along with an iron/aluminum pyrophosphate frit.

That waste would be combined/coprocessed with that generated by the reactor's off gas cleanup and uranium recovering/recycling systems because they would all consist of fluoride salts (mole-wise mostly those of the alkali metals) that readily form durable phosphate-based glass/glass ceramic waste form materials [21, 22]. The exact composition of the latter two waste streams is unknown, but since the MSFR's offgas system is similar to that described in ORNL 3791 [23], it is likely to comprise ~8 g mol of miscellaneous FP mixed with roughly 200 g mol of NaF and ~9 mol of MgF$_2$ per day.

Combined, a year's worth of those wastes could contain roughly 150,000 g mol of alkali metals, which, because they generally limit the waste loading of radwaste-type glasses,[30] correspond to the generation of ~4 m^3 of repository-ready glass/glass ceramic waste forms/GW$_e$-year.[31] This is far less than the 35–45 m^3 of glass-bonded sodalite (a.k.a., "ceramic waste form") required to immobilize the salt-wastes

Table 2. Three thorium-fueled molten salt breeder reactors.

	ORNL 1 GW$_e$ 2-salt (ORNL 3791)	ORNL 1 GW$_e$ 1-salt "MSBR" (Bettis, *NucTech*, 1970)	1.5 GW$_e$ MSFR (Florina Ph.D. thesis)
Reactor characteristics			
Type	C-moderated	C-moderated	"Fast"
Moderator	34 tonnes graphite	295 tonnes graphite	None
Energy density (kW/L)	91	22	330
Fuel salt characteristics			
Volume (m^3)	23	49	18
Content (metric ton): ^{233}U	0.76	2.3	5
^{7}Li	5.3	12.7	5.01
Be	3.1	3.6	N/A
Th	N/A	68	41
Processing rate (L/day)	409	26,350	~6
Blanket salt characteristics			
(Volume) m^3	63	N/A	7.3
Content (metric ton): Th	150		18
^{7}Li	15		1.95
Processing rate (L/day)	2860		~2

generated by the pyroprocessing of a GW$_e$-year's worth of IFR spent fuel rods [21].

Every 2–4 years those glass/glass ceramic canisters would be accompanied by a roughly half cubic meter metallic waste form generated by hot-pressing the reactor's "worn out" super alloy[32] core/blanket barrier wall plus the nickel wool FP scum adsorbent into a "brick" like that invoked for the IFR's fuel hull/anode sludge wastes [2].

A key difference between this scenario's waste forms and the raw LWR spent fuel assemblies currently constituting the USA's preferred high-level waste (HLW) waste form is that the former would contain much less long-lived TRU which would significantly simplify their ultimate disposition.

Table 3 compares a GW$_e$-yr's worth of radwaste generated by this MSFR scenario with that generated by the direct disposal of 37 GWd/THM (tonne heavy metal) LWR and 80 GW$_t$d/THM pebble bed-type VHTR (a.k.a., PBMR) reactor fuels.[33]

First, please note that the total amount of TRU-type waste discarded by this MSFR implementation scenario is much less than that in an equivalent amount of spent LWR or VHTR/PBMR fuel[34] and that most of its plutonium would be short-lived, nonfissile ^{238}Pu – not the more politically problematic ^{239}Pu. Second, recycling only the uranium in its "reprocessed" fuel rather than all of its actinides virtually eliminates the "higher" minor actinides (Am, Cm, etc.).[35] Third, the MSFR's high thermal-to-

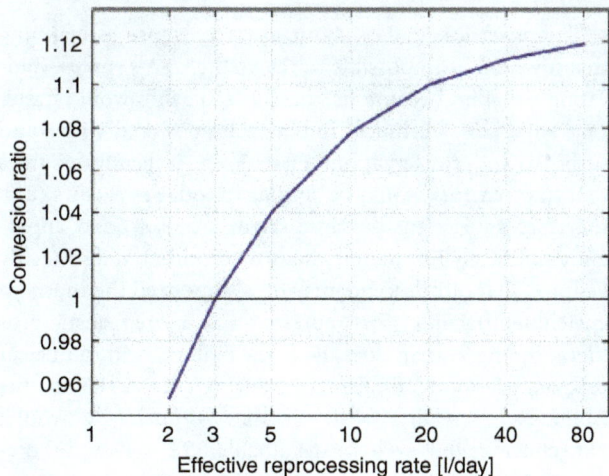

Figure 8. CR versus Reprocessing rate for MSFR Th breeder (JEFF3.1 data base, from Ref. [16]).

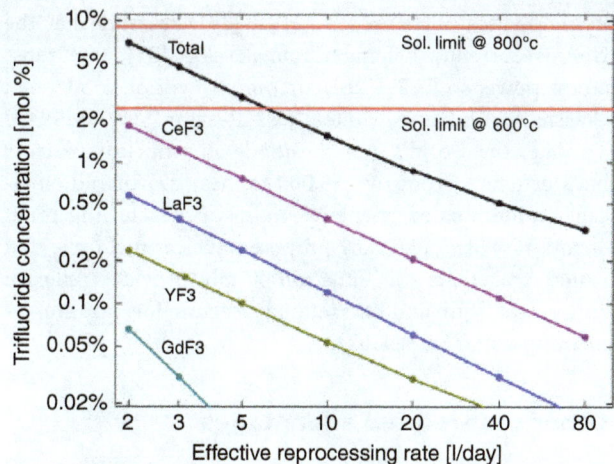

Figure 9. Rare Earth (REE) FP buildup as a function of reprocessing rate (from Ref. [16]).

Table 3. 1 GW$_e$-year's worth of LWR, PBMR, and this-scenario's MSFR wastes.

TRU isotope	kg waste/GW$_e$-year		
	MSFR	LWR	PBMR
Np			
237	3.67	9.88	NA
238	0.003	NA[1]	NA
239	4.5E-4	NA	NA
Pu			
238	1.301	3.46	3.9
239	0.159	109.00	75.2
240	0.027	63.00	56.2
241	0.003	27.00	37.0
242	0.000	7.80	32.1
Am			
241	0.002	0.98	NA
242	1.3E-6	0.01	NA
243	7.9E-5	1.95	NA
Cm			
242	1.5E-4	0.50	NA
243	1.5E-05	0.01	NA
244	2.4E-5	0.50	NA
Total TRU kg	5.16	224	204
Other radwastes			
FP	762	1154	952
U	0	~20,000	~12,000
"Hot" metallic	~700	~3000	0
"Hot" moderator	0	0	377,000

[1]NA means that my source document did not mention that isotope.

electrical power conversion efficiency translates to significantly less FP/GW$_e$.[36] Finally, it would not generate many tonnes of highly radioactive "spent" moderator per GW$_e$-year as would ORNL's classic MSBR, US/Russian cold-war "production reactors", gas-cooled, graphite-moderated reactors (acronyms include AVG, HTGR, VHTR, PBMR, NGNP, etc.), the molten salt-cooled, solid-fueled,[37] graphite-moderated (FHR or PB-AHTR [24]) reactors that the USA is currently helping China develop [25], or Trans-atomicpower's (TAP's [26]) zirconium hydride-moderated "Waste-Annihilating Molten Salt Reactor" (WAMSR).[38] To date, the world's graphite-moderated nuclear reactors have generated roughly 250,000 tonnes [27] of radiologically contaminated graphite, most of which lingers in "temporary" storage – any proposed nuclear fuel cycle that would exacerbate this situation simply provides reflexive anti-nukes with another rationale/excuse for <u>not</u> implementing a nuclear renaissance.

Other cost-related advantages

There are a host of additional reasons why MSFRs should be relatively cheap. Several are due to the characteristics of their molten salt working fluids:

(1) Their low viscosity and high heat capacity means that relatively small pumps and heat exchangers are needed.

(2) Their extremely low vapor pressure at the reactor's working temperature means that its pipes, tanks, and heat exchangers would not have to be as strong (heavy/expensive) as those of LWRs or any sort of gas-cooled reactor.

(3) Their ability to solubilize both fissile and fertile materials means that they obviate virtually all of the costs (manufacture, shipping, loading, "shuffling", unloading, storage, dissolution, separation, refabrication, more shipping, etc.) incurred with the solid fuels invoked for all of the other GIF reactor candidates.

Another reason for a relatively low cost is that the MSFR's active core region is so small (3 m^3/GW$_t$) that it should be possible to design an optimized version in which components subject to especially high neutron flux (primarily the tank wall(s) separating the blanket and fuel salt streams) would be relatively simple to replace (this should be a primary goal of future MSFR development work). Because a 2-cm thick, 9 m^3, spherical core tank would weigh roughly 4 tonnes and the real-world price of the super alloy [28] it is apt to be made of is roughly \$5/kg,[39] its cost should be quite low.[40] This is a vitally important point because a system's durability and affordability is as much determined by its maintainability as by its frequency of failure.

Issues, Arguments, Quibbles, and Excuses

The "proliferation" problem

> "Your scenario is impossible because its fissile isn't denatured."

This assertion makes about as much sense as claiming that it would be impossible to revive the USA's space exploration program because its current leadership would prefer that any "new" technical initiative be powered with wind turbines and conservation. First, since it tacitly assumes that new reactors would be similar to today's reactors (and therefore subject to the same set of arbitrary, and apparently politically immutable, man-made rules), it also tacitly assumes that our descendants will always need the uranium enrichment facilities that represent a far more realistic proliferation threat than does the fissile material within nuclear reactors. Second, diluting/denaturing the ^{233}U in the MSFR's salt stream(s) with ~8 times as much ^{238}U would: (1) render its fuel cycle unsustainable (CR < 1) and therefore defeat one of the key reasons for implementing a nuclear renaissance with them; (2) greatly complicate operation rendering them more difficult/dangerous to run and increase their electricity's cost; and (3) turn them into just another large-scale TRU-type radwaste generator. The fact remains that the US federal government – a signatory and vociferous proponent of the Nuclear Non-Proliferation Treaty – has been operating many HEU (Highly Enriched Uranium)-fueled (mostly naval) reactors for decades and is likely to continue to do so. There have been no "diversions" of their fissile by terrorists and it is unrealistic to assume that the fissile within nuclear reactors sited in any "First World" country would be either. Like war, proliferation is a political issue and its solution, if there is one, will be found in that arena. It has nothing to do with civilian reactor designs, which means that attempts to leverage it to claim superiority for a particular GEN IV technology do no favor to either nuclear power or humanity's future prospects.

Materials related issues

"Your scenario is impossible because it discards ~1 kg of isotopically pure ^{7}Li per GW$_e$-day and the USA doesn't possess lithium isotope separation capability."

This argument ignores the fact that the USA generated the ~half-tonne of "pure" ^{7}Li required to operate ORNL'S MSRE over fifty years ago. The reason why it has lost that capability is that the process employed at that time was "dirty" and it therefore became more convenient to outsource the production of both ^{6}Li and ^{7}Li to other countries.[41] Today, the USA possesses roughly twice as many tax/ratepayers as it did then and could now take advantage of the fact that China's entrepreneurs could quickly supply whatever is needed. To them ^{7}Li would probably be considered the byproduct because the more valuable isotope would go to making better (lighter) Li-ion batteries for the millions of Chinese-made electric cars rendered practical by a cheap nuclear renaissance.

"The nickel–based alloys required to build your reactor are damaged by neutrons, which means that we must wait until someone discovers something better (unobtanium?)."

This is just another excuse for more foot-dragging because the primary reason that the nuclear industry is concerned about nickel embrittlement is that it weakens thick-walled SS pipes/tanks that are supposed to withstand 1000–2200 psi pressure differentials for several decades. In a MSFR, those pressure differentials would be far lower,[42] which means that the reason for using nickel is that it renders metallic surfaces inert to fluoride/fluorine-induced corrosion. Since corrosion is a surface phenomenon, it would be reasonable to make the bulk of the core/blanket barrier wall out of a probably cheaper, low nickel alloy and plate its surfaces with a thin layer of pure Ni [29]. Another comforting fact is that the ~15% nickel D9 stainless steel-

clad fuel utilized by ANL's exhaustively studied IFR pilot plant (EBR II) experienced similar neutron fluxes while safely achieving burn ups of over 170 GWd/MTHM Ref. [2, Table 6–1]. Additionally, since a properly designed MSFR's core tank wall could be replaced as often as wished, it is unreasonable to assert that its durability constitutes a show stopper. In any case, definitive answers to such questions can only be generated by open minds willing to suggest/authorize/perform realistic tests.

"Molten salts are too corrosive/reactive/dangerous, etc."

Dry fluoride ion-based salts (not hydrofluoric acid or fluorine gas) are only corrosive to materials that are more electropositive (reactive to fluorine) than are the metals (in this case lithium and thorium) from which the cations accompanying that fluoride were derived. Since most of the materials used in/around nuclear reactors (steel, concrete, etc.) are less electropositive (reactive) than is either elemental thorium or lithium, there is little/no driving force for corrosion or any other sort of reaction if/when such contact occurs. Spilled molten salt could cause a "burn" in the same sense that hot cooking oil can but cannot actually burn (or explode) as would an IFR/SFR's molten sodium if it were to contact air, water, or human sweat.[43]

"Since the startup of each new MSFR would require about 5 tonnes of fissile, wouldn't it be impossibly expensive to get a significant number of them built?"

Before describing several ways to address this question,[44] I should point out that the startup fuel of any of the nuclear industry's proposed full-sized (not "modular"), GEN III LWRs would also contain/require ~5 tonnes of "new" fissile (^{235}U).

The first scenario is that assumed by most of the western world's NE R&D experts for any sort of fast reactor; that is, that it/they would be both started up and run with TRU "pyroprocessed" from spent LWR reactor fuel (e.g., GNEP). While that proposal is theoretically feasible and serves a *political* purpose,[45] such fuel would be far more expensive and dangerous to use than is "fresh" ^{235}UF$_4$ and thereby apt to stifle a US nuclear renaissance.[46] Consequently, I will not attempt to assign numbers to it.

My second scenario assumes that MSFRs would be started with ^{235}U (as $\geq$80% ^{235}UF$_4$) generated by a 20% increase in the USA's current investment in the uranium enrichment "separative work units" (SWUs) utilized to produce fuel for its ~100 existing LWRs. Since the nuclear power industry's spokespersons contend that its total fuel costs (roughly 25% of which goes for enrichment) are "low",[47] a temporary 20% increase in the USA's total commitment to enrichment should also constitute a "low" cost. If we also consider that each MSFR is apt to generate somewhat more fissile than it consumes (I've assumed 4.5%, see

Fig. 8), this scenario would double the USA's total nuclear power generating capacity (to ~200 GW$_e$) in 24 years and nearly quintuple it (to 487 GW$_e$) by the end of this century.

My third scenario is similar but assumes that rather than continue to keep the current LWR fleet running indefinitely, five of them (oldest first) would be shut down during each of the first 20 years and the SWU's devoted to feeding them applied to generating additional MSFR start up fuel instead. After that, all of the SWUs currently used to feed the USA's LWRs (~1.25E+7/year) would be applied to MSFR start up. This scheme would treble the USA's nuclear power generating capacity in 19 years and raise it to ~1985 GW$_e$ by the end of the century.

Fourth, since the USA purportedly still possesses ~600 tonnes of cold war-generated, weapons grade HEU, (http://www.pogo.org/blog/2014/08/20140825-no-more-excuses-for-failing-to-downblend.html), let's assume that its decision makers decide to devote 500 tonnes of it to MSFR startup over the next ten years[48] (build 10 MSFRs per year) while decommissioning an equal number of LWRs applying their SWUs to producing additional MSFR startup fuel. This scheme would double the USA's nuclear power generating capacity within 9 years and raise it to 2017 GW$_e$ by 2100 AD.

Finally, during the cold war, the USA's production reactors generated roughly 100 tonnes of bomb-grade plutonium (>90% ^{239}Pu), much of which is currently being stored in vaults at the Savannah River National Laboratory. Such plutonium is a much better fissile (fuel) than is the mix of TRU isotopes in spent LWR fuel and available in a form (pure PuO$_2$) that can be easily/cheaply converted to PuF$_3$. Utilizing it to start a small fleet of MSFRs (~20) could simultaneously kick-start a USA nuclear renaissance while rendering it useless for weapons manufacture.[49] For example, if the USA were to start five MSFRs with that Pu and shut down five of its oldest LWRs for each of four years and then continue to shut down another five LWRs per year during the next fifteen utilizing their enrichment capacity to make more MSFR startup fuel, its nuclear power generating capacity would double in ~19 years and reach ~1673 GW$_e$ by 2100 AD – all accomplished with the same enrichment capacity used to generate today's 100 GW$_e$'s worth of LWR power.

Any of the last three MSFR startup scenarios would obviate the USA's need/rationale/excuses for uranium enrichment after ~2100 AD.

Mission-related questions

"Why not just build "advanced" LWRs instead? (corollary: don't we have enough uranium to just keep doing what we're familiar with?)."

This question tacitly assumes that our political leadership will not choose to address climate change and the consequences of fossil fuel depletion with nuclear power; that is, that only a few new reactors will be built during the next several decades.[50] If that is the case then there is indeed "plenty of uranium." (Another drawback is that it would be "burning our descendant's seed corn"; that is, when they finally do decide to switch to a sustainable nuclear fuel cycle, the world won't contain as much "cheap" fissile to start it with.) However, if our leaders do decide to face up to those problems, their fix will have to be implemented with breeder-type reactors. The following ball-park calculation demonstrates the reason for this:

- Current EIA estimates peg the world's total current energy use at ~500E+18 J/a
- Since a 1 GW$_e$ LWR has an thermal-to-electrical energy conversion efficiency of about 33% and uses about 200 metric tonnes (t) of raw uranium per year...
- J$_{heat\ energy}$/t U = 1/0.33 × 1E + 9 J/s × 3600 sec/h × 24 h/d × 365 y/a/200 tU = 4.73E + 14
- The uranium industry's latest Redbook (http://www.world-nuclear-news.org/ENF-U...07127.html) states that "total identified U resources at a 'reasonable' (currently <$260/kg) price" is 7,096,600 tU.
- If the world's total current energy needs were to be met with today's reactors fueled with "reasonably priced" uranium, how long would it last?
- Since tonnes U/a = 500E + 18/4.73E + 14 = 1.06E + 6 Time 'til gone = 7.096E + 6/1.06E + 6 = 6.71 years.

If the same Redbook's 10,400,500 tU of "undiscovered resources" (*expected to exist based on existing geological knowledge but requiring significant exploration to confirm and define them*) were to be found/used too, it could fuel that scenario's clean/green LWR-powered world for another ~9.8 years.

The same report goes on to say that, "The increase in the resource base is the result of concerted exploration and development efforts. Some $2 billion was spent on uranium exploration and mine development in 2010, a 22% increase on 2008 figures..." This means that even with today's still relatively concentrated uranium ores (certainly with respect to seawater's ~3 ppb U), extending that industry's "resource base" is currently costing its customers a great deal of money.[51]

"Why is this/your MSFR implementation scenario especially 'sustainable'?"

The answer is that "it represents a permanent solution to humanity's energy problems". The following ball-park calculation supports this contention[52] :

- @ ~2.7 g/cc, the mass of the Earth's crustal landmass (not under its oceans) to 1 km depth (i.e., "readily accessible rock") $\approx$4.2E + 17 tonnes
- Total fossil fuel $(CH_x) = \sum$coal + shale kerogen + petroleum + natural gas) reserves $\approx$1638E + 9 tonnes (843 + 500 + 170 + 125 gigatonnes - recent EIA estimates) – consequently, weight fraction CH_x in readily accessible rock is $\approx$3.9 ppm ($1638E+9/4.2E+17$)
- The combustion of one gram of "average" CH_x generates ~37,000 joules of heat energy *plus about 3 grams of CO_2*
- Mankind's current fossil fuel consumption rate (~500 exajoules/a) represents about 1% (*or 500E+18/ ($1638E+9*1E+6*37000$)*) of its total CH_x reserves
- @ 12 ppm, thorium in the Earth's crustal landmass $\approx$4655 gigatonnes[53]
- @ 200 Mev/atom, the fission of one gram of ^{233}U via MSFR produces 8.3E+10 Joules of heat energy *and <u>no</u> "greenhouse" gases*
- Since this scenario would discard ~80% of the Th,
- MSFR energy/fossil energy $\approx$1.3E+6 (*1-0.8)*(4655/ 1638)*(8.3E+10/3.7E+4*)
- Therefore, @ humanity's current total energy demand, "readily accessible" thorium could power us for ~1.3E+8 years (*1.3E+6*1/about 1%*).

By circa 1.3E+8 AD, the FP accompanying the discarded thorium in most of this scenario's waste forms would have decayed thereby converting their repository to an extremely rich (~24% Th and 3% ^7Li) ore body that could go on powering our descendants for another half-billion years. After that has been consumed, they could then begin to extract thorium from the next kilometer-thick layer of the Earth's crustal landmass.

Conclusions

To summarize, a properly designed (readily maintainable) MSFR isobreeder represents today's "best" Gen IV option because:

- Its compact size and simplicity relative to alternatives invoking solid fuels and/or moderators (all of the other GIF candidates) means that it should be relatively cheap to both build and operate (less metal needed to fabricate/maintain and no initial fuel fabrication, handling, durability, shuffling, transport, reprocessing, or fuel refabrication costs).
- Its fuel cycle is genuinely sustainable – no fuel shortages "forever".
- Radwaste management should be relatively simple/cheap.
- Operation would neither generate nor require huge amounts of TRU.

- Its ~700°C operating temperature and high heat capacity working fluid translates to higher electricity generation efficiencies and more direct process heat applications.
- Its nonreactive, high-boiling, working fluid reduces both the probability and consequences of accidents (spills, etc.).
- When steady state is attained (~100 years) they would obviate the need/rationale for either uranium enrichment or uranium mining.
- Fueling them would generate far less environmental impact (e.g., mine tailings, etc.) than would any of the nonbreeder Gen IV reactor concepts.

This scenario's primary drawback is that it would require virtually everyone involved with researching, implementing, regulating, or "helping" the USA's nuclear power industry to embrace a massive paradigm shift. The reasons for this include:
- The nuclear industry's current business model is already profitable, firmly established, and primarily cost-plus[54] which means that most of its leadership resists change.[55]
- The USA's political system strongly caters to established (moneyed) interests and has become virtually incapable of addressing any politically sensitive and/or nonimmediate national problem (e.g., climate change) regardless of how important it might be from a "technical" point of view.[56]
- Its leadership has supported (via earmarks) a series of long-winded, multi $billion, boondoggles (e.g., the Savannah River Site's MO_x plant, LLNL's "National Ignition Facility", SANDIA et. al.'s Yucca Mountain studies, Hanford's "Waste Treatment Plant", the Idaho National Laboratory's "steam reformer", etc.) that have served to convince many people that any sort of US nuclear renaissance is apt to be risky, environmentally impactful, and prohibitively expensive.
- The leadership of the USA's national laboratories is no longer able to make tough decisions[57] or authorize the sorts of "messy" real-world experimentation required to develop an unfamiliar reactor concept and then demonstrate its viability (currently, if a proposed project can't be almost 100% proven/completed with "paper studies", it won't be undertaken). The Nuclear Regulatory Commission possesses the same mindset.
- The leadership of most of the world's environmental organizations does not realize that a properly implemented nuclear renaissance represents the most reasonable way to serve their cause and therefore continue to resist anything that might lead to one.

The most sensible way to implement a US nuclear renaissance would be to build clusters of MSFRs (a total of 1500–3000 GW_e's worth) both where today's LWRs are located and at US DOE's already radiologically compromised nuclear facilities. These "power parks" would include a centralized reprocessing/waste treatment facility to recover useful materials (e.g., ^{99}Mo,^{58} rhodium, and palladium) from radwastes before they are vitrified. They would be surrounded by energy intensive manufacturing facilities (e.g., water desalination,^{59} aluminum, steel, fertilizer/synthetic fuel (most of which could be ammonia, http://nh3fuelassociation.org/), titanium, lithium, rare earths, cement plants...) which would provide millions of high-quality jobs and thereby address other problems generated by the USA's post-Vietnam War policy shifts.

The USA currently spends about 300 times as much to "maintain its nuclear deterrence" as it does on the "Advanced Reactor Concept" R&D which could address the root causes of conflict.[60] At this point in time the MSFR is just an especially reasonable paper reactor that wasn't "discovered" by the USA and is therefore unknown to its political leadership and most of its nuclear engineers. However, since the EU has demonstrated that a nuclear renaissance implemented with them could likely address both climate change and the otherwise inevitable social/economic consequences of fossil fuel depletion in a uniquely affordable and environmentally correct fashion, the people responsible for implementing the USA's NE R&D programs should be encouraged/enabled to do the scientific research, design work, and pilot plant testing necessary to turn it into a practical (both maintainable and affordable) reactor.

Acknowledgments

I thank Kirk Sorensen, whose "energy from thorium" blogsite reminded me of why I had decided to become a scientist in the first place, and then provided the technical information/tools required to pursue what's become my retirement hobby. Second, I thank Professor MaryLou Dunzik-Gougar (NE ISU), who paid my registration fee at the conference (GLOBAL 2013) where I first learned about MSFR. Finally, I thank Carlo Fiorina, whose presentations at that conference revealed that the European Union's EVOL program had finally addressed the chief weakness of Weinberg's "chemist's reactor" and then volunteered to perform the calculations which "proved" it.

Conflict of Interest

None declared.

Notes

[1] Global warming/climate change, ocean acidification, air pollution, biofuel-driven food cost escalation, water shortages/pollution, relentless cost of living increases, widespread poverty, and political

impasses up to and including outright war (see http://www.cna.org/sites/default/files/MAB_2014.pdf).
[2] The USA's ~5% of the world's population uses ~20% (~100 exajoules/year) of its total energy: 100 exajoules/year corresponds to the thermal energy output of ~1000 1 GW$_e$ nuclear reactors (or 5555 of B&W's currently front-running "mPower" small modular reactor (SMR) concept - see http://en.wikipedia.org/wiki/B%26W_mPower).
[3] Conversion ratio (CR) ≡ fissile generated/fissile consumed: if CR < 1 the reactor is a "converter" (and also unsustainable); if CR ≡ 1.00, it's an "isobreeder"; if CR > 1.00, it's a "breeder".
[4] ^{235}U/(all U+Th) ≈ 0.002.
[5] Witness the perpetual brouhaha generated by Iran's uranium enrichment facility.
[6] Unstated reasons include the fact that molten salt breeder reactors (MSBRs) are much better suited to "burning" thorium than is any solid-fueled reactor. This is important because the earth's crust contains ~4 times as much thorium as uranium and its "combustion" generates far less long-lived TRU waste.
[7] The original six GIF candidates included three fast reactors (gas-cooled [GFR], sodium cooled [SFR], and lead cooled [LFR]) and three thermal reactors (graphite moderated/molten salt [MSR], gas cooled/graphite moderated/very high temperature [VHTR], and supercritical water-cooled/moderated [SCWR]). The seventh added later is the subject of this paper, the molten salt fast reactor (MSFR). See https://www.gen-4.org/gif/upload/docs/application/pdf/2014-06/gif_2013_annual_report-final.pdf.
[8] "Cans" and "coulds" are italicized because there are many ways to implement MSRs, only some of which exhibit the characteristic in question (see D. Holcomb et. al., "Fast Spectrum Molten Salt Reactor Options", ORNL/TM 2011/105 R (July 2011).
[9] ORNL's MSR/MSBR program cancellation in 1973 was accompanied by the downsizing of its long-time Director, Alvin Weinberg (see Ref. [1]). The reason stated for the AEC's actions was that the USA "could no longer afford" to support two breeder reactor programs (this is ironic because in 1972, ANL's LMFBR development work had cost taxpayers 26 times as much ($123.2M/$4.8M) as had ORNL's MSBR program – see L. Cohen's "The Technology Pork Barrel", Brooking Institution Press, 1991, p. 234). Overall, total US LMFBR R&D spending has been ~150 times that devoted to MSR-related work.
[10] "Two-fluid" rather than "two-salt" because most of ORNL's MSBR modeling work assumed that the reactor's fuel salt would contain thorium as well as fissile. Such systems are now generally called "1½ salt" reactors. "Two-salt" is now generally reserved for reactor concepts in which 100% of the thorium (and therefore fissile generation) is in the surrounding blanket.
[11] "Poisons" are materials that scavenge/absorb neutrons that would otherwise serve a useful purpose.
[12] Dr. McPherson's development of boron-free graphite (boron is a powerful neutron poison) had previously rendered Hanford's war-winning ^{239}Pu production reactors possible.
[13] The technical reason for moderating a MSR's core (with graphite, zirconium hydride or BeO) is that doing so would increase fission cross sections thereby allowing it to operate with a smaller fissile inventory – a characteristic deemed to be of paramount importance when ORNL was pursuing its MSR studies. One of the downsides of moderation is the fact that parasitic neutron absorption cross sections become much higher which means that far more "reprocessing" (greater degree of FP removal) is required to achieve break-even fissile regeneration.
[14] http://www.ecolo.org/documents/documents_in_english/Rickover.pdf.
[15] That breakthrough invoked a ~20 stage countercurrent extraction/back extraction system utilizing a molten bismuth solvent containing

an electrochemically generated metallic reductant (lithium and/or thorium) to selectively recover key species (U, Th, REE, and Pa) from a molten salt slipstream [8].

[16]The periphery of the MSBR's core possessed larger salt channels than did its center (less graphite/cc) which would have simultaneously suppressed fission and enhanced conversion (breeding) therein.

[17]"Degree" because a great deal of reprocessing (>1200 L/day) would have been required to achieve break-even fissile generation: "kind" because that system would have to be able to remove ^{233}Pa (a readily transmuted neutron poison), store it ex situ until it had decayed to ^{233}U, and then return it to the fuel salt loop. Two-fluid MSRs can isobreed without a ^{233}Pa isolation/hold system.

[18]"Safer" because EVOL's studies indicate that a MSFR would possess much larger negative temperature and void reactivity coefficients than would the classic MSBR.

[19]The basic concept was described over 50 years ago when US physicists realized that an unmoderated MSBR would likely be simpler to both build and operate. See L.G. Alexander, "MSFRs", *Proceedings of Breeding Large Fast Reactors*, ANL-6792 (1963). Unfortunately, it was subsequently essentially forgotten until the EVOL's MSR researchers decided to resurrect it.

[20]All else being equal, a chloride-based MSR is "faster" than its fluoride analog because the halide atoms comprising ~two-thirds of those present in either of their salt streams are heavier.

[21]"Durability" because the blanket salt would intercept the fast-moving neutrons that would eventually damage the reference MSFR's axial reflectors.

[22]The key difference between the MSFR and the 8-ft core diameter, two-fluid reference reactor described in Chapter 14 of Ref. [6], is that ORNL's modelers chose to limit the latter's maximum fuel-side salt's thorium concentration to 7 mole % because (to them) its fissile inventory would become "excessive" above that figure (that was before today's centrifuges rendered HEU *relatively* "cheap"). The addition of thorium to a MSR's fuel salt "hardens" its core's neutron spectrum rendering it "faster".

[23]Xenon and krypton (roughly 16 wt% of all FP) comprise the majority of inert gas FP. Nonsoluble FP comprise elements (e.g., Pd, Ru, Re, Te, Ag, and Mo) too "noble" to exist as cations in its redox-controlled fuel salt stream and which therefore tend to accumulate as a metallic scum at the gas/liquid interface within the sparge gas disengaging system where they could be adsorbed onto nickel "wool".

[24]Descriptions of MSFR reprocessing systems still invariably invoke >20 stage counter current liquid--liquid extraction systems like those assumed for ORNL's MSBR; for example, see slide 50 of Ref. 15.

[25]Uranium separation/recycling is the best-proven of the technologies proposed/investigated for MSR salt clean-up [7]: elemental fluorine bubbled through the molten salt – either blanket or fuel – selectively oxidizes its uranium to gaseous UF_6 that is then adsorbed upon a cool filter comprised of NaF salt granules. It would then be transferred to a second molten fuel salt slip stream by heating that filter and bubbling the re-evaporated UF_6 through it along with sufficient hydrogen gas to convert (reduce) it back to salt-soluble UF_4.

[26]"Potentially" because solid-fuel reprocessing/recycling schemes for the LFR and GFR haven't been worked out yet and that proposed for the IFR is apt to be too expensive (see p. 124 of http://www.pdfdrive.net/fast-reactor-development-in-the-united-states-e261 1848.html.)

[27]A typical "deep burn" MSR scenario (e.g., ORNL/TM- 7207, available gratis at http://energyfromthorium.com/ornl-document-repository/) invokes a huge (~320 tonnes of fuel salt containing both thorium and ^{238}U-denatured fissile in a BeF_2/^{7}LiF solvent moderated with ~1200 ton of graphite) one-fluid reactor that is to be operated for thirty years with no reprocessing other than gas sparging and noble metal scum removal. Fresh fissile (typically 20% ^{235}U enriched uranium) is periodically added to compensate for the fact that although it exhibits "high" conversion (~0.8), it can't generate enough of its own fissile (^{233}U and ^{239}Pu) to continue operating without it. At the end of thirty years, everything including the graphite is to be "reprocessed" by whoever owns it at that time. Other than "simplicity", the chief driver/rationale for this concept is greater proliferation resistance: imaginary terrorists who have managed to "divert" some of its intensely radioactive fuel salt would have to chemically isolate the uranium and/or plutonium in it and implement an equally surreptitious/successful isotopic separation process in order to generate bomb-grade fissile (some decision makers and most anti-nukes seem to take such ridiculous scenarios *very* seriously).

[28]The reason for this is that the fissile generated in its ^{233}Pa isolation/hold tank would contain insufficient ^{232}U to discourage imaginary terrorists.

[29]The assumed 6 liter/day reprocessing/discard rate is determined by the fuel salt's REE FP solubility limit (Fig. 9), not achieving a CR of exactly 1.000 (see Fig. 6). Because there is still some uncertainty in ^{233}U's fast neutron fission-to-capture ratio, a reprocessing/discard rate of up to 9–10 L/per day *might* eventually prove to be necessary.

[30]Alkali metals generate "non bonded" oxygen atoms in glasses which lower their water leach resistance.

[31]This figure assumes recycle of most of the fluorine as makeup ThF_4 ^{7}LiF, NaF, etc., see Ref. [22].

[32]The most promising super metal candidates tested to date are Hastelloy EM 721 and HN80MTY.

[33]This scenario's MSFR TRU-waste generation figures were calculated by Carlo Fiorina using an "extended-EQL3D" program and the JEFF3.1 nuclear data base. LWR TRU-waste figures are based upon a recent analysis of 37 GWd/MTU Fukishima fuel rods (ORNL TM2010/286 http://info.ornl.gov/sites/publications/files/Pub27046.pdf). HTGR Pu-waste/GW_e figures were calculated from the data in Julian Lebenhaft's MS thesis (MIT 2001) http://hdl.handle.net/1721.1/28288 which source did not mention/discuss the minor actinides (MA).

[34]And also less TRU waste than that apt to be generated by a sustainable version of the SFR (IFR). The reason for this is that any U/Pu based, solid-fueled breeder would build up a huge inventory of TRU (typically 5–15 tonnes/GW_e) which, in that case, would be contained within ~100,000 steel-clad, sodium-containing, fuel pins all of which would have to be repeatedly dissolved/reprocessed/refabricated via an intrinsically "arty" batch-type pyrochemical process. The low TRU loss figures (typically <0.1%) usually attributed to the IFR (by far the most exhaustively studied of GIF's proposed fast reactor fuel cycles) ignores the fact that such low losses have never actually been achieved: see slide 16 of http://energy.gov/sites/prod/files/NEA-C_Rev5.pdf.

[35]Actinide discard eliminates most MA precursors and most of that which is produced would be burned to FP by this scenario's combination of ~3000 day in-core residence time and extremely high neutron flux (the MSFR is also a "deep burn" reactor).

[36]The reasons for the MSFR's superior heat-to-electrical conversion efficiency are that it operates at a higher temperature than do either LWRs (~300°C) or LMFBRs (~500°C) and would also likely be coupled to more efficient turbines (supercritical CO_2/Brayton instead of steam/Rankine).

[37]It is *almost* impossible (too expensive, difficult, and "dirty") to reprocess graphite-based solid reactor fuels, which means that it is *almost* impossible to implement a genuinely sustainable nuclear renaissance with reactors that require them.

[38]Analysis of the numbers revealed by Ref. [26] suggests that WAMSR's core will contain about 47 tonnes of zirconium hydride encased within metallic cladding subject to the same conditions that a MSFR's core tank wall would experience. If it also lasts for 4 years, this translates to generating ~22 tonnes of a probably pyrophoric metallic radwaste/GW_e-year's worth of WAMSR power.

[39]GOOGLE "Hastelloy N" and peruse ALIBABA's price quotes for "large lots."

[40]that is, probably under $50,000 worth of super metal/replacement. To put this into perspective, at 6 cents per kWhr, the electricity generated by the reference MSFR each day would be worth $2.16 million.

[41]The USA's LWRs currently consume ~1000 kg of ^{7}LiOH per year, 100% of which is imported. There is also currently very little demand for the pure ^{6}Li that some of its fusion bomb warhead designs call for. Mole-wise, lithium is as common as chlorine in the Earth's crust and most of it is ^{7}Li.

[42]Peak pressure anywhere within a MSFR's fuel salt system would be <100 psi (6.8E5 Pa) [16].

[43]Most of chapter 6 of James Mahaffey's latest book, "Atomic Accidents", is devoted to the "events" –mostly sodium leaks and fires – that have plagued most of the world's sodium cooled reactors.

[44]Assumptions: (a) all scenarios begin immediately (2014); (b) each LWR consumes 0.685 tonne of ^{235}U per year (CR~0.4); (c) MSFR CR = 1.045; (d) all of the reactors generate ~3 GW_t (consume ~3.13 kg fissile/day); and, (e) enrichment of natural uranium to 80% ^{235}U rather than to 4.5% requires ~50% more SWUs (see http://www.wise-uranium.com/nfcue.html).

[45]The dry-cask storage of today's huge backlog of spent LWR fuel is a fully developed, genuinely safe, and affordable technology. On the other hand, implementing a HLW repository for it (disposal) constitutes a "transcientific" (political) problem which means that NE R&D scientists/engineers can (and do) only "study" proposed solutions such as GIF's waste-burning reactor scenarios.

[46]Since cost-related risk dominates decision making in the "Free World", its electrical utility CEO's will be reluctant to employ reactors that would commit them to utilizing a fabulously expensive, tough-to-handle, low-quality fuel obtained via the "pyroprocessing" (by whom?) of spent LWR fuel, and then continuously separating/partially recombining/partially discarding everything (^{7}Li, TRU, U, ^{233}Pa, Th, and misc. FP) in that fuel in order to operate them. As had happened during the course of the USA's "Clinch River" LMFBR boondoggle, they are apt to believe that the adoption of an unnecessarily complex nuclear fuel cycle is unlikely to benefit them or their customers.

[47]At $142/SWU, the enrichment cost of the fuel currently feeding the USA's LWRs adds ~$1.8 billion per year to its citizens' utility bills (http://www.eia.gov/uranium/marketing/).

[48]DOE's current management plan for that HEU is to down blend it with natural or depleted uranium to render it less attractive to the horde of imaginary terrorists seeking to "divert" fissile from the USA's civilian reactors. Its implementation would waste an already-made $5 billion SWU investment.

[49]DOE's current management plan for that plutonium is to substitute it for ^{235}U in "mixed oxide fuel" (MOX) destined for use by the USA's civilian LWRs (http://www.huffingtonpost.com/project-on-government-oversight/budget-for-mox-program-cu_b_2662552.html). That progam is currently 300% over budget (expenditures to date, ~ $7.7 billion), a decade behind schedule, and has sparked zero interest from its proposed customers (MOX is far more radioactive than is their usual fuel and would therefore require extensive changes).

[50]It also tacitly assumes that the USA will neither choose to reindustrialize itself (windmills and solar panels are too unreliable to power modern factories) nor address the root cause of climate change in time to head off probable disaster (i.e., by circa 2100 AD, see http://www.worldbank.org/en/topic/climatechange/publication/turn-down-the-heat.)

[51]To get some idea of what the uranium industry's definition of "reasonable" could become if the whole world were to try to power itself with "advanced" LWRs, see... http://www.foe.org.au/anti-nuclear/issues/oz/u/cartel.

[52]Readers are encouraged to GOOGLE the figures used in my examples and repeat the calculations with whatever they come up with - any *reasonable* set of different inputs will support the contentions.

[53]This figure is ~190,000 times greater than that of the ^{235}U in the earth's oceans.

[54]Electrical utilities are natural monopolies regulated in a way that guarantees a "reasonable" profit. The US Federal Government's employees (both direct and contractor) enjoy a similar monopoly on NE R&D research requiring the use and/or generation of other than trace levels of radionuclides and/or radiation.

[55]"...nuclear engineering is to engineering as modern Islam is to religion", (James Mahaffey, "Atomic Awakening: A New Look at the History and Future of Nuclear Power," Pegasus Books, June 2009, p. XVI). This cultural characteristic plus a tight technical/academic job market inhibits innovation by lower level employees.

[56]For example, it is incapable of siting a HLW repository anywhere within~640 million acres of federal land some of which (e.g., the Nevada Test Site) is both otherwise useless and already contaminated.

[57]In contrast, Admiral Rickover had an unambiguous mission combined with both the will and technical talent required to make the decisions required to keep his project on track. He could not have succeeded if he had been either technically clueless or forced to embrace "all of the above" (unfocused).

[58]Any one of the MSFRs in those parks could supply 100% of the world's ^{99}Mo requirement (it's the radioisotope in the "cows" milked to generate the ^{99}Tc used for medical imaging). Because its production has been outsourced to other countries, it now costs US consumers ~$350 million/year.

[59]Two 1.5 GW_e MSFRs coupled to a reverse osmosis-based seawater desalination plant could generate the ~5 million acre ft/year of water required to revive California's already climate change impacted agriculture industry. A few more of them might be able to save Texas' cattle industry.

[60]https://www.armscontrol.org/act/2012_06/Resolving_the_Ambiguity_of_Nuclear_Weapons_Costs points out that DOE/NNSA currently spends ~$6.9 billion/a to maintain the USA's stockpile of nuclear weapons & that total (DOD+DOE) nuclear weapons-related expenditures are 20–40 billion/a. In 2014, total expenditures for "advanced nuclear reactor concept R&D" were $0.12 billion.

References

1. Goeller, H. E., and A. M. Weinberg. 1976. The age of substitutability. Science 191: 683–689, also available as OSTI 5045860

2. Till, C., and Y. I. L. Yang. 2011. Plentiful energy: the story of the integral fast reactor. Create Space 182–188.

3. Forsberg, C. A., P. F. Peterson, and H. H. Zhao. 2004. An advanced molten salt reactor using high technology. ICAPP 04, Pittsburg, PA, Je13-17.

4. 2008. Review of DOE's nuclear energy research and development program. Appendix A, NAP Press.

5. Akerlund, I., and J. Freed. Nuclear energy renaissance set to move on without US. Third Way, Available at http://content.thirdway.org/publications/851/Third_Way_Report_-_Nuclear_Energy_Renaissance_Set_to_Move_Ahead_Without_U.S.pdf (A comprehensive description of what has been happening with USA's NE R&D programs) (accessed 14 August 2014).

6. James, Lane (AEC). Fluid fuel reactors. Addison-Wesley (1958) (this book can be accessed gratis at http://energyfromthorium.com/ornl-document-repository/)

7. MacPherson, H. G. 1985. Molten salt reactor adventure. Nucl. Sci. Eng. 90: 374–380.

8. Whatley, M. E., L. E., McNeese, W. L., Carter, L. M., Ferris, and E. L., Nicholson. 1970. Engineering development of the MSBR fuel cycle. Nucl. Appl. Technol. 8: 170–178. (can be accessed gratis at http://energyfromthorium.com/ornl-document-repository/)

9. Mathieu, L., D. Heuer, R. Brissot, C. Garzenne, C. Le Brun, Lecarpentier, et al. 2009. Possible configurations for the TMSR and advantages of the fast non moderated version. Nucl. Sci. Eng. 161:78–89.

10. Mathieu, L., D. Heuer, E. Merle-Lucotte, R. Brissot, C. Le Brun, E. Liatard, et al. 2006. The thorium molten salt reactor: moving on from the MSBR. Prog. Nucl. Energy 48:664–679.

11. Mourogov, A., and P. M. Bokov. 2006. Potentialities of the fast spectrum molten salt reactor concept: REBUS-3700. Energy Convers. Manage. 47:2761–2771.

12. Ignatiev, V., O. Feynberg, I. Gnidoi, A. Merzlyakov, V. Smirnov, A. Surenkov, et al. 2007. Progress in development of Li,Be,Na/F Molten SaltActinide Recycler & Transmuter Concept. Proc. ICAPP 2007, May 13–18, Nice, France.

13. EVOL Project. 2012. Evaluation and viability of liquid fuel fast reactor systems. Available at: http://www.li2c.upmc.fr/.

14. Aufiero, M., and O. Geoffrey. 2013. A few comments on the MSFR safety and design optimization. EVOL Meeting, Grenoble, France 26–28 June 2013.

15. E Merle-Lucotte. Introduction to the Physics of the MSFR. Thorium Energy Conference 2013 (ThEC13) – Cern, Geneva, Available at http://indico.cern.ch/getFile.py/access?contribId=36&sessionId=9&resId=1&materialId=slides&confId=222140 (accessed 9 Jan 2015).

16. Aufiero, M., et al. 2013. An extended version of the SERPENT-2 Code to investigate fuel burnup and core evolution in the molten salt fast reactor. J. Nucl. Mater. 441:473–486.

17. Fiorina, C. The molten salt fast reactor as a fast-spectrum candidate for thorium implementation. Doctoral dissertation, POLITECNICO DI MILANO, 2013, Available at https://www.politesi.polimi.it/bitstream/10589/74324/1/2013_03_PhD_Fiorina.pdf (accessed 9 Jan 2015).

18. Merle-Lucotte, E., D. Heuer, M. Allibert, M. Brovchenko, N. Capellan, and V. Ghetta. 2011. Launching the thorium fuel cycle with the molten salt fast reactor. Proc. ICAPP 2011, May 2–5, Nice, France.

19. Merle-Lucotte, E., D. Heuer, M. Allibert, V. Ghetta, and C. Le Brun. 2008. Introduction to the physics of molten salt reactors. Materials issues for generation IV systems. NATO Sci. Peace Security Ser. B Phys. Biophys. 2008:501–521.

20. Olson, L. C. Materials corrosion in molten LiF-NaF-KF eutectic salt. Section 2–8, Ph.D. Dissertation, UWM, 2009, Available at http://allen.neep.wisc.edu/docs/dissertation-olson-luke.pdf.

21. Siemer, D.. 2012. Improving the integral fast reactor's proposed salt waste management system. Nucl. Technol. 178:341–352.

22. Siemer, D.. 2014. Molten salt breeder reactor (MSBR) waste management. Nucl. Technol. 185:101–108.

23. Scott, C. D., and W. L. Carter. Preliminary design study of a continuous fluorination-vacuum distillation system for regenerating fuel and fertile streams in a molten salt breeder reactor. ORNL-3791, UC-80-ReactRt Technology, TID-4500, 1966

24. Forsberg, C. W., P. F. Peterson, and R. A. Kochendarfer. Design options for the advanced high temperature reactor. Proceedings of ICAPP '08, Anaheim CA, Je 8-12, 2008.

25. Thorium-Fueled Molten Salt Reactors Weinberg Foundation Je2013 p. 14, http://www.the-weinberg-foundation.org/wp-content/uploads/2013/06/Thorium-Fuelled-Molten-Salt-Reactors-Weinberg-Foundation.pdf

26. TRANSATOMICPOWER. Technical white paper, V 1.01. Available at http://transatomicpower.com/white_papers/TAP_White_Paper.pdf (accessed March 2014).

27. Progress in radioactive graphite waste management. IAEA TECDOC 1647, 2010 http://www-pub.iaea.org/MTCD/Publications/PDF/te_1647_web.pdf

28. Serp, J., and H. Boussier. Molten salt reactor system 2009–2012 status. Available at http://www.iaea.org/Nuclear Power/Downloadable/Meetings/2013/2013-02-28-03-01-INPRO-GIF/11.anzieu1.pdf (accessed 9 Jan 2015).

29. Olson, L. C. Materials Corrosion. (Ref 20, Section 4–8).

14

Biochar: a new promising catalyst support using methanation as a probe reaction

Lingjun Zhu, Shi Yin, Qianqian Yin, Haixia Wang & Shurong Wang

State Key Laboratory of Clean Energy Utilization, Zhejiang University, Hangzhou 310027, China

Keywords

Activation, biochar, catalyst support, methane, syngas

Abstract

Biochar (BC), a by-product from the fast pyrolysis of rice husk, was activated by chemical and physical process. The activated biochar (ABC) showed graphite-like morphology and had a large amount of random pores with BET surface area of 1058 m^2/g. The biochar properties observed by further characterization revealed its feasibility to be used as a catalyst support. Syngas methanation as a probe reaction was utilized to characterize biochar supported Ru catalyst. The catalytic performance of the Ru/ABC catalyst in methanation was superior or comparable to the conventional activated carbon (AC) supported Ru catalyst. A high CH_4 selectivity of 98% and a CO conversion of 100% were obtained under the proper reaction conditions over the Ru/ABC catalyst.

Introduction

Biochar (BC), a residual by-product from fast pyrolysis of lingo-cellulosic biomass for bio-oil production, is rich in carbon, up to 60 wt%, and typically accounts for 15–40 wt% of the biomass feedstock [1]. At present, the majority of BC is discarded, utilized as process fuel or sold commercially for soil amendment and carbon sequestration [2–4]. In general, BC is similar to graphite. It is mainly composed of the aromatic carbon rings linked together. However, these rings are irregularly stacked and randomly arranged [5]. As with activated carbon (AC), the surface chemistry of BC can be varied using chemical methods. As

a consequence, BC, which is an abundant and low-cost renewable carbon source, has great potential to be used as a catalyst or catalyst support. In addition, BC can be easily separated from catalysts by oxidation to recover precious metals [6]. Therefore, the utilization of BC as catalyst will not only enhance its utilization but also promote the development of a variety of catalysts. Yan et al. [7] applied BC in the synthesis of carbon-encapsulated iron nanoparticles, and its catalytic activity on the conversion of synthesis gas to liquid hydrocarbons was evaluated. It was found that the CO conversion was about 95% and the selectivity of liquid hydrocarbon was as high as 68%. Dehkhoda et al. and Yu et al. [5, 8] reported the similar results that the BC, activated with KOH and sulfonated using fuming H_2SO_4, had a high activity on the transesterification of canola oil with methanol because of its high surface area and large acid density. To date, there have been few reports on the utilization of activated biochar (ABC) as a support for a heterogeneous catalyst.

It is known that a well-developed pore structure and high surface area, which are likely to favor the uniform dispersion of active components and to stabilize these against sintering, are essential for a good catalyst support. Since the original BC has a low density of pores, an activation process is required before it can be used as a support [9].

The methanation reaction of carbon monoxide and hydrogen has received considerable attention since it was first reported by Sabatier and Senderens [10]. By methanation process, syngas, derived from coal and biomass gasification and other kinds of methods such as coke oven gas, can be converted to substitute natural gas (SNG). Besides used as fuel gas, SNG in liquefied or compressed form can be used in transport vehicles as a substitute for diesel and gasoline. In the past decades, the selective methanation of CO from syngas was developed and has been widely applied in the gas purification process, where the content of CO in a syngas can be reduced to a very low level [11, 12]. As for the many fuels and chemicals synthesis from syngas, methanation is one of the reactions which have simple products. Syngas can be converted to methanol by methanol synthesis process and further converted to gasoline using methanol-to-gasoline process. Syngas is also used as a feed in producing synthetic petroleum such as gasoline and diesel via Fischer–Tropsch process. Herein, methanation was studied as a probe reaction for the catalyst support application of ABC.

Over recent decades, considerable effort has been made to develop efficient catalysts for syngas methanation [13, 14]. Carbonaceous materials, such as activated carbon (AC) and carbon nanotubes, have recently received a great deal of attention as catalysts or supports for the production of hydrocarbons. They show excellent catalytic activity, due to their unique magnetic and electric properties and their high mechanical and chemical stabilities [1, 15, 16].

The Ru-based catalyst is the most active for syngas methanation [11, 13, 17]. Compared to cobalt and iron, ruthenium catalysts have some unique features in the methanation process, and are suitable for fundamental research [18]. In this study, the BC, produced by rice husk pyrolysis, was activated to make it a suitable catalyst support. Methanation of syngas was selected just as a probe reaction to investigate the properties of ABC for the catalyst support application. And methanation over AC supported Ru catalyst was studied for the comparison.

We hope to develop the utilization of biochar in catalytic field in our future work based on this exploration.

Experimental

BC activation and catalyst preparation

The original BC was produced by the fast pyrolysis of rice husk at 773 K. It was activated before being used as a support, in order to increase the surface area and porosity. Under stirring, original BC was impregnated in 5 mol/L KOH aqueous solution for 4 h at ambient temperature, and then filtered, and dried at 383 K overnight. The sample was calcined in a tube furnace for 2 h under a flow of N_2, while the temperature was raised to 973 K at 5 K/min. After the activation, the sample was thoroughly washed with deionized water, and impregnated with 0.1 mol/L HCl aqueous solution for 3 h. Finally, it was filtered and washed with deionized water to remove soluble salts until the neutral pH value was achieved, and then dried at 383 K overnight to obtain ABC.

The as-prepared ABC and commercial AC supports were separately refluxed in the 30% aqueous solution of HNO_3 for 6 h at 373 K. Then, the mixture was filtered, washed, and dried as described above. The pretreated support (5 g) was added to 20 mL aqueous solution of $RuCl_3$ with a concentration of 0.124 mol/L. The suspension was agitated ultrasonically for 1 h, and then stirred for another 5 h at ambient temperature. Subsequently, the samples were filtered and dried at 383 K overnight, to give the Ru/ABC and Ru/AC catalysts with a Ru loading of 5 wt%.

Catalyst characterization

The pore structures of the samples were determined from nitrogen adsorption–desorption isotherms obtained at 77 K using a Quantachrome Quadrasorb SI apparatus (Quantachrome Instruments, Boynton Beach, FL, USA). The specific surface area was calculated based on the multipoint BET method. Powder X-ray diffraction (XRD) analysis of the catalysts was carried out using a PANalytical X'Pert PRO X-ray diffractometer with a Cu Kα radiation source operating at 40 kV and 30 mA. The morphology of the samples was observed using a transmission electron microscope (TEM), Philips-FEI Tecnai G^2F30 (Philips-FEI Corp., Amsterdam, Netherlands). Temperature-programmed reduction of hydrogen (H_2-TPR) was conducted on a Micromeritics AutoChem II 2920 instrument (Micromeritics Instrument Corp., Norcross, GA, USA). The catalysts were pretreated with Ar at 473 K for 60 min and then cooled to 323 K. The TPR process was carried out by heating to 1173 K at a rate of 10 K/min in a mixture of 10% H_2/Ar. Temperature-programmed desorption of carbon monoxide

(CO-TPD) was also conducted on the AutoChem II 2920 instrument. Dispersion of Ru was calculated from the moles of CO desorbed by assuming a CO/Ru stoichiometry of 1.

Test of catalytic activity

The methanation reactions were carried out in a continuous-flow fixed-bed reactor. In each experiment, 3 mL (0.65 g for Ru/ABC, 1.95 g for Ru/AC) of catalyst was sandwiched in quartz sand, which was then packed into a steel tube reactor with an inner diameter of 8 mm. Before the reaction, the catalyst was activated with pure H_2 at 723 K for 5 h, ramping at a rate of 3 K/min, then cooling to the reaction temperature. The flow rate of simulated syngas was controlled with a precise mass flowmeter (Brooks 5850E, Brooks Instrument, Hatfield, PA, USA). The reaction temperature was monitored with a K-type thermocouple inserted into the center of the catalyst bed. Typically, the methanation reaction was carried out at 2 MPa, H_2/CO mole ratio of 3, and gas hourly space velocity (GHSV) of 1200 h^{-1}, unless otherwise specified.

The products were analyzed using an on-line gas chromatograph (GC) (Agilent 6820, Agilent Technologies, Santa Clara, CA, USA). H_2, CO, CO_2 were separated using a TDX-01 packed column (0.5 m) and then detected by a thermal conductivity detector. The hydrocarbons were separated by a Propack-Q packed column (1 m) and then detected by a flame ionization detector.

The CO conversion and product selectivity were calculated relative to the number of moles of C, according to the following equations:

$$X_{CO}(\%) = (n_{CO_{in}} - n_{CO_{out}})/n_{CO_{in}} \times 100\%$$
$$S_{C_m}(\%) = m \times n_{C_m}/(n_{CO_{in}} - n_{CO_{out}}) \times 100\%$$

where X_{CO} and S_{C_m} refer to the CO conversion and the product selectivity. $n_{CO_{in}}$ and $n_{CO_{out}}$ represent the number of moles of CO in the feed and vent gas, respectively, and m represents the carbon number of the product C_m. n_{C_m} refers to the number of moles of C_m.

Control experiments were carried out with simple ABC and AC using the condition 613 K, 2 MPa, H_2/CO mole ratio of 3, and gas hourly space velocity (GHSV) of 1200 h^{-1}. The conversion of CO over the ABC and AC was zero. Therefore, pure ABC or AC has no catalytic effect on the methanation.

Results and Discussion

Characterization of the support and catalyst

Nitrogen adsorption—desorption measurements were used to evaluate the textural properties of the samples. The BET surface area, pore volume and the average pore size of the BC before and after activation, the AC, as well as the Ru/ABC and Ru/AC catalysts are presented in Table 1.

The original BC showed a small BET surface area (40 m²/g), which was in accordance with the reported value [19]. It is indicated that only a few pores were generated during the fast pyrolysis, because of the incomplete pyrolysis of biomass in a short residence time and the entrapment of tar-like materials within the pores. During the chemical activation, further disruption of the carbon lattice and the release of the volatiles in the BC facilitated the enlargement of the existing pores and the generation of new ones [5]. The ABC showed a BET surface area of 1058 m²/g, some 25 times higher than that of the original BC. The average pore diameter of BC decreased from 5.5 to 2.3 nm after activation, which indicated that most of the new generated pores had smaller size than the original ones. In addition, it can be inferred from Table 1 that the surface area of the micropores accounts for ~78% of the total surface area in the ABC, and the value is 95% in the commercial AC. It indicated that microporous and mesoporous structures coexisted in ABC and its supported catalysts, while the AC and the Ru/AC catalyst showed the microporous structure. For a catalyst support, the coexistence is better than having a solely microporous structure, since any diffusion limitation in the reaction can be relieved and the mesopores can act as passages,

Table 1. Texture properties of the biochar and the Ru/ABC catalyst[1].

Catalyst	S_{BET} (m²/g)	$S_{External}$ (m²/g)	S_{Micro} (m²/g)	V_{Pore} (cm³/g)	D_{Pore} (nm)	d_{Ru} (nm)[2]	Ru dispersion (%)[3]
BC	40	30	10	0.06	5.5	–	–
ABC	1058	225	833	0.59	2.3	–	–
Ru/ABC	806	240	566	0.58	2.9	3.3	36
AC	653	31	622	0.35	2.3	–	–
Ru/AC	558	24	534	0.34	2.4	1.9	41

[1]The $S_{External}$, S_{Micro} and V_{Micro} were determined by the t-method.
[2]Determined by TEM measurement.
[3]Determined by CO-TPD.

allowing reactant and products to pass in and out of the interior of the catalyst [19, 20]. The BET surface area of the Ru/ABC and Ru/AC catalyst was lower than that of the ABC and AC, respectively, suggesting that some of the pores of the support had been blocked by active components. In addition, the BET surface area of the micropores was significantly reduced, while that of the mesopores remained almost unchanged. It is deduced that the size of the active components was small enough to enter the micropores of the support [21].

The XRD patterns of the BC before and after activation, the AC, as well as the catalysts before and after reduction are depicted in Figure 1. All the XRD patterns showed two wide diffraction peaks corresponding to the characteristic peaks of carbon. The strongest diffraction peaks, located at $2\theta = 20°-30°$, are indicative of the

amorphous carbon structure. The weaker diffraction peak at around $2\theta = 40°-50°$ is assigned to the reflection from graphite. The presence of more obvious graphite-like structure in the ABC indicates that this is the more stable state for the carbonaceous material [5, 22]. No diffraction peaks of active components were detected on the Ru/ABC and Ru/AC catalysts before reduction. After reduction, the diffraction peaks emerged at $2\theta = 38.4°$, $42.2°$, and $44.0°$ over the Ru/ABC catalyst. These are assigned to the crystal planes of Ru (1 0 0), (0 0 2), and (1 0 1), respectively, indicating that the RuCl$_3$ precursor has been transformed to metallic Ru [16]. However, there was no Ru diffraction peak observed over the reduced Ru/AC catalyst, mostly because of their small particle size and uniform distribution. The stronger Ru diffraction peaks on

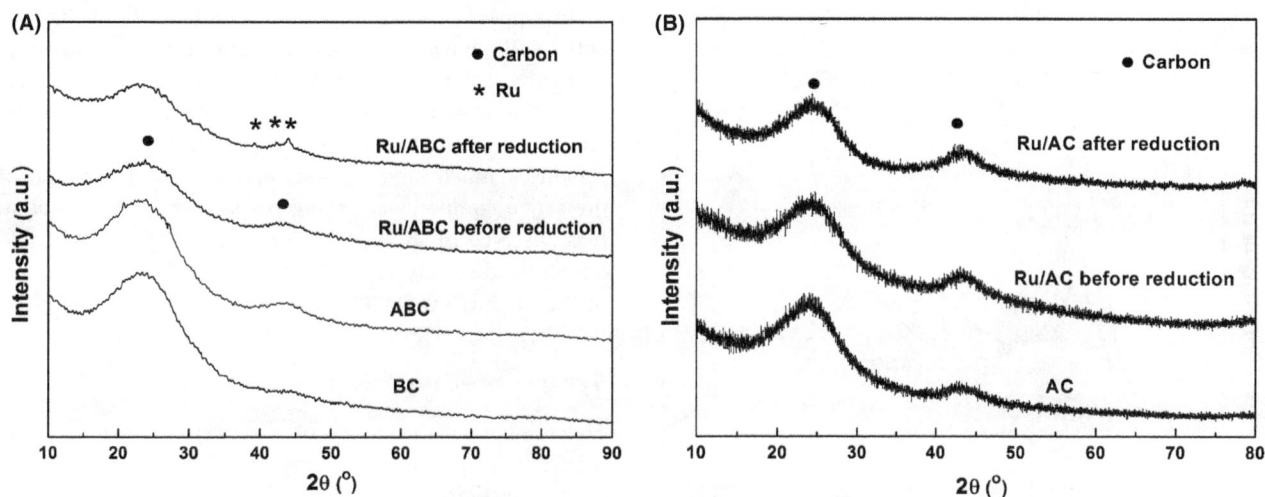

Figure 1. XRD patterns of (A) BC and BC supported catalysts and (B) AC and AC supported catalysts.

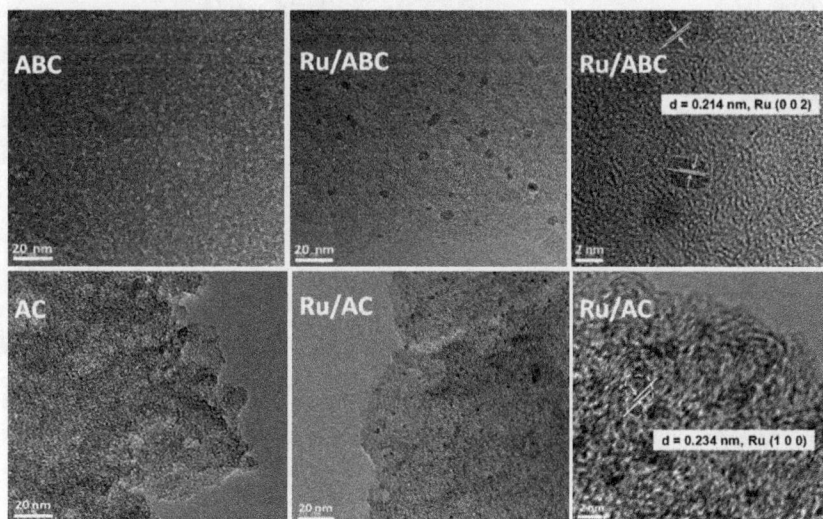

Figure 2. TEM and HRTEM images for ABC, Ru/ABC, AC, and Ru/AC.

the reduced Ru/ABC catalyst than that on the Ru/AC catalyst indicated the larger Ru crystallite size on the ABC.

The structure of the supports and the distribution of active components were analyzed by TEM, shown in Figure 2. TEM images showed that large numbers of random pores were observed on the ABC, owing to the release of volatiles and the reaction between the BC and the activation agent [23]. The presence of the metallic Ru, Ru (0 0 2), and Ru (1 0 0) on the Ru/ABC and Ru/AC catalysts, respectively, was verified by its interplanar crystal spacing on the high-resolution transmission electron microscope (HRTEM) images. Ru particles were uniformly distributed on both of the reduced Ru/ABC and Ru/AC catalysts. The average particle size of the metallic Ru was 3.3 nm for the Ru/ABC catalyst, while a smaller value of 1.9 nm was observed on the Ru/AC catalyst (Table 1). This was calculated from the TEM images by

$d_{TEM} = \sum n_i d_i^3 / \sum n_i d_i^2$, where n_i is the number of particles having a characteristic diameter d_i (within a given diameter range) [24].

The TPR profiles of the Ru/ABC and the Ru/AC catalyst are depicted in Figure 3. The wide reduction peaks at about 873 K and 800 K for Ru/ABC and Ru/AC catalysts, respectively, were caused by the gasification of carbon [25]. The intense reduction peaks at 523 K was assigned to the reduction of the Ru species on both of the catalysts [26]. It was reported that the reduction temperature of Ru species on Ru/Al$_2$O$_3$ catalyst was 380 K [27]. The higher reduction temperature in this work suggests the presence of a relatively strong interaction between Ru and the supports, which might decrease the possibility of sintering [28].

Based on the above results, a possible mechanism of the BC activation and the Ru/ABC catalyst preparation was proposed (Fig. 4). Firstly, the original BC was treated with KOH aqueous solution. Then, most of the pore structure of the BC was developed during activation, widening the existing pores and generating new ones, partly forming graphite-like morphology. Finally, the ABC can provide a much larger surface area for the dispersion of the active components during preparation of the catalyst. It is observed in Table 1 that the Ru/ABC catalyst showed a high Ru dispersion of 36%, which was close to the value of Ru/AC catalyst (41%).

Syngas methanation

The catalytic performance of the prepared catalysts was carried out in the syngas methanation, to evaluate the catalytic activity of the Ru/ABC catalyst, compared with the Ru/AC catalyst. The effect of different reaction conditions on the catalytic activity was investigated.

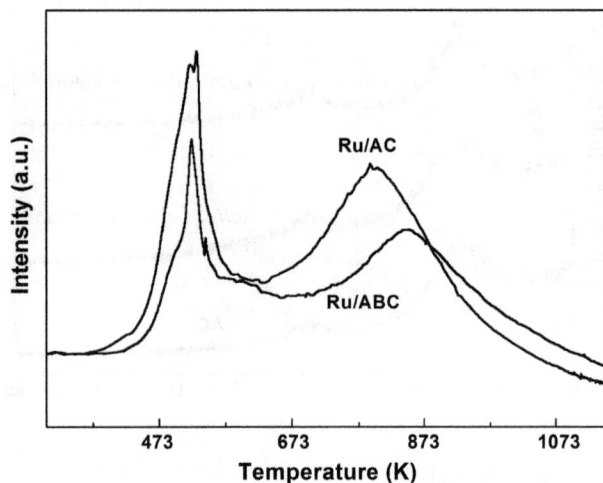

Figure 3. TPR profiles of the Ru/ABC and Ru/AC catalysts.

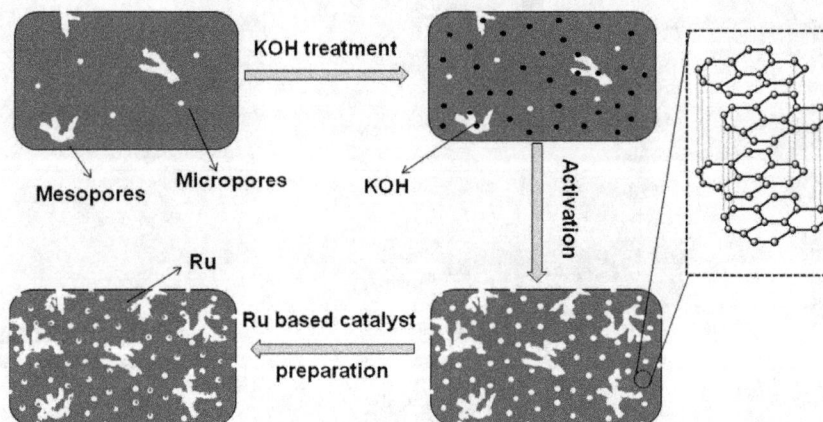

Figure 4. A schematic illustration of the preparation of ABC and Ru/ABC catalyst.

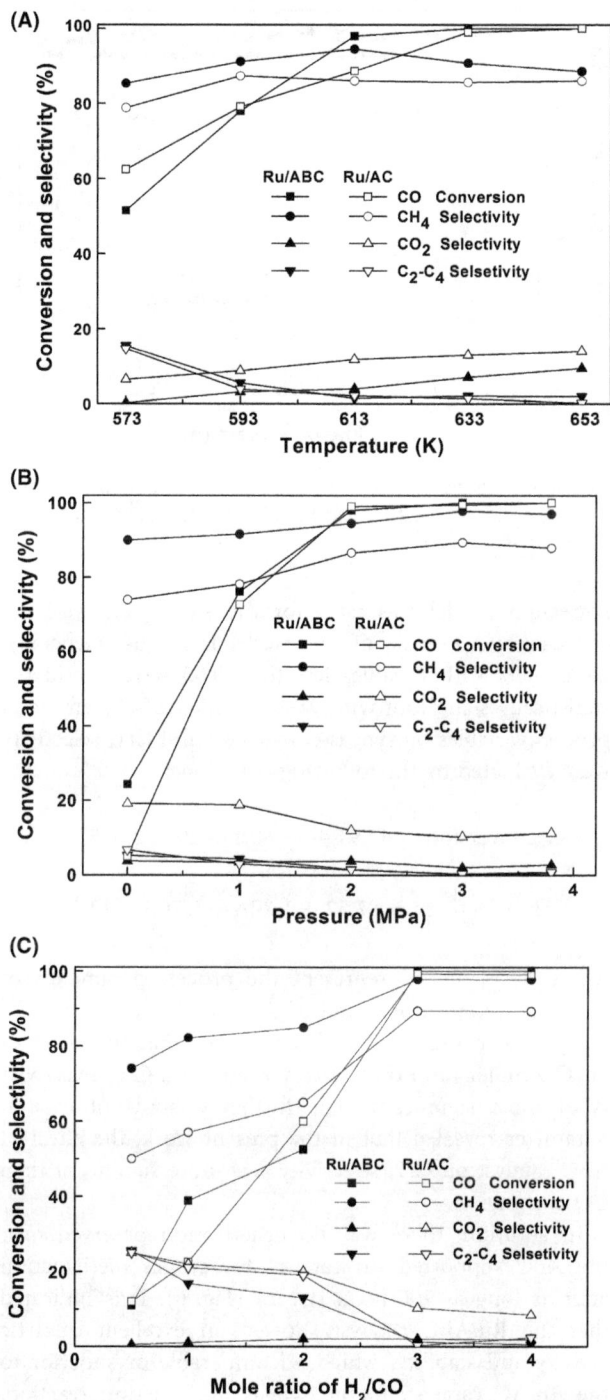

Figure 5. Catalytic performance of the Ru/ABC and Ru/AC catalysts: (A) effect of reaction temperature; (B) effect of reaction pressure; (C) effect of H_2/CO mole ratio.

The effect of reaction temperature on the activity of Ru/ABC and Ru/AC catalysts was conducted under the conditions of $H_2/CO = 3$, $P = 2$ MPa, gas hourly space velocity (GHSV) = 1200 h^{-1}, and the results are pre-

sented in Figure 5A. The CO conversion increased significantly with elevated temperature, because the mobility of the H atom on the catalyst surface was promoted at higher temperature [29]. The Ru/ABC catalyst showed a high CO conversion of 98% at 613 K, while a higher temperature of 633 K was needed to obtain the same CO conversion over the Ru/AC catalyst. For the two catalysts, the main different properties are dispersed Ru crystallite size, the support pore structure, and may the electronic state. The Ru particle size is 3.3 nm for the Ru/ABC catalyst and 1.9 nm for the Ru/AC catalyst, as shown in Table 1. The effect of Ru crystallite size on CO and CO_2 hydrogenation reactions has been reported in several works. The extent of the impact was depended on the nature of support. It was found that the TOF of CO over Ru/TiO_2 catalyst increased by a factor of 40, with increasing d_{Ru} from 2.1 to 4.5 nm. While on Ru/Al_2O_3 catalyst, the TOF value increased only by a factor of 3.5, with increasing d_{Ru} from 1.3 to 13.6 nm [30]. Highly dispersed Ru–Zr/CNTs catalyst in amorphous form with the particle size ranging from 4 to 8 nm exhibited more active in CO methanation than the corresponding catalyst with the larger particle size in crystallized form [31]. The effect of Ru crystallite size on the catalytic activity and product selectivity was found be insensitive in CO_2 methanation over SiO_2 supported catalyst [32]. The catalytic behavior of Ru can be modified by the metal oxide supports through electron transfer. CO concentration decreased to below 10 ppm over ZrO_2 promoted CNTs supported Ru catalyst, since Zr has high electronegativity and its charge transfer weaken C=O bond of the adsorbed CO molecules [31, 33]. The coexistence of mesopores and micropores over the Ru/ABC catalyst was the superiority to the Ru/AC catalyst in terms of mass transfer, especially in the reaction of large molecule synthesis. Therefore, the slightly high catalytic activity of ABC supported catalyst may be ascribed to the relative large Ru crystallite size, the special structure, and the probably appropriate electron state of catalyst.

The selectivity of CH_4 reached the maximum value of 95% and a CO conversion of 98% at 613 K over the Ru/ABC catalyst. However, further increase of the temperature reduced the selectivity of CH_4. The CO conversion reached 100%, and the selectivity of CH_4 decreased to 91% at 633 K over the Ru/ABC catalyst. While the selectivity of CH_4 over the Ru/AC catalyst remained stable at around 87% at the temperature higher than 613 K, which was lower than that over the Ru/ABC catalyst. The results indicate that the structure and electron state in ABC supported catalyst are beneficial for methanation while inhibiting the side reaction, as the physical effect, such as Ru particle size, governs the selectivity for CO methanation [34]. CO hydrogenation over Al_2O_3 supported Ru catalyst was reported by Tajammul HS and his coworker. The promotion of Mn led to increase methane selectivity from 67% to 89% [35]. The selectivities of methane between 49.2% and 79.6% were obtained in CO methanation on Ni/Al$_2$O$_3$ catalysts with transition metals and rare-earth metals improved [36].

CO$_2$ and C$_2$–C$_4$ hydrocarbons are the main by-products of the methanation reaction. It is known that CO$_2$ is the product of water–gas shift (WGS) reaction in CO hydrogenation. At higher CO conversion, the increased production of water provided higher water concentration for WGS reaction [37]. Small amount of C$_2$–C$_4$ hydrocarbons was produced, and their selectivity fell close to zero at temperatures higher than 613 K. Thus, 613 K and 633 K appeared to be the optimum temperatures for syngas methanation using the Ru/ABC and Ru/AC catalysts, respectively, considering both of the high CO conversion and CH$_4$ selectivity.

Figure 5B illustrates the effect of reaction pressure on the catalytic performance of the Ru/ABC and Ru/AC catalysts under their optimal temperatures. It is observed that the CO conversion increased rapidly with increasing pressure and maintained at 100% at the pressure higher than 3 MPa over both of the catalysts. Syngas methanation is a stoichiometric number reducing reaction, so high pressure would facilitate the reaction, leading to high CO conversion [13]. The selectivity of CH$_4$ maintained the values higher than 90% at all the testing pressures, and reached the maximum value of 98% at 3 MPa over the Ru/ABC catalyst. However, the Ru/AC catalyst showed lower CH$_4$ selectivity. In addition, further increase in the reaction pressure had little effect on either the CO conversion or CH$_4$ selectivity.

Under the conditions of 3 MPa, 613 K, and 633 K for the Ru/ABC and Ru/AC catalysts, respectively, the effect of H$_2$/CO mole ratio was studied and the results are displayed in Figure 5C. The CO conversion was as low as 11% at the H$_2$/CO ratio of 0.5, while it sharply increased to 100% at a H$_2$/CO ratio of 3 and the selectivity to CH$_4$ simultaneously increased to its highest value of 98% over the Ru/ABC catalyst. For the Ru/AC catalyst, CO conversion increased from 12% to 99% with increasing H$_2$/CO from 0.5 to 3, companied by the increasing of CH$_4$ selectivity. The high concentration of H$_2$ favors the formation of CH$_4$ and might suppress the WGS reaction, resulting in a reduced selectivity for CO$_2$ with the increase of the H$_2$/CO ratio. Thus, a H$_2$/CO mole ratio of 3 is proper for the complete conversion of CO, with high selectivity of CH$_4$.

The statistical analysis on process parameters was performed using design expert 8.0 software with response surface methodology to evaluate the effect of reaction conditions on the methanation [38]. According to the preliminary analysis, the process variables multiple linear

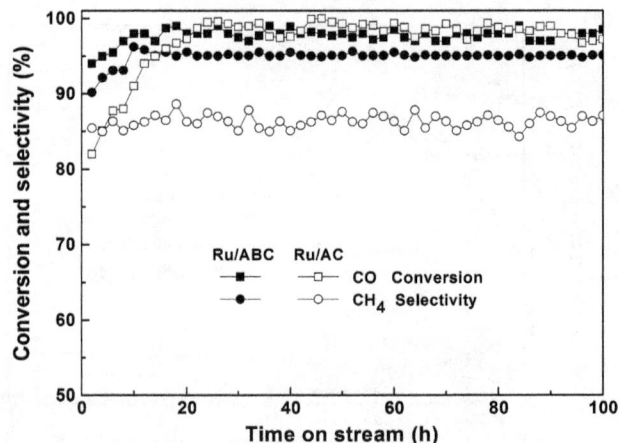

Figure 6. Catalytic stability of the Ru/ABC (613 K, 2 MPa, H$_2$/CO = 3, GHSV = 1200 h^{-1}) and Ru/AC catalysts (633 K, 2 MPa, H$_2$/CO = 3, GHSV = 1200 h^{-1}).

regression model was used for the subsequent variance analysis, and P-value of the model is 0.0004. Generally, parameters with P-values less than 0.05 were considered statistically significant with 95% confidence. The effects of process variables on syngas conversion and CH$_4$ selectivity were evaluated by the following equations:

$$CO\ conversion = 62.94 + 23.80A + 20.72B + 52.37C$$

$$CH_4\ selectivity = 87.52 + 1.20A + 2.70B + 12.29C$$

where A, B, and C represent the process parameters of temperature, pressure, and H$_2$/CO molar ratio, respectively. The parameter coefficients show that the effect of H$_2$/CO molar ratio on CO conversion and CH$_4$ selectivity is of most significant. The further analysis of catalyst parameter revealed that in the present work, the effect of ABC catalyst on CH$_4$ selectivity was more significant than CO conversion.

In addition, there was no deactivation observed over the ABC supported Ru catalyst for syngas methanation after as long as 100 h on stream (Fig. 6). It is indicated that the Ru/ABC catalyst showed an excellent catalytic activity and stability, which is comparable or superior to the Ru/AC catalyst, in the syngas methanation reaction. Furthermore, syngas methanation over ABC supported Ru catalyst was a gas–solid heterogeneous catalysis reaction. It is quite different with the liquid phase reaction. Hanson et al. pointed out that the catalysts would be deactivated due to the leaching of active sites in the sulfonated carbon during the esterification of free fatty acids for biodiesel production [39]. However, the leaching of active metals could not be observed in the gas–solid phase reaction, as was also confirmed by the fact that no deactivation was observed over the ABC supported Ru catalyst during the 100-h catalytic stability test. There is reason to believe that the BC is a promising substitute to the AC in the syngas methanation reaction. Furthermore, using the insights herein, BC can also be used as catalyst or support into other kind of reactions.

Conclusions

The porosity of BC was significantly developed with the BET surface area increasing from 40 to 1058 m^2/g after activation, and the structure of which became more graphite-like, which made the ABC feasible to be used as catalyst support. The catalytic activity of Ru/ABC catalyst was evaluated in syngas methanation compared with the Ru/AC catalyst. The proper Ru particles and the coexistence of microporous and mesoporous structure of the Ru/ABC catalyst showed advantages over the Ru/AC catalyst with smaller Ru particles and solely microporous structure in syngas methanation. The Ru/ABC catalyst exhibited excellent catalytic activity in syngas methanation reaction that CO conversion and the selectivity of CH_4 were 100% and 98%, respectively. Additionally, the Ru/ABC catalyst showed excellent stability for a period of at least 100 h on stream.

Acknowledgments

We acknowledge financial support from the National Natural Science Foundation of China (51276166), the National Science and Technology supporting plan through contract (2015BAD15B06), and the International Science and Technology Cooperation Program of China (2011DFR60190).

Conflict of Interest

None declared.

References

1. Brown, T. R., M. M. Wright, and R. C. Brown. 2011. Estimating profitability of two biochar production scenarios: slow pyrolysis vs fast pyrolysis. Biofuels, Bioprod. Biorefin. 5:54–68.
2. Abdullah, H., K. A. Mediaswanti, and H. Wu. 2010. Biochar as a fuel: 2. Significant differences in fuel quality and ash properties of biochars from various biomass components of Mallee trees. Energy Fuels 24:1972–1979.
3. Kim, Y., and W. Parker. 2008. A technical and economic evaluation of the pyrolysis of sewage sludge for the production of bio-oil. Bioresour. Technol. 99:1409–1416.
4. Bird, C. M., C. M. Wurster, P. H. Paula Silva, A. M. Bass, and de Nys R. 2011. Algal biochar – production and properties. Bioresour. Technol. 102: 1886–1891.
5. Yu, J. T., A. M. Dehkhoda, and N. Ellis. 2011. Development of biochar-based catalyst for transesterification of canola oil. Energy Fuels 25:337–344.
6. Xiong, H, M. A. M. Motchelaho, M. Moyo, L. L. Jewell, and N. J. Coville. 2011. Correlating the preparation and performance of cobalt catalysts supported on carbon nanotubes and carbon spheres in the Fischer–Tropsch synthesis. J. Catal., 278:26–40.
7. Yan, Q., C. Wan, J. Liu, J. Gao, F. Yu, J. Zhang, et al. 2013. Iron nanoparticles in situ encapsulated in biochar-based carbon as an effective catalyst for the conversion of biomass-derived syngas to liquid hydrocarbons. Green Chem. 15:1631–1640.
8. Dehkhoda, A. M., A. H. West, and N. Ellis. 2010. Biochar based solid acid catalyst for biodiesel production. Appl. Catal. A 382:197–204.
9. Azargohar, R. and A. K. Dalai 2006. Biochar as a precursor of activated carbon. Appl. Biochem. Biotechnol. 131:762–773.
10. Kopyscinski, J., T. J. Schildhauer, F. Vogel, S. M. A. Biollaz, and A. Wokaun. 2010. Applying spatially resolved concentration and temperature measurements in a catalytic plate reactor for the kinetic study of CO methanation. J. Catal. 271:262–279.
11. Chen, A., T. Miyao, K. Higashiyama, H. Yamashita, and M. Watanabe. 2010. High catalytic performance of ruthenium-doped mesoporous nickel–aluminum oxides for selective CO methanation. Angew. Chem. Int. Ed. 49:9895–9898.
12. Kimura, M., T. Miyao, S. Komori, A. Chen, K. Higashiyama, H. Yamashita, et al. 2010. Selective methanation of CO in hydrogen-rich gases involving large amounts of CO_2 over Ru-modified Ni–Al mixed oxide catalysts. Appl. Catal. A 379:182–187.
13. Zhang, J., Z. Xin, X. Meng, and M. Tao. 2013. Synthesis, characterization and properties of anti-sintering nickel incorporated MCM-41 methanation catalysts. Fuel 109:693–701.
14. Liu, Z., B. Chu, X. Zhai, Y. Jin, and Y. Cheng. 2012. Total methanation of syngas to synthetic natural gas over Ni catalyst in a micro-channel reactor. Fuel 95:599–605.
15. Malek Abbaslou, R. M., J. Soltan, and A. K. Dalai. 2011. Iron catalyst supported on carbon nanotubes for Fischer–Tropsch synthesis: effects of Mo promotion. Fuel 90:1139–1144.
16. Wu, Z., Y. Mao, X. Wang, and M. Zhang. 2011. Preparation of a Cu-Ru/carbon nanotube catalyst for hydrogenolysis of glycerol to 1,2-propanediol via hydrogen spillover. Green Chem. 13:1311–1316.
17. Panagiotopoulou, P., D. I. Kondarides, and X. E. Verykios. 2008. Selective methanation of CO over supported noble

metal catalysts: effects of the nature of the metallic phase on catalytic performance. Appl. Catal. A 344:45–54.

18. Kang, J., K. Cheng, L. Zhang, Q. Zhang, J. Ding, W. Hua, et al. 2011. Mesoporous zeolite-supported ruthenium nanoparticles as highly selective Fischer-Tropsch catalysts for the production of C5–C11 isoparaffins. Angew. Chem. 123:5306–5309.

19. Özçimen, D., and A. Ersoy-Meriçboyu. 2010. Characterization of biochar and bio-oil samples obtained from carbonization of various biomass materials. Renew Energ 35:1319–1324.

20. Surisetty, V. R., A. K. Dalai, and J. Kozinski. 2011. Influence of porous characteristics of the carbon support on alkali-modified trimetallic Co–Rh–Mo sulfided catalysts for higher alcohols synthesis from synthesis gas. Appl. Catal. A 393:50–58.

21. Muthu Kumaran, G., S. Garg, K. Soni, M. Kumar, L. D. Sharma, G. Murali Dhar, et al. Effect of Al-SBA-15 support on catalytic functionalities of hydrotreating catalysts: I. Effect of variation of Si/Al ratio on catalytic functionalities. Appl. Catal. A, 2006, 305: 123–129.

22. Patrick, J. W. 1995. Pp. 1–48 in Porosity in carbons: characterization and applications. Halsted Press, John Wiley & Sons, New York.

23. Yang, K., J. Peng, C. Srinivasakannan, L. Zhang, H. Xia, and X. Duan. 2010. Preparation of high surface area activated carbon from coconut shells using microwave heating. Bioresour. Technol. 101:6163–6169.

24. Wang, S., Q. Yin, J. Guo, B. Ru, and L. Zhu. 2013. Improved Fischer-Tropsch synthesis for gasoline over Ru, Ni promoted Co/HZSM-5 catalysts. Fuel 108:597–603.

25. Xiong, J., X. Dong, and L. Li. 2012. CO selective methanation in hydrogen-rich gas mixtures over carbon nanotube supported Ru-based catalysts. J. Nat. Gas Chem. 21:445–451.

26. Perkas, N., J. Teo, S. Shen, Z. Wang, J. Highfield, Z. Zhong, et al. 2011. Ru supported catalysts prepared by two sonication-assisted methods for preferential oxidation of CO in H_2. Phys. Chem. Chem. Phys. 13:15690–15698.

27. Rynkowski, J., T. Paryjczak, A. Lewicki, M. Szynkowska, T. Maniecki, and W. Jóźwiak. 2000. Characterization of Ru/CeO_2-Al_2O_3 catalysts and their performance in CO_2 Methanation. React. Kinet. Catal. Lett. 71:55–64.

28. Li, X., S. Wang, Q. Cai, L. Zhu, Q. Yin, and Z. Luo. 2012. Effects of preparation method on the performance of Ni/Al_2O_3 catalysts for hydrogen production by bio-oil steam reforming. Appl. Biochem. Biotechnol. 168:10–20.

29. Trépanier, M., and A. Tavasoli, A. K Dalai, and N. Abatzoglou. 2009. Fischer-Tropsch synthesis over carbon nanotubes supported cobalt catalysts in a fixed bed reactor: Influence of acid treatment. Fuel Process. Technol. 90:367–374.

30. Panagiotopoulou, P., D. I. Kondarides and X. E. Verykios 2009. Selective methanation of CO over supported Ru catalysts. Appl. Catal. B 88:470–478.

31. Xiong, J., X. Dong, and L. Li. 2012. CO selective methanation in hydrogen-rich gas mixtures over carbon nanotube supported Ru-based catalysts. J. Nat. Gas Chem. 21:445–451.

32. Scir'e, S., C. Crisafulli, R. Maggiore, S. Minic'o, and S. Galvagno. 1998. Influence of the support on CO_2 methanation over Ru catalysts: an FT-IR study. Catal. Lett., 51:41–45.

33. Perkas, N., J. Teo, S. Shen, Z. Wang, J. Highfield, Z. Zhong, et al. 2011. Ru Supportedcatalysts prepared by two sonication-assisted methods for preferential oxidation of CO in H_2. Phys. Chem. Chem. Phys. 13:15690–15698.

34. Eckle, S., M. Augustin, and H.-G. Anfang, and R. J. Behm. 2012. Influence of the catalyst loading on the activity and the CO selectivity of supported Ru catalysts in the selective methanation of CO in CO_2 containing feed gases. Catal. Today 181:40–51.

35. Tajammul, H. S., and A. M. Ashraf. 1997. Carbon monoxide hydrogenation over Ru-Mn-K/Al2O3 catalysts. Turk. J. Chem. 21:77–83.

36. Hwang, S., J. Lee, U. G. Hong, J. C. Jung, D. J. Koh, H. Lim, et al. 2012. Hydrogenation of carbon monoxide to methane over mesoporous nickel-M-alumina (M=Fe, Ni, Co, Ce, and La) xerogel catalysts. J. Ind. Eng. Chem. 18:243–248.

37. Abbaslou, R. M. M., J. Soltan, and A. K. Dalai. 2010. Effects of nanotubes pore size on the catalytic performances of iron catalysts supported on carbon nanotubes for Fischer-Tropsch synthesis. Appl. Catal. A 379:129–134.

38. Bezerra, M. A., R. E. Santelli, E. P. Oliveira, L. S. Villar, and L. A. Escaleira. 2008. Response surface methodology (RSM) as a tool for optimization in analytical chemistry. Talanta 76:965–977.

39. Deshmane, C. A., M. W. Wright, A. Lachgar, M. Rohlfing, Z. Liu, J. Le, et al. 2013. A comparative study of solid carbon acid catalysts for the esterification of free fatty acids for biodiesel production. Evidence for the leaching of colloidal carbon. Bioresour. Technol. 147:597–604.

15

Energy efficient inactivation of *Saccharomyces cerevisiae* via controlled hydrodynamic cavitation

Lorenzo Albanese[1], Rosaria Ciriminna[2], Francesco Meneguzzo[1] & Mario Pagliaro[2]

[1]Istituto di Biometeorologia, CNR, via Caproni 8, 50145 Firenze, Italy
[2]Istituto per lo Studio dei Materiali Nanostrutturati, CNR, via U. La Malfa 153, 90146 Palermo, Italy

Keywords

Energy efficiency, hydrodynamic cavitation, pasteurization, *Saccharomyces cerevisiae*, yeast

Abstract

We investigate hydrodynamic cavitation to inactivate commonly employed *Saccharomyces cerevisiae* yeast strains in an aqueous solution using different reactors and hydraulic circuit selected to demonstrate the process feasibility on the industrial scale. The target to achieve an useful lethality of the yeast at lower temperature when compared with standard thermal and even with other cavitation processes was achieved, with 90% yeast strains lethality at lower temperature (6.3–9.5°C), and about 20% lower energy input. A separate model simulating the combined thermal and cavitational effects on yeast lethality allows to accommodate the data into a comprehensive framework providing a tool to design further targeted experiments and to predict results when changing the process parameters.

Introduction

Saccharomyces cerevisiae (SC) is the yeast which is most commonly used in the food industry for the fermentation of wine and beer, as well as it is responsible for spoilage in fruit juices and milk [1]. Its inactivation is traditionally performed mostly by means of thermal pasteurization, even though extensive research was aimed at developing alternative methods to achieve a sufficient lethality of SC while keeping temperatures as low as possible [2].

Lower pasteurization temperatures would lead at least in principle to a double advantage: preserving superior nutritive and organoleptic qualities of the food liquids, as well as saving thermal energy, the latter provided that the alternative techniques are sufficiently energy effective.

Beer production, for example, which need a costly thermal pasteurization stage after fermentation in order to avoid further fermentation after bottling as well as to obtain a safe product before release to the market, could benefit from the adoption of new more efficient pasteurization techniques.

Unfortunately, improving lethality and inactivation of SC over standard thermal methods using lower temperatures is quite challenging due to the intrinsic resilience of such microorganisms eventually stopping their reproduction and remaining in a latent state until nutrients restore or environmental conditions turn again favorable [3, 4].

Interest in cavitation processes in liquid media, particularly for food treatment and water processing, has been growing in recent years due to unique features such as scalability, stability and localized release of high-density energy [5, 6]. A myriad of microscopic *hot spots* characterized by high to extreme density thermal energy, with temperatures up to 10,000 K, and mechanical energy, with pressure waves up to 5000 bar activate or accelerate specific chemical reactions as well as advanced oxidation processes by means of the generation of extremely reactive species such as hydroxyl radicals that can lead to degradation of organic and inorganic chemicals and pollutants [7–9].

Proposed applications of cavitation in liquid media cover a wide range of technological fields, with serious perspectives for industry in the area of biochemical engineering and biotechnology, including microbial cell disruption and disinfection [10]. Coupling with other methods and components stimulating or enhancing oxidation processes, such as Fenton chemistry, hydrogen peroxide, etc., was proven to be advantageous in industrial wastewater treatment applications [11–15].

Nevertheless, in applications concerning liquid foods, oxidation is not desirable as it is harmful to the quality of the food [2]. Remarkably, the shock waves generated by bubbles implosion in hydrocavitation were identified as the dominant factor responsible the disruption of microorganisms [16], which represents a significant motivation to further investigate the applicability of cavitation techniques to yeast inactivation in liquid foods and its feasibility at the industrial scale.

A further practical advantage of cavitation in liquids is the heating occurs directly within the liquid without exchange with hot surfaces or pipes, thereby dramatically reducing heat losses due to internal friction, pressure loss downstream of nozzles, curves, and so forth [17]. As a consequence, in the cavitation treatment of liquid food, thermal gradients are minimized and undesired phenomena such as sugar caramelization are prevented; moreover, the cavitation-induced turbulence enhances the liquid homogenization, thereby accomplishing a further task normally required in liquid food treatment.

Research on cavitation processes aimed at improving sanification of liquid foods has focused on the two main cavitation sources, namely, acoustic (ultrasonic) sources and hydrodynamic (mechanical) reactors. Gogate and coworkers have repeatedly shown that hydrodynamic cavitation (HC) allows to achieve up to two orders of magnitude higher energy efficiency when compared to acoustic cavitation [18], with clear advantages as reduced energy consumption, achievement of microbial lethality at reduced temperatures, and maintenance of fresh-like product quality during processing [5].

Recently, the outperformance of HC over acoustic cavitation and other conventional methods by about two orders of magnitude in terms of energy yield was convincingly proved in the technical field of the synthesis of biodiesel based on the interesterification of waste cooking oil [19], thereby suggesting that such figure could be a broad rule.

This study is mainly aimed at assessing the comparative advantages of the synergistic addition of HC processes to the usual thermal treatment to inactivate the SC in terms of processing temperatures and energy saving. A set of measurements is identified, allowing to perform a significant diagnosis of cavitation yield which is practically feasible in industrial processing, as well as a model is provided to enable the assessment and prediction of the benefits of HC processes on the basis of measurements of bulk properties of a liquid sample and of cavitation parameters. The model is general and can be extended to other yeasts, microorganisms, bacteria, fungi, and spores, as well as to other liquid foods.

The specific substance used in the experiments, namely, a water–sugar solution, makes the results directly applicable to the brewing processes and in particular to beer pasteurization, representing a novelty for both the specific application area and the demonstration of feasibility at the industrial scale.

Theoretical and Experimental Background

HC reactors invariably include some nozzles in order to locally accelerate the liquid, thus lowering its pressure due to the Bernoulli equation [20] and in turn creating countless void, plasma- and vapor-filled bubbles that, under the average pressure of the downstream undisturbed medium, collapse after few milliseconds [21].

The cavitation process is generally represented in terms of bubble density as well as collapse intensity by means of the cavitation number, hereinafter indicated as CN or σ:

$$\sigma = (P_0 - P_\mathrm{v})/(0.5 \cdot \rho \cdot u^2) \tag{1}$$

where P_0 is the downstream average pressure, P_v is the liquid vapor pressure, in turn a function of the average temperature for any given liquid, ρ is the liquid density, and u is the flow velocity [22].

The relationships between features of the cavitation bubbles and the CN σ, in particular the inverse relationship of their density and the direct relationship of the intensity of their collapse, has been known for a long time [23]. More recently, the strongly nonlinear relationship between HC efficiency and CN was clearly shown in terms of production of the hydroxyl radicals [9].

Static HC reactors crossed by the liquid flow are easy to assemble with commonly available commercial electromechanical components, generally with two configuration: orifice plates and Venturi tubes [5, 6]. Venturi tubes show an obvious practical advantage over orifice plates with liquids including solid residues and/or high viscosity components potentially occluding smaller holes [24]. On the other hand, orifice plates allow more flexibility resulting from the different possible geometric arrangement, number, and morphology of the holes, increasing the chance to enhance the cavitation process due to the downstream turbulent interaction of the various jets. The latter effect was translated into a modified CN σ_{max} [23]:

$$\sigma_{\text{mod}} = \frac{\sigma}{\sum p_{\text{h}}/p_{\text{mp}}} \quad (2)$$

where $\sum p_{\text{h}}$ is the sum of perimeters of the plate holes and p_{mp} is the inner perimeter of the main pipe downstream of the orifice plate.

Assessment of yeast concentration from bulk properties of processed samples

SC yeast consumes sugar contained in a liquid substance according to the general alcoholic fermentation reaction $C_6H_{12}O_6 \Longrightarrow 2\ C_2H_5OH + 2\ CO_2$ [25], therefore the evolution of sugar concentration is a clear marker of the yeast concentration and vitality.

To quantitatively assess the SC concentration, we developed and applied to the postprocessing phase (i.e., after any thermal and/or cavitation treatment) a simple *bulk* model built upon basic principles of ecological resource–population dynamics [26, 27], the resource being sugar in solution and population being the SC yeast.

The sugar concentration depletion rate at any given time is assumed to be proportional to the product of yeast and sugar concentrations at the same time, so that, in the event the yeast concentration were constant, the sugar concentration would decrease exponentially over time:

$$\frac{dC_S(t)}{dt} = -k \cdot N_{\text{SC}}(t) \cdot C_S(t) \quad (3)$$

where $C_S(t)$ is the sugar concentration, $N_{\text{SC}}(t)$ is the SC (yeast) concentration, k is a constant that, in the event N_{SC} were constant over time, after its multiplication by N_{SC} can be interpreted as the reciprocal of the time constant in the exponential decay function describing the

sugar concentration tendency. Since $N_{\text{SC}}(t)$ is actually changing with time, generally equation (3) can be solved only numerically.

In turn, the yeast concentration $N_{\text{SC}}(t)$ is assumed to change as a combined result of growth by self-duplication [4] and decay produced by the depletion of sugar. Rigorously, a balance of growth with death rate induced by the toxic metabolites produced during the multiple duplications should be taken into account, but the latter process is assumed to work effectively only over longer time scales, when the liquid has been eventually spoiled. Moreover, although different choices are possible, an arctangent function of the ratio between sugar and yeast concentration, normalized between the extreme values -1 and 1, was deemed to be sufficiently flexible to allow for an accurate description of both yeast's growth and decay rates. The relevant equation reads as follows:

$$\frac{dN_{\text{SC}}(t)}{dt} = \lambda \cdot N_{\text{SC}}(t) \cdot \frac{\arctan\left[a_{\text{SC}} + b_{\text{SC}}\frac{C_S(t)}{N_{\text{SC}}(t)}\right]}{\pi/2} \quad (4)$$

where λ is a constant that, in the event the arctangent function equaled 1 (high values of the sugar to the yeast concentration ratio), can be interpreted as the reciprocal of the time constant in the exponential growth function describing the yeast concentration tendency; a_{SC} and b_{SC} are further constants allowing to describe the yeast concentration decay at low sugar concentration and to appropriately scale the ratio of sugar to yeast concentration, respectively.

All parameters k, λ, a_{SC}, and b_{SC} were estimated on the basis of the first and most unaffected liquid sample extracted at the beginning of the only experiment carried out without any cavitation reactor (hereinafter "*blank experiment*"), as described in section "Experiments and estimation of model parameters." The parameter estimation method along with the respective assumptions, as well as the solution method for the coupled equations (3) and (4), is discussed in the Appendix.

SC inactivation rate from combined thermal and cavitation treatments

The second model is aimed at quantitatively relating the lethality and permanent inactivation induced on SC by a thermal or combined thermal and cavitation process.

First, a model is proposed to simulate the lethality rate induced by a purely thermal treatment, that is, by heating the liquid including the yeast. It should take into account both temperature and residence time; however, the large thermal inertia of the industrial scale experimental installation allows to discard such residence time as a controlling parameter.

Although different choices are possible, in analogy with the *bulk* model described in section "Assessment of yeast concentration from bulk properties of processed samples," we represent the ratio between the SC concentration at some time during the process to its initial value by means of the inverse of an arctangent function of the temperature, normalized between the extreme values 0 and 1. The relevant equation reads as follows:

$$N_{SC}(t_p) = N_{SC}(0) \cdot \{F_T[T(t_p)]\}^{-1}$$
$$F_T(T) = \{\arctan[A_T \cdot (T - T_c)] + \pi/2\}/\pi \quad (5)$$

where t_p is the heating process time starting from $t_p = 0$, $T(t_p)$ is temperature at time t_p during the process, A_T is a constant regulating the "sharpness" of the arctangent function, T_c is the temperature value at which $F_T = 1/2$. It is worth noting that the function F_T cannot be allowed to decrease in the course of a process, regardless of possible short transient cooling phases due to technical needs such as sampling, because inactivation or lethality are irreversible on such short time scales.

The parameters A_T and T_c were estimated from the *blank* experiment, when the liquid treatment was limited to heating, by matching the ratios $N_{SC}(t_p)/N_{SC}(0)$ derived from equation (5) with the respective values computed as explained in the Appendix. Moreover, normalizing the F_T function in equation (5) enables to fit the interval between null and maximum lethality. It should be noted that the same matching could be performed with observed values of the ratios of the SC concentration. The estimated values of A_T and T_c are shown in the Appendix.

The model represented by equation (1) is now generalized to include the additional effect of HC upon the yeast's lethality. In order to keep the model as simple as possible, the additional lethality induced by cavitation is accounted for by multiplying the "thermal" function $F_T(T)$ in equation (5) by the integral over the process time of a further "cavitation" function F_C; such function F_C depends on the CN σ, as per equation (1) or, in the modified form, equation (2), on the frequency of cavitation processes, that is, the number of passages of each fluid parcel through the cavitation reactor per unit time, and again on the temperature, due to the observed strongly synergistic effect of temperature and cavitation [1].

The dependence of F_C on the CN σ was chosen such that, all other quantities held constant, it produces up to two local maxima at some values $\sigma = \sigma_{1-max}$ and $\sigma = \sigma_{2-max}$ with $0 < \sigma_{1-max} < \sigma_{2-max}$, and falls to zero at both $\sigma = 0$ and $\sigma \to \infty$, where no cavitation occurs. The choice to allow up to two local maxima derives from the consideration that in low CN cavitation regimes more bubbles are generated and their collapse is moderate, while in higher CN cavitation regimes less bubbles are

generated and their collapse is more violent; a square dependence on the CN was chosen in order to get more pronounced local maxima in the range of CN of interest.

The relevant equation reads as follows:

$$F_C(t) = B_c \cdot e^{T(t)-T_c} \cdot N_{cav}(t) \cdot$$
$$\left[\left(\frac{\sigma(t)}{\sigma_{1-max}}\right)^2 \cdot e^{-\left(\frac{\sigma(t)}{\sigma_{1-max}}\right)^2} + \left(\frac{\sigma(t)}{\sigma_{2-max}}\right)^2 \cdot e^{-\left(\frac{\sigma(t)}{\sigma_{2-max}}\right)^2}\right]$$
$$(6)$$

where B_c is a "scaling" constant factor, N_{cav} is the frequency of cavitation processes, that is, the number of passages of each fluid parcel through the cavitation reactor per unit time. In other words, the dependence of F_C on the cavitation frequency is kept linear, while an exponential growth function of the temperature is used to represent the observed sharp changes of the cavitation efficiency with temperature itself.

The parameter T_c is chosen as the same temperature value as in equation (5) in order to avoid the introduction of a new parameter. Usually, the quantities N_{cav} and σ change during any process due to the observed weak dependence of the liquid flow rate and its speed on the actual temperature, which in turn affects the liquid internal friction, therefore their average values are computed.

The equation describing the synergistic effects of the thermal and cavitation processes on the yeast concentration reads as follows:

$$N_{SC}(t_p) = N_{SC}(0) \cdot \left\{F_T(t) \cdot \left[\int_0^{t_p} F_C(t)dt + 1\right]\right\}^{-1} \quad (7)$$

The parameters B_c, σ_{1-max}, and σ_{2-max} in equation (6) are estimated from the combined thermal and cavitation experiments by matching the ratios $N_{SC}(t_p)/N_{SC}(0)$ derived from equation (7) with the respective values computed as in the Appendix, corresponding to samples extracted at different stages of the process.

Again, the same value matching could be performed with observed values of the ratios of the SC concentration, should such measurements be available. The estimated values of B_c, σ_{1-max}, and σ_{2-max} are displayed in the Appendix.

Materials and Methods

Experimental stand

The experimental stand consisted of a closed hydraulic circuit equipped with a main centrifugal pump of nominal mechanical power 4 kW (Lowara SHE 40-160/40); a filling tank where the mixture of water, sugar, SC yeast, and nutrients was prepared before entering the circuit with the help a smaller pump; an expansion tank used to regulate the hydraulic pressure downstream of the cavitation reactor; the cavitation reactor; a secondary circuit to the heat exchanger supplied with a small 100 W circulation pump; On/Off valves to manage filling; and a secondary circuits as well as expansion tank (Fig. 1).

The main hydraulic circuit including the cavitation reactor consists of a food quality "AISI 304" stainless steel pipe with internal diameter 97.6 mm, and of a vertical extent slightly taller than 5 m starting from the main pump exit (volume = 90 L). The secondary circuit consists of a one inch (about 25.4 mm) diameter steel pipe connected to the heat exchanger, with a volume about 25 L. The total volume including both the main and the secondary circuit was therefore about 115 L.

The only energy source is electricity powering the main pump's motor. Any heating of the circulating liquid is the result of the conversion of the pump's rotor mechanical energy into heat in the closed hydraulic circuit. Moreover, the HC-induced bulk heating is an energy-intensive process, especially when a vertical geometry of the hydraulic circuit is used [17].

1. Filling tank
2. Expansion tank
3. Cavitation reactor (if any)
4. Main centrifugal pump
5. On/Off valve
6. Heat exchanger

T = Thermometer

P = Manometer

⊘ = Circulation pump

Ⓥ = Air/vapor degassing valve

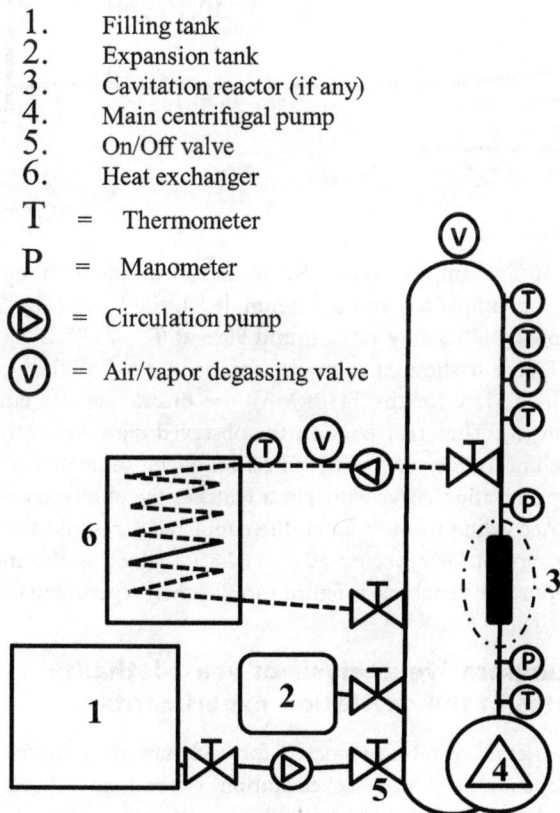

Figure 1. Experimental stand: general scheme with identification of main components.

The cavitation reactors used for the experiments were an orifice plate equipped with 156 holes (each having an internal diameter 2 mm) and a Venturi tube (Fig. 2A and B). Both are upscale models of effective configurations used in laboratory scale experiments by researchers in India [23, 28]. The orifice plate total opening was 490 mm^2 (6.55% of the main pipe's inner section). The corresponding value for the opening of the Venturi tube's nozzle was 452 mm^2 (6.05% of the pipe's section).

The preparation of the yeast before each experiment, as well as the measurement methods, is described in the Appendix.

Experiments and estimation of model parameters

The main structural and operational features of the performed experiments are listed in Table 1.

Figure 3 shows the time series of the liquid average temperature for all the performed yeast lethality experiments.

The temperature curves for the VENTURI test #3_flash and VENTURI test #5_flash differ from all the others because those were the only processes carried out without heat exchange, that is, semi-adiabatically (unless unavoidable heat losses due to practical limitations to the thermal insulation). The last phase of the experiment VENTURI test #4 was carried out without heat exchange too, with most of the temperature rise occurring in that phase. The estimated values of the parameters included in the models described in sections "Assessment of yeast concentration from bulk properties of processed samples" and "SC inactivation rate from combined thermal and cavitation treatments" are shown in Table A1 in the Appendix.

Results and Discussion

As explained in section "Assessment of yeast concentration from bulk properties of processed samples," the initial postprocessing SC yeast concentration for each sample of any experiment listed in Table 1, indicated with $N_{SC}(t_0)$, can be retrieved after matching the respective observed sugar concentration time series with the same series predicted on the basis of the model represented by equations (3) and (4) with the parameter values shown in Table A1 in the Appendix.

THERMAL test: the benchmark experiment

Due to its own relevance as the reference experiment for the calibration of model parameters as well as a "benchmark" for all other experiments, the results for the THERMAL test are shown in full detail. Figure 4 shows

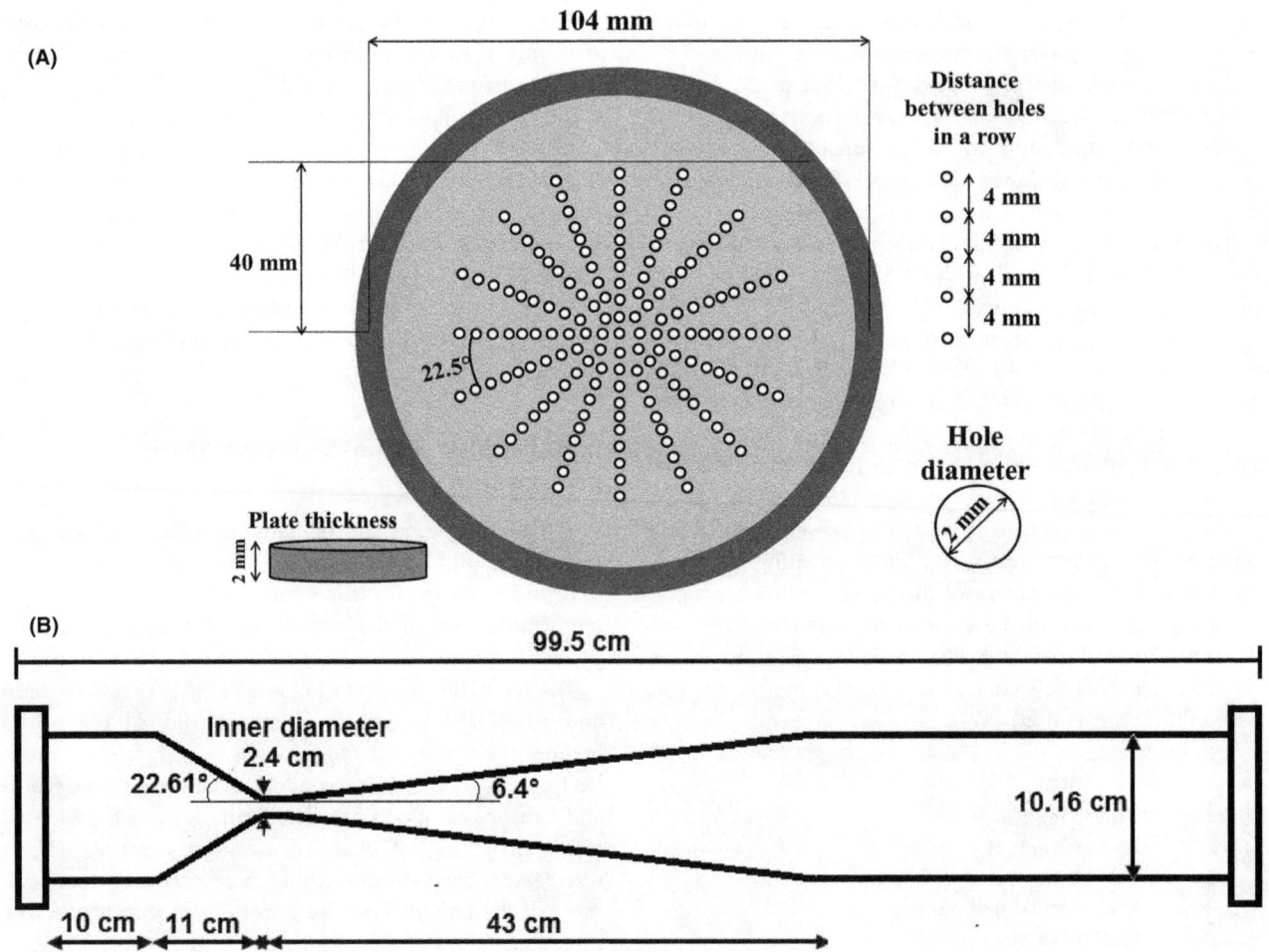

Figure 2. Cavitation reactors: (A) orifice plate and (B) Venturi tube.

the sugar concentration time series for all the samples extracted during the process.

Given the measurement error on the order at most of 5 g/L, the differences among the tendencies shown by the time series associated with the first three samples (S0, S1, and S2), the fourth (S3), and the fifth (S4) in Figure 4 are quite significant, showing that a relevant effect upon the yeast activity occurs at 54°C < T < 61°C. A small drop in the sugar concentration occurs even in the S4 sample (heating up to 73.9°C) about 48 h after its extraction.

Charts in Figure 5 show the observed and simulated sugar concentration time series for all samples, along with the predicted yeast concentration time series associated with the best simulated sugar concentration curve.

As explained in the Appendix, the identification of the simulated sugar concentration curve best matching the observed one was performed computing the respective average square distances during the earliest 72 h after sampling.

The initial postprocessing SC yeast concentration $N_{SC}(t_0)$ appears to change from the preprocessing value

of 50 for samples S0 and S1, to 35 for sample S2, then to 15 for sample S3, and 5 for sample S4, the latter (Fig. 5E) showing a lethality rate around 90% at $T = 73.9°C$.

Figure 6 shows the model reconstruction of the yeast lethality rate for the THERMAL test on the basis of equation (5). The error bars on the observed data derive from the uncertainty of the identification of the simulated sugar concentration curve which best matches the observed one.

According to the model, the simulated threshold lethality rate of 90% occurs at $T = 62.8°C$, which is the most important benchmark figure for all other experiments.

Comparative analysis of yeast lethality rates in the cavitation experiments

The lethality rates estimated for each sample collected in the course of the six cavitation experiments listed in Table 1 are compared with the benchmark values of the same quantity retrieved from the THERMAL test in section "THERMAL test: the benchmark experiment".

Table 1. Main features of the performed experiments.

Experiment ID	Cavitation reactor	Volume (L)	Temperature range and samples' extraction temperature (°C)	Range of cavitation number[1]	Range of cavitation frequency[2] (min^{-1})	Processing time[3] (min)
THERMAL test[4]	None	115	**26.0–73.9**	N/A	N/A	440
			S0–32.8			
			S1–41.7			
			S2–53.7			
			S3–61.5			
			S4–73.9			
PLATE test	Orifice plate	115	**20.2–74.3**	0.14–0.21[5]	7.0–8.6	484
			S0–25.5	0.44–0.67[1]		
			S1–41.0			
			S2–52.0			
			S3–62.0			
			S4–74.3			
VENTURI test #1	Venturi tube	115	**26.3–62.0**	0.64–0.72	9.3–10.2	263
			S0–33.5			
			S1–45.8			
			S2–52.1			
			S3–57.1			
			S4–62.0			
VENTURI test #2	Venturi tube	115	**25.5–51.8**	0.94–1.18	8.8–9.8	159
			S0–35.6			
			S1–51.8			
VENTURI test #3_flash[6]	Venturi tube	90	**25.9–72.0**	0.84–1.02	11.6–13.4	75
			S0–30.5			
			S1–53.0			
			S2–62.5			
			S3–72.0			
VENTURI test #4	Venturi tube	115	**27.8–62.5**	0.87–1.15	9.7–11.8	211
			S0–30.8			
			S1–42.0			
			S2–55.5			
			S3–62.5			
VENTURI test #5_flash[6]	Venturi tube	90	**22.8–61.1**	0.29–0.44	11.7–14.4	58
			S0–32.6			
			S1–51.7			
			S2–57.3			
			S3–61.1			

[1]Cavitation number computed as per equation (1).
[2]Number of passages of any liquid parcel per minute through the cavitation reactor.
[3]Length of the thermal or thermal and cavitational process, in minutes.
[4]Also referred to as "blank" experiment in section "Assessment of yeast concentration from bulk properties of processed samples".
[5]Modified cavitation number computed as per equation (2).
[6]Only tests with free heating, all others equipped with secondary circulation to an heat exchanger at different cooling rates.

Figure 7 shows the whole set of lethality rates, as well as a focus over the temperature range with the steepest increase of the lethality rate, that is, 50–63°C.

Figure 7A shows that all lethality rates data points derived from the cavitation tests lie below the THERMAL test lethality curve when both the following conditions are fulfilled: temperature higher than 50°C and lethality rate greater than 15%; moreover, the 90% threshold of the lethality rate is achieved, within the represented uncertainty, at temperatures up to 10°C lower than in the THERMAL test, that is, around 52°C.

Figure 7B, focused on the temperature range where changes of the lethality rate are steepest, shows as well the CN associated with samples collected during all processes, computed at the time of sampling. Although the picture is quite complex, it appears that the best result, namely, the lowest temperature at the 90% lethality rate is achieved with the highest CN (CN = 1.18 in VENTURI test #2).

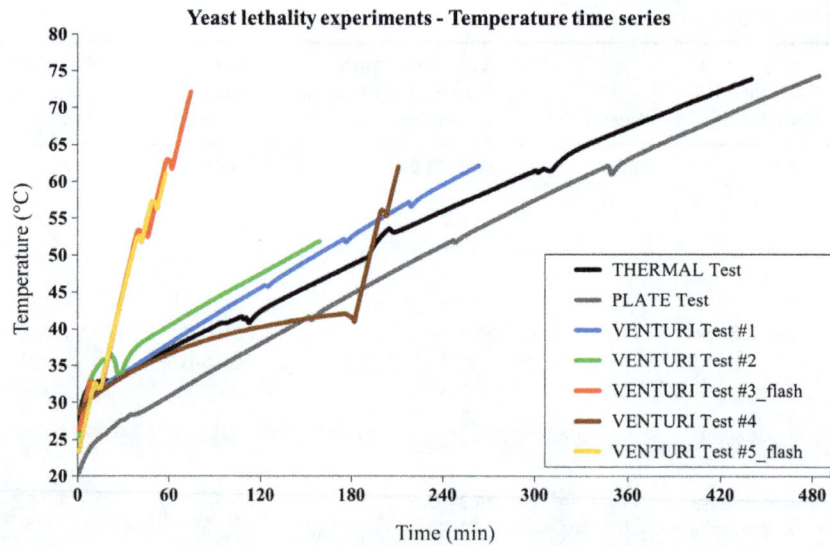

Figure 3. Liquid average temperature time series for all the performed yeast lethality experiments.

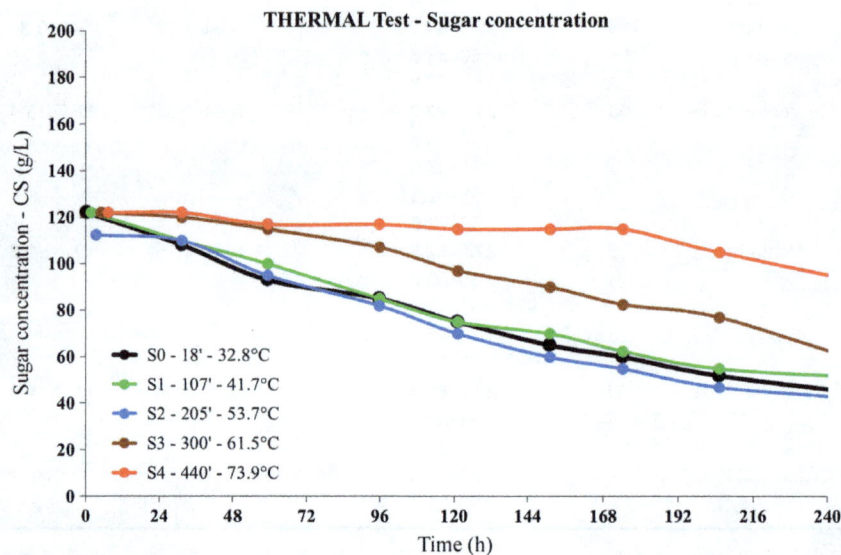

Figure 4. THERMAL test: postprocessing sugar concentration versus time series for all samples.

The semi-adiabatic tests, that is, VENTURI test #3_flash and VENTURI test #5_flash, show approximately the same values for the cavitation frequency and the processing time at any temperature (Table 1), but very different CN (CN = 0.84 and CN = 0.29 for the considered samples extracted from the VENTURI test #3_flash and the VENTURI test #5_flash, respectively, allowing to conclude that the lethality rate is greater when the CN is smaller, that is, in VENTURI test #5_flash).

The other two tests carried out by means of Venturi tube reactor, namely, VENTURI test #1 (CN = 0.71) and VENTURI test #4 (CN = 1.09), the first one with a little longer processing time (Table 1), show a comparable

behavior, close to the best of the two flash tests but after much longer processing times.

A tentative conclusion from Venturi tube experiments, within the limits of the cavitation regimes under study, could be that the impact of cavitation processes on the lethality rate shows a local peak at low CNs, around CN = 0.29, decreases as the CN increases up to a little more than 1, after that it increases very sharply, therefore adding confidence to the "two-peak" yeast lethality model represented by equation (6).

Last, the PLATE test, showing a modified CN = 0.14 and CN = 0.46, shows the worst efficiency in terms of lethality, this conclusion being supported by its processing

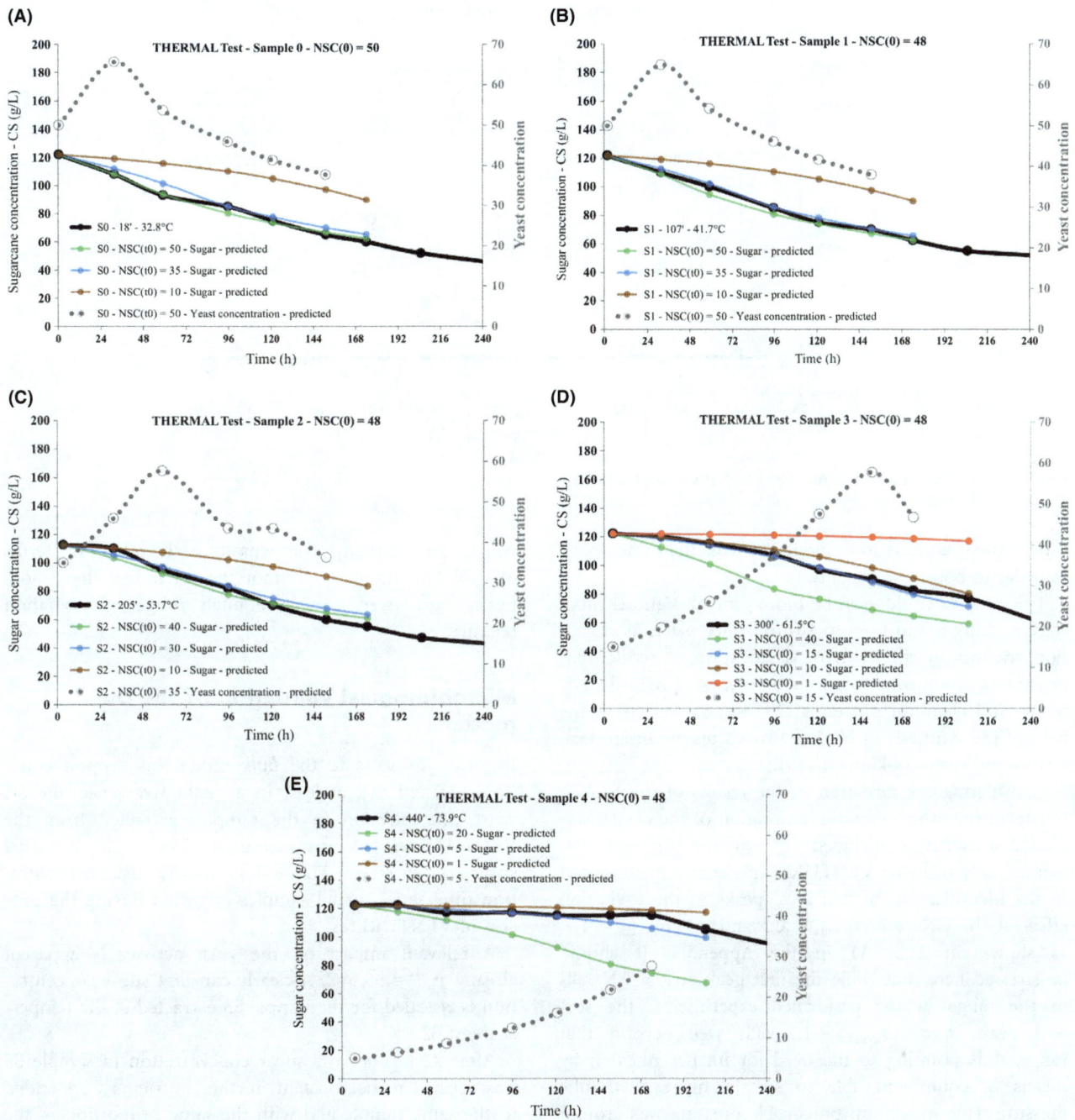

Figure 5. THERMAL test: observed (black) and model-simulated (colored) sugar concentration, model predicted yeast concentration (dotted gray) versus time series for all samples, each chart (A-E) concerning a specific sample. Time and temperature of any sample during the process are indicated. The postprocessing initial yeast concentration (first point in its time series) is indicated too.

time which is by far the longest among the other cavitation experiments (and close to the processing time of the THERMAL test). Whether this is due to the values of its CN, modified CN, lower cavitation frequency, or other specific features of the reactor configuration, remains so far unclear.

Assuming that the cavitation regime of the PLATE test corresponds to an effective CN somewhere between 0.14 and 0.46, therefore close to that of VENTURI test #5_flash, the important recommendation arises that, in addition to the practical advantages discussed in section "Theoretical and Experimental Background," the Venturi

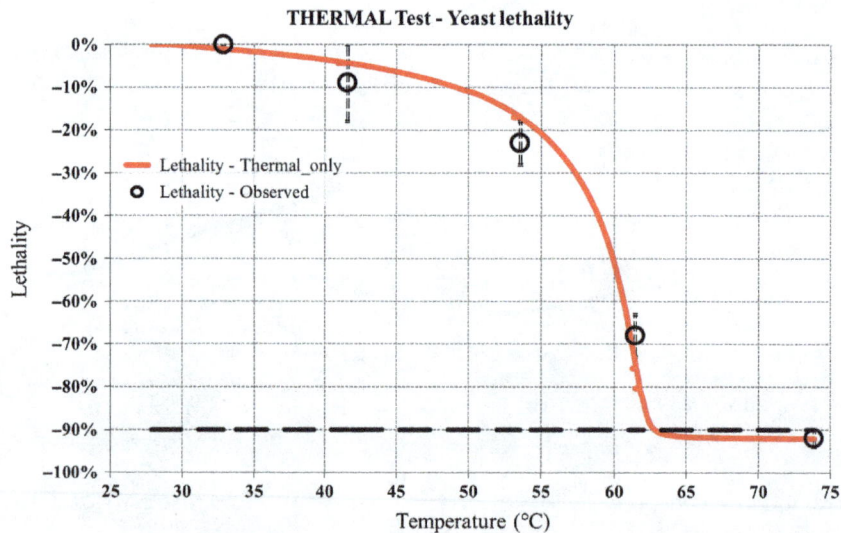

Figure 6. THERMAL test: model reconstruction of the yeast lethality rate.

tube configuration is to be preferred over the orifice plate in order to boost the lethality rate.

This finding could even be more general: Venturi tubes were recently found to outperform orifice plates as cavitation reactors in the fields of degradation of recalcitrant pollutants in aqueous solutions by means of hybrid cavitation and chemical processes [29, 30], as well as in the field of the synthesis of biodiesel based on the interesterification of waste cooking oil [19].

Considering the deviation of the results of the PLATE test from the other ones, the calibration of the cavitation-related parameters included in equation (6) was performed only over the VENTURI experiments, turning out in the identification of two local peaks of the cavitation effect at the CNs $\sigma = \sigma_{1\text{-max}} = 0.3$ and $\sigma = \sigma_{2\text{-max}} = 1.7$, as shown in Table A1 in the Appendix. It should be stressed here that while the first peak at low CN falls in the range of the performed experiments, the second peak at $\sigma = \sigma_{2\text{-max}} = 1.7$ falls well beyond that range, thus pointing to the need for further research by means of equipment able to sustain higher hydraulic pressures (the maximum practicable pressure was around 7.5 bar).

Complex nonlinear behavior of the HC efficiency was found in other works; for example, using an orifice plate as cavitation reactor, the degradation of dichlorvos in aqueous solution showed a sharp peak at high inlet pressure and low temperature [31]; similar results were found in other works using Venturi tubes and dealing with different pollutants such as rhodamine B and p-nitrophenol [29, 30]. The problem in those works is that, increasing the inlet pressure, both liquid circulation velocity and downstream recovery pressure increase,

which can partially compensate each other in terms of CN; therefore, it cannot be excluded that more peaks exist over a wide enough range of cavitation regimes.

Microbiological validation of the bulk model

In order to validate the *bulk* model built upon equations (3) and (4), at least in a qualitative sense, the SC yeast concentration in the samples extracted from the VENTURI test #4 was measured along with the usual sugar concentration. Figure 8 shows the sugar concentration time series for all samples extracted during the process of VENTURI test #4.

A relevant impact on the yeast activity is apparent already at $T = 55.5°C$. A clearly constant sugar concentration is revealed for the sample S3 extracted at the temperature of 62.5°C.

After 72 h, once the sugar concentration in sample S3 was verified to be constant, further nutrients were added to the same sample and with the same proportion of the liquid initially loaded into the circuit, in order to investigate possible recovery of the yeast cells. The same sample was renamed as "S3N" after the addition of nutrients.

On the basis of the sugar concentration time series shown in Figure 8 for the S3N sample, the yeast recovery was quite slow and limited in extent.

Selected portions of each sample extracted during the VENTURI test #4 process were used to measure the yeast cells concentration according to the methods described in the Appendix.

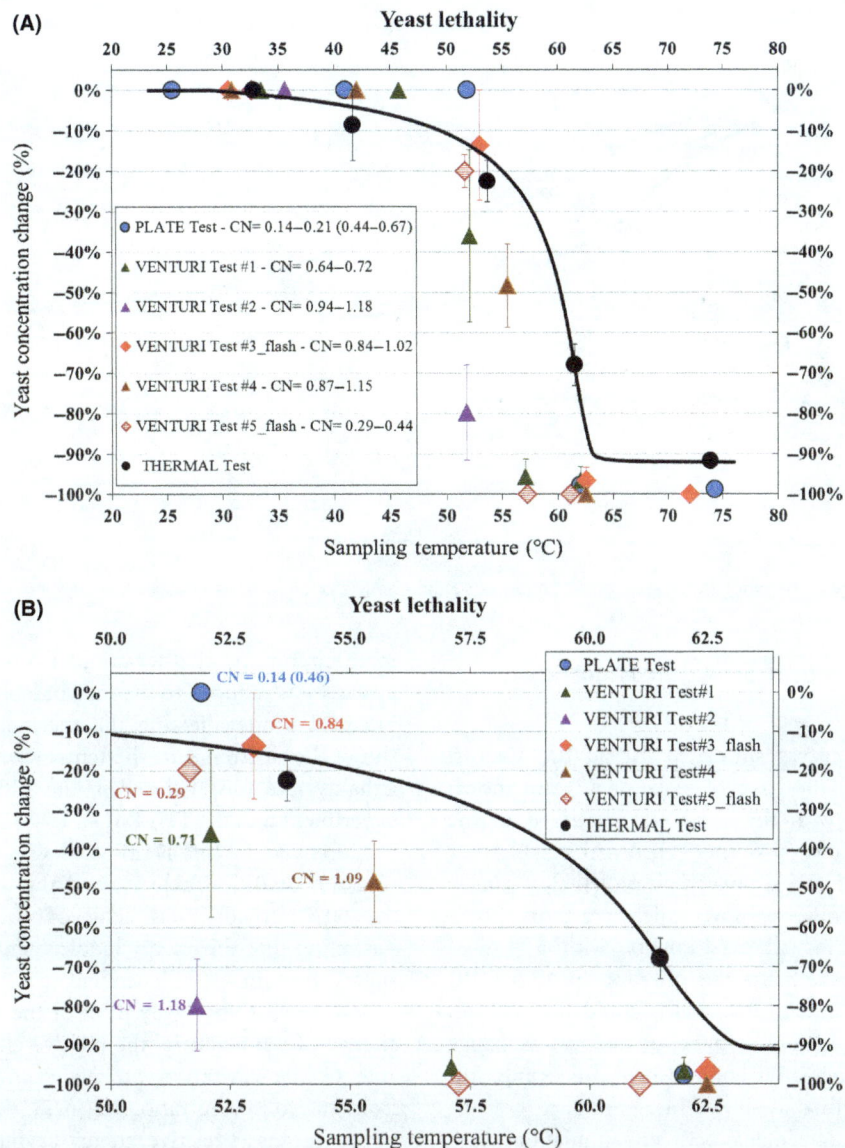

Figure 7. Whole set of lethality rates (A) and focus over the temperature range with the steepest increase of the lethality rate, including labels highlighting the cavitation number occurred before any sampling (B); for the PLATE test, the modified CN is shown, along with the ordinary CN in brackets.

Charts in Figure 9 show the direct comparison of the observed and model-simulated yeast concentration time series, the latter associated with the best matching of the observed and simulated sugar concentration curves. Error bars representing the standard deviations are based upon the analysis of at least five microbiological samples at all the observed data points except the first two, for which only one sample was available.

The simulated steep fall of the initial postprocessing yeast concentration (i.e., the quantity $N_{SC}(t_0)$) going from sample S1 (Fig. 9B) to sample S2 (Fig. 9C) and the even larger decrease going from S2 (Fig. 9C) to S3 (Fig. 9D)

are qualitatively very well reproduced by the experimental data. The overall evolution of the postprocessing yeast concentration is also reproduced with fair accuracy at least up to 96 h after sampling, especially for samples S0 (Fig. 9A) and S1 (Fig. 9B).

Therefore, the microbiological analysis offers a qualitative validation of the proposed model and of the overall results of the research, even though the discrepancy of the predicted and observed peak concentration values in samples S2 and S3, as shown in Figure 9C and D, respectively, could suggest that the yeast tendency model as represented by equation (4) is oversimplified and needs further development.

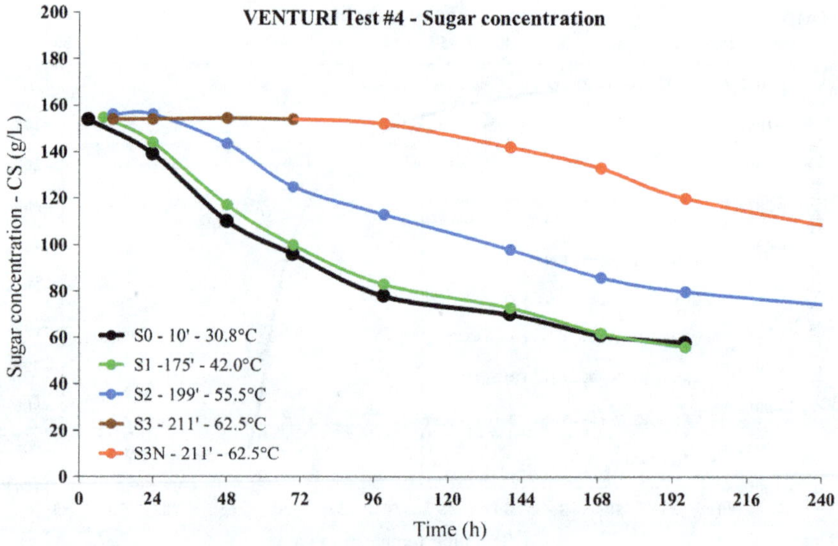

Figure 8. VENTURI test #4: postprocessing sugar concentration versus time series for all samples

Summary results and energy efficiency considerations

The yeast lethality curves simulated for all the Venturi tube experiments on the basis of equation (7) are shown in Figure 10, along with the same curve derived in the absence of any cavitation process (THERMAL test).

The agreement of the simulated yeast lethality curves with the available observations, which are not shown here, is very good for all experiments, with a limited overestimation of the lethality rate only for the VENTURI test #4 (blue curve in Fig. 10), which could derive from its unique features such as the break of the heat exchange regime (slow warming) after about 3 h of processing, followed by a *flash* heating, as shown in Figure 3.

Relying on this agreement between simulations and observations, both the temperature associated with the 90% yeast lethality threshold and the difference with regard to the THERMAL test can be inferred for any VENTURI experiment; these results are shown in Table 2.

The 90% lethality threshold temperature differences span the range −6.3°C to −9.5°C; limited to the *flash* tests, the best result is −7.4°C, achieved within the VENTURI test #5_flash.

The achievement of the 90% threshold of the yeast lethality rate at lower temperatures can lead both to a significant improvement of the organoleptic and nutritional qualities of the food liquid, and to some energy saving arising from the reduction of the energy requirements for heating.

The experiment VENTURI test #5_flash will be considered for the assessment of the energy saving because it was performed in semiadiabatic conditions, that is, the heat exchanger shown in Figure 1 was disconnected from the

main circuit and the deviations from an adiabatic heating process was reduced to the unavoidable heat loss due to the practical limits affecting the thermal insulation. Table 2 shows the difference of the temperature at which the 90% lethality rate was achieved in the VENTURI test #5_flash experiment and the THERMAL test, namely, −7.4°C.

Moreover, in the THERMAL test, the initial temperature was 26.0°C (Table 1), and the temperature at which the 90% lethality was achieved was 62.8°C (Table 2). Assuming that the energy requirements to heat the water–sugar solution is independent of the temperature and proportional to both the mass of the substance and to the change of temperature, the relative energy saving attributable to the cavitation process can be simply estimated from the following ratio: 7.4/(62.8 − 26.0) = 0.20 (20%), which means a relative energy saving per unit temperature decrease with regard to the purely thermal treatment on the order of 2.7%/°C.

The digital Watt-meter mentioned in the Appendix provided the direct measurements of the grid electricity consumed to bring the water–sugar solution from the initial temperature of VENTURI test #5_flash experiment, that is, 22.8°C (Table 1), to the temperature at which the 90% yeast lethality rate is achieved, namely, 55.4°C (Table 2), resulting in approximately 3.1 kWh (11,160 kJ).

Given the water–sugar solution volume of 90 L, a specific energy consumption in the VENTURI test #3_flash equal to 11,160/[90 × (55.4 − 22.8)] = 3.804 kJ/kg °C is estimated, therefore the absolute specific energy saving can be computed as follows: 7.4 × 3.804 = 28.15 kJ/kg.

It should be noted that, in comparison with previous work dealing with cavitation-induced lethality on SC by

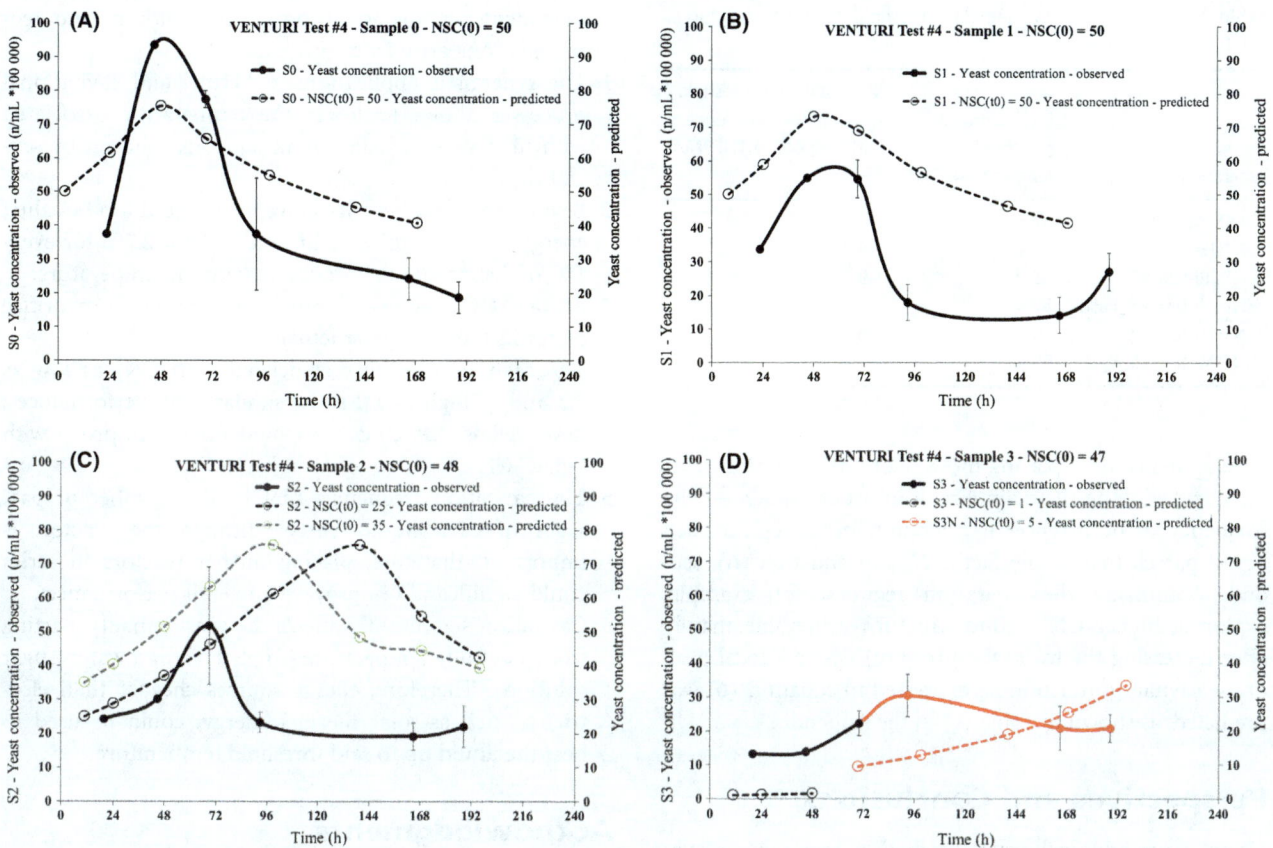

Figure 9. VENTURI test #4: observed and model-simulated yeast concentration time series, each chart (A-D) concerning a specific sample. The observation data points are in units of number of yeast cells per mL, divided by 10^5.

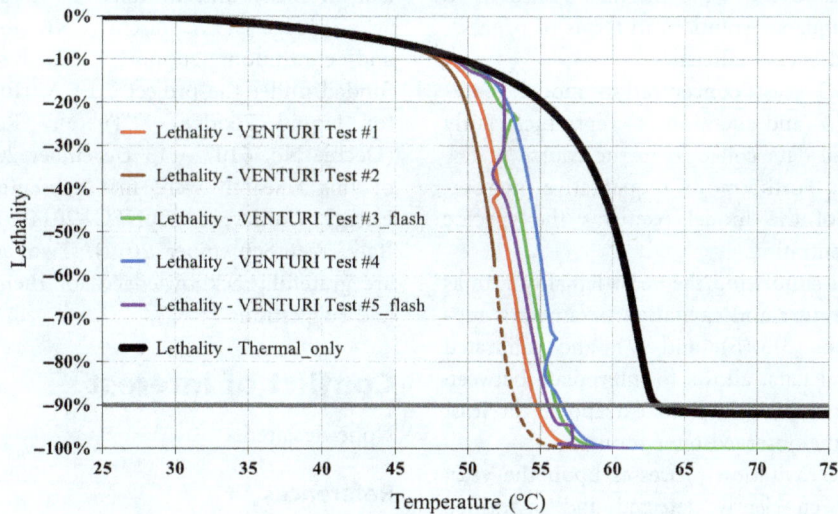

Figure 10. Yeast lethality curves for all VENTURI tests; the dashed portion of the brown curve highlights the extrapolation beyond the maximum temperature achieved in the VENTURI test #2; the thick black curve represents the yeast lethality simulated in the absence of any cavitation process.

means of laboratory-scale devices [1, 32], the values of energy consumption reported in this study are lower by nearly one order of magnitude.

Achieving even lower temperatures corresponding to the 90% yeast lethality rate, therefore energy savings well above 20%, is thought to be feasible by at least two different

Table 2. 90% Comparative lethality threshold temperatures for all VENTURI experiments.

Experiment ID	90% Lethality threshold temperature (°C)	90% Lethality threshold temperature difference with respect to THERMAL test (°C)
THERMAL test	62.8	
VENTURI test #1	54.7	−8.1
VENTURI test #2	53.3	−9.5
VENTURI test #3_flash	56.4	−6.4
VENTURI test #4	56.5	−6.3
VENTURI test #5_flash	55.4	−7.4

developments; first, placing more than one cavitation reactor in series along the main circuit in order to increase the frequency of occurrence of cavitation processes for each liquid parcel, that is, the factor N_{cav} in equation (6); second, optimizing the cavitation regimes, for example, exploring higher CN regimes (in turn achievable mainly after increasing the hydraulic pressure) where a local peak of the cavitation function F_C expressed in equation (6) was predicted, as shown in Table A1 in the Appendix.

Perspectives and Conclusions

Using a preindustrial scale installation and relying over basic and general principles of resource–population dynamics, we have assessed the comparative advantages of adding HC processes to the usual thermal treatment to inactivate SC in an aqueous solution, in terms of processing temperatures and energy efficiency.

The sugar and SC yeast concentration model represented by equation (3) and equation (4) reproduces fairly well the observational data collected in the course of few targeted experiments. Furthermore, a qualitative microbiological validation of the model confirms the specific trends of yeast concentration.

An original model simulating the yeast lethality rate as a function of temperature and cavitation parameters, represented by equations (5), (6), and (7), and calibrated over the experimental data, allows to interpolate between the sparse data points as well as to extrapolate at least within the range of the observed temperatures.

The impact of the cavitation processes upon the yeast cells lethality has been clearly detected and quantified beyond the purely thermal resistance effects, resulting in a remarkable decrease of the maximum process temperature for the same high lethality threshold. Moreover, the existence of a local peak for the impact upon the yeast lethality at a cavitation regime far beyond the observational data was predicted, pointing to the need for further experiments by means of new equipment.

In conclusion, the following general findings and recommendations arise from this study:

1 The synergistic application of thermal and cavitational processes allows to lower the temperature associated with high yeast lethality in an aqueous solution by several degrees.

2 Beyond the clear benefits in terms of liquid food quality, energy saving is quite significant: at least 2.7% for every 1°C of decrease of the process maximum temperature.

3 Among HC reactors, Venturi tubes outperform orifice plates in terms of yeast lethality.

4 The cavitation regimes associated with CNs as low as 0.3 and as high as 1.2 show similar peak performances; nevertheless, the efficiency could further improve with higher CNs.

5 The cavitation frequency, that is, the number of passages of each liquid parcel through the reactor, is important; therefore, placing further reactors in series could significantly improve the overall performance.

6 Cavitation significantly affects the yeast lethality starting from a definite temperature (51 ± 1°C for a water–sugar solution). Therefore, energy sources cheaper than electricity, such as solar thermal energy, could be used to heat the liquid up to said threshold temperature.

Acknowledgments

This article is dedicated to the memory of Francesco Foresta (1965–2014) for all he has done to advance journalism in Sicily and in Italy. L. A. and F. M. gratefully acknowledge Dr. A. Raschi (CNR) for fruitful discussions and continuous support. This research was partially funded under the project T.I.L.A. (Innovative Technology for Liquid Foods) of Tuscany Regional Government (Decree No. 6107 – 13 December 2013). Selected results of this research were first presented at the Sun New Energy Conference SuNEC 2014 (Santa Flavia, Sicily, Italy, 8–9 September 2014). Two anonymous reviewers are gratefully acknowledged for their valuable comments and suggestions.

Conflict of Interest

None declared.

References

1. Milly, P. J., R. T. Toledo, W. L. Kerr, and D. Armstead. 2008. Hydrodynamic cavitation: characterization of a novel design with energy considerations for the inactivation of *Saccharomyces cerevisiae* in apple juice. J. Food Sci. 73: M298–M303.

2. Ngadi, M. O., M. B. Latheef, and L. Kassama. 2012. Emerging technologies for microbial control in food

processing. Pp. 363–411 in J. I. Boye and Y. Arcand, eds. Green technologies in food production and processing. Springer, Boston, MA.

3. Guth, E., T. Hashimoto, and S. F. Conti. 1972. Morphogenesis of ascospores in Saccharomyces cerevisiae. J. Bacteriol. 109:869–880.

4. Neiman, A. M. 2005. Ascospore formation in the yeast Saccharomyces cerevisiae. Microbiol. Mol. Biol. Rev. 69:565–584.

5. Gogate, P. R. 2011. Hydrodynamic cavitation for food and water processing. Food Bioprocess Technol. 4:996–1011.

6. Gogate, P. R., and A. B. Pandit. 2001. Hydrodynamic cavitation reactors: a state of the art review. Rev. Chem. Eng. 17:1–85.

7. Batoeva, A. A., D. G. Aseev, M. R. Sizykh, and I. N. Vol'nov. 2011. A study of hydrodynamic cavitation generated by low pressure jet devices. Russ. J. Appl. Chem. 84:1366–1370.

8. Capocelli, M., M. Prisciandaro, A. Lancia, and D. Musmarra. 2014. Hydrodynamic cavitation of p-nitrophenol: a theoretical and experimental insight. Chem. Eng. J. 254:1–8.

9. Capocelli, M., D. Musmarra, and M. Prisciandaro. 2014. Chemical effect of hydrodynamic cavitation: simulation and experimental comparison. AIChE J. 60:2566–2572.

10. Gogate, P. R., and A. M. Kabadi. 2009. A review of applications of cavitation in biochemical engineering/biotechnology. Biochem. Eng. J. 44:60–72.

11. Bagal, M. V., and P. R. Gogate. 2014. Wastewater treatment using hybrid treatment schemes based on cavitation and Fenton chemistry: a review. Ultrason. Sonochem. 21:1–14.

12. Chand, R., D. H. Bremner, K. C. Namkung, P. J. Collier, and P. R. Gogate. 2007. Water disinfection using the novel approach of ozone and a liquid whistle reactor. Biochem. Eng. J. 35:357–364.

13. Gogate, P. R., S. Mededovic-Thagard, D. McGuire, G. Chapas, J. Blackmon, and R. Cathey. 2014. Hybrid reactor based on combined cavitation and ozonation: from concept to practical reality. Ultrason. Sonochem. 21:590–598.

14. Gogate, P. R. 2007. Application of cavitational reactors for water disinfection: current status and path forward. J. Environ. Manage. 85:801–815.

15. Chakinala, A. G., D. H. Bremner, P. R. Gogate, K.-C. Namkung, and A. E. Burgess. 2008. Multivariate analysis of phenol mineralisation by combined hydrodynamic cavitation and heterogeneous advanced Fenton processing. Appl. Catal. B Environ. 78:11–18.

16. Yusaf, T., and R. A. Al-Juboori. 2014. Alternative methods of microorganism disruption for agricultural applications. Appl. Energy 114:909–923.

17. Baurov, Y. A., F. Meneguzzo, A. Y. Baurov, and A. Y. J. Baurov. 2012. Plasma vacuum bubbles and a new force of

nature, the experiments. Int. J. Pure Appl. Sci. Technol. 11:34–44.

18. Gogate, P. R., and A. B. Pandit. 2011. Cavitation generation and usage without ultrasound: hydrodynamic cavitation. Pp. 69–106 in D. S. Pankaj and M. Ashokkumar, eds. Theoretical and experimental sonochemistry involving inorganic system. Springer, Dordrecht–Heidelberg–London–New York.

19. Maddikeri, G. L., P. R. Gogate, and A. B. Pandit. 2014. Intensified synthesis of biodiesel using hydrodynamic cavitation reactors based on the interesterification of waste cooking oil. Fuel 137:285–292.

20. Braeutigam, P., M. Franke, Z.-L. Wu, and B. Ondruschka. 2010. Role of different parameters in the optimization of hydrodynamic cavitation. Chem. Eng. Technol. 33:932–940.

21. Gogate, P. R. 2011. Theory of cavitation and design aspects of cavitational reactors. Pp. 31–68 in D. S. Pankaj and M. Ashokkumar, eds. Theoretical and experimental sonochemistry involving inorganic system. Springer, Dordrecht–Heidelberg–London–New York.

22. Gogate, P. R. 2011. Application of hydrodynamic cavitation for food and bioprocessing. Pp. 141–173 in H. Feng, G. Barbosa-Canovas, and J. Weiss, eds. Ultrasound technologies for food bioprocessing. Springer Science + Business Media, LLC, New York, NY.

23. Vichare, N. P., P. R. Gogate, and A. B. Pandit. 2000. Optimization of hydrodynamic cavitation using a model. Chem. Eng. Technol. 23:683–690.

24. Arrojo, S., Y. Benito, and A. M. Tarifa. 2008. A parametrical study of disinfection with hydrodynamic cavitation. Ultrason. Sonochem. 15:903–908.

25. Carrascosa Santiago, A. V., R. Munoz, and R. G. Garcia. 2011. Molecular wine microbiology. Academic Press, London, UK.

26. Gallopin, G. C. 1971. A generalized model of a resource-population system. Oecologia 7:382–413.

27. Gallopin, G. C. 1971. A generalized model of a resource-population system II. Stability analysis. Oecologia 7:414–432.

28. Kumar, P. S., and A. B. Pandit. 1999. Modeling hydrodynamic cavitation. Chem. Eng. Technol. 22:1017–1027.

29. Pradhan, A. A., and P. R. Gogate. 2010. Removal of p-nitrophenol using hydrodynamic cavitation and Fenton chemistry at pilot scale operation. Chem. Eng. J. 156:77–82.

30. Mishra, K. P., and P. R. Gogate. 2010. Intensification of degradation of Rhodamine B using hydrodynamic cavitation in the presence of additives. Sep. Purif. Technol. 75:385–391.

31. Joshi, R. K., and P. R. Gogate. 2012. Degradation of dichlorvos using hydrodynamic cavitation based treatment strategies. Ultrason. Sonochem. 19:532–539.

32. Yusaf, T. 2013. Experimental study of microorganism disruption using shear stress. Biochem. Eng. J. 79:7–14.

33. Quarteroni, A., and A. Valli. 2008. Numerical approximation of partial differential equations. Springer, Berlin/Heidelberg.

Appendix

Bulk Model: Parameter Estimation Method

The parameter estimation method for the *bulk* model described in section "Assessment of yeast concentration from bulk properties of processed samples," along with the respective assumptions, is discussed in the following:

- a series of values for $C_S(t)$ are known from measurements taken at different times, for example, at the initial time t_0 and at later times t_1, t_2, and so forth;
- a value of $N_{SC}(t_0)$ just after the sample extraction is assumed arbitrarily, because the ratio of initial yeast concentrations for different experiments is known;
- the value of parameter k is computed from equation (3), according to the following equation:

$$k = -\frac{C_S(t_1) - C_S(t_0)}{t_1 - t_0} \cdot \frac{1}{N_{SC}(t_0) \cdot C_S(t_0)} \quad (A1)$$

- again from equation (3), any value for $N_{SC}(t > t_0)$ can be computed assuming the value for parameter k as per equation (A1) as well as measurements for $C_S(t > t_0)$, according to the following equation:

$$N_{SC}(t_i) = -\frac{C_S(t_{i+1}) - C_S(t_i)}{t_{i+1} - t_i} \cdot \frac{1}{k \cdot C_S(t_i)} \quad (A2)$$

- at the initial time $t = t_0$, it is assumed that the ratio $C_S(t_0)/N_{SC}(t_0)$ is high enough so that equation (4) simplifies to: $\frac{dN_{SC}(t)}{dt} \approx \lambda \cdot N_{SC}(t)$, hence the parameter λ can be retrieved as follows:

$$\lambda = \frac{N_{SC}(t_1) - N_{SC}(t_0)}{t_1 - t_0} \cdot \frac{1}{N_{SC}(t_0)} \quad (A3)$$

- the parameters a_{SC} and b_{SC} of the arctangent function in equation (4) can finally be estimated from the same equation since the other parameters k and λ are known from equations (A1) and (A3), respectively, $C_S(t)$ is

measured at any time and $N_{SC}(t)$ is computed by means of equation (A2); using for $C_S(t)$ and $N_{SC}(t)$ the first three available values following the sample extraction in order to minimize any deviation of the respective tendencies from the herein described model, the following simple equation system can be written:

$$\begin{cases} a_{SC} + b_{SC} \frac{C_S(t_0)}{N_{SC}(t_0)} = \tan\left[\frac{\pi}{2} \cdot \frac{1}{\lambda \cdot N_{SC}(t_0)} \cdot \frac{N_{SC}(t_1)-N_{SC}(t_0)}{t_1-t_0}\right] \\ a_{SC} + b_{SC} \frac{C_S(t_1)}{N_{SC}(t_1)} = \tan\left[\frac{\pi}{2} \cdot \frac{1}{\lambda \cdot N_{SC}(t_1)} \cdot \frac{N_{SC}(t_2)-N_{SC}(t_1)}{t_2-t_1}\right] \end{cases} \quad (A4)$$

The estimation of parameters k, λ, a_{SC}, and b_{SC} was therefore fairly objective, although based upon an only liquid sample from the THERMAL test described in Table 1.

Bulk Model: Solution Method

The main objective of the model described in section "Assessment of yeast concentration from bulk properties of processed samples" is the assessment of the initial value of the yeast concentration $N_{SC}(t_0)$ in any sample extracted during a specific process, as a fraction of the yeast concentration before that same process. When dealing with any experiment different from the *blank* one, the initial value of the yeast concentration before a process is expressed as a fraction (lower or greater than unity, depending on the volume of the processed liquid substance) of the arbitrary value attributed to the first liquid sample extracted at the beginning of the *blank* experiment, multiplied by the latter value, while the sugar concentration $C_S(t)$ is measured at any time.

However, since the value of $N_{SC}(t_0)$ is unknown, only the first value of the sugar concentration $C_S(t_0)$ is assumed to be known, then different first values of the yeast concentration $N_{SC}(t_0)$ are used to produce a bundle of curves for $C_S(t)$ by solving the coupled equations (3) and (4) with a simple finite differences "forward in time" scheme [33], which, for the first two time steps, reads as follows:

$$\begin{cases} C_S(t_1) = C_S(t_0) - k \cdot N_{SC}(t_0) \cdot C_S(t_0) \cdot (t_1 - t_0) \\ N_{SC}(t_1) = N_{SC}(t_0) + \lambda \cdot N_{SC}(t_0) \cdot \frac{\arctan\left[a_{SC}+b_{SC}\frac{C_S(t_0)}{N_{SC}(t_0)}\right]}{\pi/2} \cdot (t_1 - t_0) \\ C_S(t_2) = C_S(t_1) - k \cdot N_{SC}(t_1) \cdot C_S(t_1) \cdot (t_2 - t_1) \\ N_{SC}(t_2) = N_{SC}(t_1) + \lambda \cdot N_{SC}(t_1) \cdot \frac{\arctan\left[a_{SC}+b_{SC}\frac{C_S(t_1)}{N_{SC}(t_1)}\right]}{\pi/2} \cdot (t_2 - t_1) \end{cases}$$

$$(A5)$$

Finally, the series of measured values for sugar concentration are compared with the respective computed values for the same quantity based on the system of equation (A5), in order to identify the first value of the yeast concentration $N_{SC}(t_0)$ originating the computed series, which shows the best matching with the observed tendency during the earlier 72 h after sampling (such time period being chosen as a compromise between the need to avoid significant effects from yeast cells death rate as well as contamination by different microorganisms, on one hand, and allow for recovery of SC cells temporarily inactivated during the experimental processes). The ratio of such $N_{SC}(t_0)$ to the initial value of the yeast concentration before a specific process finally gives the measure of the lethality induced by that process on the *Saccharomyces cerevisiae* (SC).

SC Inactivation Model: Parameter Estimation Method

The parameters A_T and T_c of the SC inactivation rate model described in section "SC inactivation rate from combined thermal and cavitation treatments" were estimated from all the five samples extracted in the course of the THERMAL test too, according to the best qualitative matching of the lethality curves reconstructed from equation (5) with the respective values computed as explained above in this Appendix.

Finally, the parameters B_c and σ_{max} in equation (6) were estimated the same way as A_T and T_c and assuming the values of the latter as known, except that the reference experiments were the five ones involving a cavitation reactor as described in Table 1 and equation (7) was used.

Preparation of the Yeast

The SC yeast was prepared before each experiment by supplying nutrients along with water heated to about 40°C.

Nutrients are composed of a mix of APA, vitamins and minerals; the APA, composed of ammoniacal nitrogen, amino acids and polypeptides, is instrumental for cell multiplication and for the biosynthesis of proteins. The vitamins (thiamin, biotin, and pantothenic acid) regulate and limit the production of sulfur compounds and fatty acids, as well as are instrumental in the biosynthesis of amino acids and proteins and are important in cell proliferation and resistance to stress. The mineral salts (manganese, magnesium, zinc, copper, potassium, calcium) regulate cell growth, the formation of alcohols and esters, tolerance to ethanol and the temperature also act as enzyme cofactors.

Measurement Methods

The temperature of the circulated liquid was measured every 10 sec and recorded as minute averages by means of a set of five thermocouples in contact with the liquid itself, from which the average value is directly computed. Pressure gauges (manometers) both upstream and downstream of the cavitation reactor, if any, are used for visual inspection, the first mainly for safety reasons, the second in order to check that the pressure regulated by means of the expansion tank corresponds to the downstream pressure, sometimes referred to as "recovery pressure" too, meaning the pressure "recovered" by the liquid flow shortly after passing through the reactor nozzle, which in turn modulates the collapse of the cavitation bubbles. Thermocouples and manometers are placed as shown in Figure 1.

The other relevant physical measurement concerns the water flow: it is assessed from the characteristic curves of the main centrifugal pump supplied by the manufactured, which relate the liquid flow to its power absorption, the

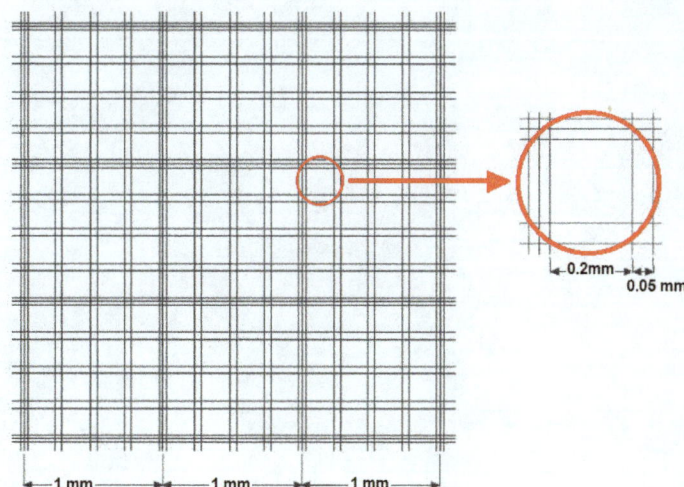

Figure A1. Buerker counting chamber

latter in turn measured by means of a commercial digital Watt-meter. Once the water flow is known, the liquid speed through the cavitation reactor nozzle is computed simply dividing the water flow itself by the total nozzle opening surface.

The sugar concentration in water solution was measured by means of a simple "Guyot" analogical gauge combining liquid density and temperature, with an estimated maximum error around 5 g/L.

The microbiological measurements were performed after diluting a small sample of processed liquid with methylene blue, which is a basic dye because the methylene chloride salt dissociates in water in the positively charged blue methylene ion and the negatively charged colorless chloride ion. Because of their chemical nature, cytoplasms of any bacterial cells have a weak negative charge therefore the microorganism shall get directly colored. Of course, such coloring process can happen before the cell gets completely destroyed, that is, only early after its death, therefore counting the absolute number of alive cells is needed for a proper balance as much as identifying the dead and alive.

Figure A1 shows the Buerker counting chamber used to count the alive (white) and dead (blue colored) cells by means of a "biological" 600× microscope equipped with a digital camera.

Estimated Values for All the Models' Parameters

All the parameters were quantitatively estimated assuming an arbitrary value equal to 50 for the initial, preprocessing yeast concentration for the THERMAL test and scal-

Table A1. Estimated values of models' parameters and the respective relevant equations.

Parameter	Estimated value[1]	Relevant equation
k	0.000073	Equation (3)
λ	0.0121	Equation (4)
a_{SC}	−10.9	Equation (4)
b_{SC}	5.9	Equation (4)
A_T	1.6	Equation (5)
T_c	62°C	Equation (5)
B_c	50	Equation (6)
$\sigma_{1\text{-}max}$	0.3	Equation (6)
$\sigma_{2\text{-}max}$	1.7	Equation (6)

[1]Units according to the International System, unless differently indicated, and consistent with the respective equations.

ing the yeast initial concentration values for the other experiments according to the ratio of their respective volumes to the THERMAL test one, since the same absolute quantity of yeast, equal to 50 g, was used for each experiment; therefore, on the basis of the volume data in Table 1, the initial yeast concentration was equal to 50 for all experiments except for the "VENTURI test #3_flash" with about 64 g.

On the basis of the above written, remembering that for different reasons all the values of the models' parameters should be considered as *first guesses* until further experiments will allow to perform more robust assessment of their central values and uncertainties by means of objective estimation methods, the preliminary estimated values of the parameters are shown in Table A1.

16

Fuel-mix and energy utilization analysis of Port Harcourt Refining Company, Nigeria

Ismaila Badmus[1,*], Richard Olayiwola Fagbenle[1,2] & Olanrewaju Miracle Oyewola[1]

[1]Mechanical Engineering Department, University of Ibadan, Nigeria
[2]Mechanical Engineering Department, Obafemi Awolowo University, Ile-Ife, Nigeria

Keywords

Capacity utilization, fuel mix, Port Harcourt, refinery

Abstract

This study analyses the fuel-mix and energy utilization patterns in Port Harcourt Refining Company from 2000 to 2011. The average fuel mix over the study period is 43% refinery fuel gas, 0% liquefied petroleum gas (LPG), 44% low pour fuel oil (LPFO), 8% Coke, and 5% automotive gas oil (AGO). The present ratio of high-carbon fuel consumption to low-carbon fuel consumption adversely influences the specific fuel consumption by increasing it. Our proposal is that the present AGO and LPFO consumption levels are totally replaced with equivalent amounts of natural gas. This would yield the following fuel mix: 46% refinery fuel gas, 46% natural gas, and 8% coke. This, in effect, would result in a proportion of 92% low-carbon fuels and 8% coke. It would also lead to a specific fuel consumption that is averagely unaffected by high-carbon/low-carbon fuel consumption ratios. Natural gas utilization has its main advantage in its flare/waste and consequent environmental degradation reduction. Finally, the proposed fuel mix would generally lead to a reduction in specific fuel consumption, thus saving energy and reducing costs.

Introduction

The Nigerian economy depends primarily on the oil sector. Moreover, the country holds the largest natural gas reserves in Africa. Natural gas associated with oil production is mostly flared. However, regional pipeline development, liquefied natural gas (LNG) infrastructural expansion, as well as policies to ban gas flaring are expected to accelerate growth in the sector, both for export and domestic electricity generation [1]. Nigeria has four domestic refineries, with a total capacity of 445,000 barrels per day [2]. In 2001, the country had 150,000 km road network, 5900 km of pipelines for product receipts/transportation, 21 government-owned fuel depots, and numerous marketers' terminals/depots. All these were in addition to a storage capacity in excess of 7 million barrels, 6 ocean/tanker terminals, 2000 fuels tank trucks, 5100 service stations, and 50 inland terminals/depots [3]. For about a decade there was very little change, except the fact that the country reduced its pipeline network to 5000 km [4].

An assessment of the oil and petrochemical industry in Nigeria would yield a myriad of challenges. Despite the oil fortune, Nigeria is a net importer of refined petroleum products. At the top of the petroleum industry is the Federal Government–owned Nigerian National Petroleum Corporation (NNPC) that operates joint venture agreements with some foreign multinational oil companies in Nigeria to produce the nation's oil and gas. In contrast, the Venezuelan Petroleum Corporation, which was established about the same time as NNPC as a public corporation, has grown to be the third largest international conglomerate and also a net exporter of refined petroleum products [5]. As observed by Iwayemi [6], despite the fact that Nigeria has domestic refineries owned by the government, with capacity to process nearly 450,000 barrels of oil per day, the country still imports more than 75% of petroleum products requirements for local consumption.

This situation has led to very unstable petroleum product price regimes in the country, forcing poor Nigerians to look for cheaper fuel alternatives to satisfy their domestic energy needs, with the attendant negative environmental effects [7, 8]. Aigbedion and Iyayi [9] observe that: "The sub-sector has also been constrained by the unenviable state of the nation's refineries which have been producing at minimal capacities in the past few years, despite huge expenses incurred on the Turn Around Maintenance of the crisis – ridden refineries. This development has led to massive importation of petroleum products to fill demand gaps that exist in domestic consumption." Importation of refined petroleum products obviously increases their pump prices. For instance, the Petroleum Products Pricing Regulation Agency (PPPRA) pricing templates are a function of the exchange rate of Nigerian Naira to the US Dollar, landing costs, and distribution margins. The landing cost, which accounts for most of the cost, is the cost of imported products delivered into the jetty depots [10].

Virtually every Nigeria life is affected by availability or otherwise of petroleum products, primarily through transportation, but also through household energy use and industrial use, which includes operation of internal combustion engines to generate electricity due to the electrical power crisis in the country. However, before these products can be obtained from crude oil it has to pass through the refining process where it is 'cracked' into different useful components such as premium motor spirit (PMS) or gasoline, automotive gas oil (AGO) or diesel oil, dual purpose kerosene (DPK), and others. As there is no other way of obtaining these components from crude oil except through this process, the energy consumed during the process is a critical determinant of the final product price in the energy market.

There is a widespread acceptance that there is a general decline in the level of technical efficiency at which Nigerian refineries operate. Naturally, inadequate maintenance and technological obsolescence would lead to a general decline in technical efficiency of a system. Perhaps this is why even social scientists such as Bamisaye and Obiyan [11] have also advocated urgent repair of Nigeria's oil refineries.

The Port Harcourt Refining Company (PHRC) comprises an old plant which was commissioned in 1965 and a newer one which was commissioned 1989. The old plant has an installed capacity of 60,000 barrels per stream day (bpsd), whereas the new plant has a capacity of 150,000 bpsd, giving a combined capacity of 210,000 bpsd and 47% of all NNPC refining capacity.

In particular, there have been several studies on the PHRC. Isah et al. [12] discussed the importance fluidized catalytic cracking (FCC) unit in improving gasoline production rate in the PHRC. Akpabio and Ekott [13] discussed the need for incorporation of delayed coking process units into the Nigerian refineries, including

PHRC. Tonnang and Olatunbosun [14] applied a neural network controller in the analysis of a crude distillation unit (CDU), using PHRC as a case study. Akpa and Okoroma [15] carried out a pinch analysis of the heat exchanger network in the CDU of PHRC. Uzukwu and Iyagba [16] also carried out a process analysis of the new Port Harcourt refinery crude charge heater.

Jesuleye et al. [17] analyzed the energy demand of the Port Harcourt refinery, Nigeria, based on information obtained from its annual publications, backed-up by spot interviews. The analytical approach adopted for the study involves the calculation of energy intensities to determine the refinery's annual energy demand for various energy types from 1989 to 2004. The results showed that the actual energy demand per year for processing crude oil into refined products exceeded, in varying degrees, the stipulated refinery standard of 4 barrels of oil equivalent (BOE) per 100 BOE. It varied from 4.28 to 8.58 BOE per 100 BOE. In terms of energy demand efficiency, this implies very poor performance of the refinery during the 16-year period under investigation. Lack of optimal fuel utilization mix and noncompliance with the turnaround maintenance schedules were attributed to the refinery's inefficient energy demand pattern.

As observed by Badmus et al. [18] while appraising Jesuleye et al. [17], despite apparent preponderance of literature on general assessments of Nigerian refineries in the open literature, only this recent one by Jesuleye et al. [17] appears to be an in-depth energy analysis of only one of the refineries. However, although Jesuleye et al. [17] observed that "the refinery energy demand mix (by type of fuel use) did not follow the most technically efficient path", they have not stated this "most technically efficient path" explicitly. This is why the work embarked upon in this study is important.

Methodology

Data collection

The study employs a quantitative method of analysis through intensive data gathering, collation, and processing. In particular, data collection has been done for a period spanning 2000 to 2011. Two separate field visits were paid to the PHRC, Port Harcourt, for the purpose of primary data gathering and secondary data authentication. The required secondary data were obtained from annual reports of the government-owned refineries and various publications of relevant institutions such as NNPC, US Energy Information Administration and Organisation of Petroleum Exporting Countries (OPEC), Austria. In particular, NNPC Annual Statistical Bulletin editions for the respective years and the PHRC library have been consulted extensively for secondary data. The aggregated data on overall annual fuel types and utilization, total quantity of crude oil processed, and refinery capacity utilization in the refinery have been obtained for a period of 12 years. Interviews with some of the PHRC technical staff were also used to elicit information concerning data authentication. An energy analysis (based on first law of thermodynamics) of the data collected has been done and the results are presented. In addition to this, the mass conservation law has also been applied in conjunction with this energy conservation law.

Theoretical background

In applying mass conservation law, the total mass of crude oil processed must be equal to the total mass of various products.

$$\sum \dot{m}_{in} = \sum \dot{m}_{out} \tag{1}$$

In practice, as refinery fuels are derived from the crude oil itself,

Mass of crude oil processed

= Total mass of various finished products

+ Total mass of refinery fuels $\qquad$ (2)

+ Total mass of losses

From the first law of thermodynamics,

$$\sum \dot{E}_{in} + \dot{Q}_{cv} = \sum \dot{E}_{out} + \dot{W}_{cv} \tag{3}$$

As a combustion process involves no work interaction, the relevant steady flow energy equation is as follows:

$$Q = H_{products} - H_{reactants} = \Delta H. \tag{4}$$

ΔH is called enthalpy of combustion.

Heating value

In practice, the parameter used to determine the energy content of a fuel is its heating value, defined as the amount of heat released when a fuel is burned completely in a steady flow process and the products are returned to the thermodynamic state of the reactants. The heating value is thus equal to the absolute value of the enthalpy of combustion of the fuel:

$$\text{Heating value (HV)} = |\Delta H| \tag{5}$$

The heating value depends on the phase of the H_2O that must necessarily accompany the products of combustion of a hydrocarbon fuel, as the hydrogen component is oxidized to H_2O.

Higher heating value

The heating value is called the higher heating value (HHV) when the H_2O in the products is in the liquid form.

Lower heating value

This is the heating value when the H_2O in the combustion products is in the vapor form. Hence, HHV and lower heating value (LHV) are related thus:

$$\text{HHV} = \text{LHV} + (mh_{fg})_{H_2O} \text{kJ/kg} \tag{6}$$

$m \equiv$ mass of H_2O products per unit mass of fuel; $h_{fg} \equiv$ specific enthalpy of vaporization of H_2O at the specified temperature

In applying the first law of thermodynamics in this study, LHV of fuels have been used to determine the fuels' energy content. This is because this is the heating value used in practice in situations where the flue gases cannot be safely cooled below temperature values that enable utilization of enthalpy of steam condensation or cooling below dew points. Situations like these arise where it is feared that cooling below dew points will cause corrosive acid formation that can attack the system. Tables 1 and 2 show the LHVs used in this study.

Mean LHV

When a fuel is a mixture of different components with distinct LHVs, the mean LHV can be derived.

Total energy consumed based on LHV of the fuels is given by:

Table 1. Lower heating values of nigerian refinery finished fuel products [19, 20].

S/N	Fuel	LHV (MJ/kg)
1	AGO*	42.7
2	DPK	43.1
3	FO (fuel oil)*	40.19
4	LPG (liquefied petroleum gas)*	45.3
5	PMS	44.0

*These are also used as refinery fuels.
AGO, automotive gas oil; DPK, dual purpose kerosene; PMS, premium motor spirit.

Table 2. Lower heating values of other refinery fuels [19, 21].

S/N	Fuel	LHV (MJ/kg)
1	Coke (Petroleum coke)	31.00
2	Natural gas	44.95

LHV, lower heating value.

$$\sum E_i = \sum m_i \mathrm{LHV}_i \qquad (7)$$

Mean LHV is given by:

$$\overline{\mathrm{LHV}} = \frac{\sum m_i \mathrm{LHV}_i}{\sum m_i}. \qquad (8)$$

Equivalent fuel mass

It may be necessary to obtain the equivalent mass of a particular fuel that will provide the same quantity of thermal energy as the fuel mixture, as in the case of fuel substitution. For instance, in this study, a case is made for natural gas as a fuel substitute. In this case, mass of natural gas, m_{NG}, for the same energy value as in equation (5) is given by:

$$m_{NG} = \frac{\sum m_i \mathrm{LHV}_i}{\mathrm{LHV}_{NG}}. \qquad (9)$$

Capacity utilization

This is one other parameter that is used to assess the performance of systems like the refineries. For the refineries,

Capacity utilization
= (crude oil quantity actually processed
    annually/annual quantity designed to be processed)
× 100%

$$\qquad (10)$$

Specific fuel consumption

The specific fuel consumption as used in this work is ratio of fuel utilized per unit mass of crude oil processed, expressed as a percentage.

Results

Using equation (8), the mean LHV for PHRC fuel gas is 51.91 MJ/kg. It should be noted that the fuel gas is free from hydrogen sulfide, which when combusted, is environmentally deleterious.

Discussion and Conclusion

Discussion of results

Table 3 gives the details of annualized fuel consumption in PHRC over the study period.

Capacity utilization and fuel consumption

As seen in Table 3, there has been gross capacity underutilization in the PHRC within the period under consideration. The capacity utilization has ranged from 8.82% in 2009 to 60.73% in 2001. The specific fuel consumption as shown in Figure 1A is almost inversely proportional to the capacity utilization at low capacities and practically constant at high capacities. It ranges from 4.47% in 2006 to 10.57% in 2010. The specific fuel consumptions corresponding to the lowest and highest values of capacity utilization of 8.82% and 60.73% are 9.34% and 5.53%, respectively. This implies that the highest specific fuel consumptions are the values for the years 2009 and 2010. However, it should be observed that the ratios of high-carbon fuel to low-carbon fuel in 2009 and 2010 are 2.53 and 4.28, respectively (Table 3). Hence, the adverse effects of high high-carbon fuel to low-carbon fuel ratios has aggravated the effect of low capacity utilization to make the specific fuel consumption for 9.17% capacity utilization higher than that for 8.82% capacity utilization. At

Table 3. Annualized PHRC fuel mixes (metric tons).

Year	AGO	Coke	Refinery fuel gas	LPFO	Total fuel consumed	Total quantity of crude processed	Specific fuel consumption (%)	Refinery capacity utilization (%)
2000	26,056	4934	98,183	69,512	198,685	3,214,333	6.18	30.95
2001	9736	53,229	175,839	108,818	347,622	6,290,480	5.53	60.73
2002	49,619	82,710	178,895	75,357	386,581	5,403,324	7.15	52.17
2003	20,213	27,395	103,007	94,300	244,915	3,980,603	6.15	41.88
2004	0	0	111,387	113,253	224,640	3,223,894	6.97	31.04
2005	0	17,790	141,864	130,194	289,848	4,368,783	6.63	42.18
2006	0	0	89,901	142,736	232,637	5,206,407	4.47	50.26
2007	0	0	66,532	84,407	150,939	2,590,779	5.83	24.87
2008	5208	0	38,797	64,787	108,792	1,886,074	5.77	17.84
2009	0	4435	24,124	56,480	85,039	910,751	9.34	8.82
2010	9320	0	19,000	72,063	100,383	950,134	10.57	9.17
2011	11,750	0	38,890	82,952	133,592	1,580,767	8.45	15.26

AGO, automotive gas oil; LPFO, low pour fuel oil.
Sources: NNPC [22–33].

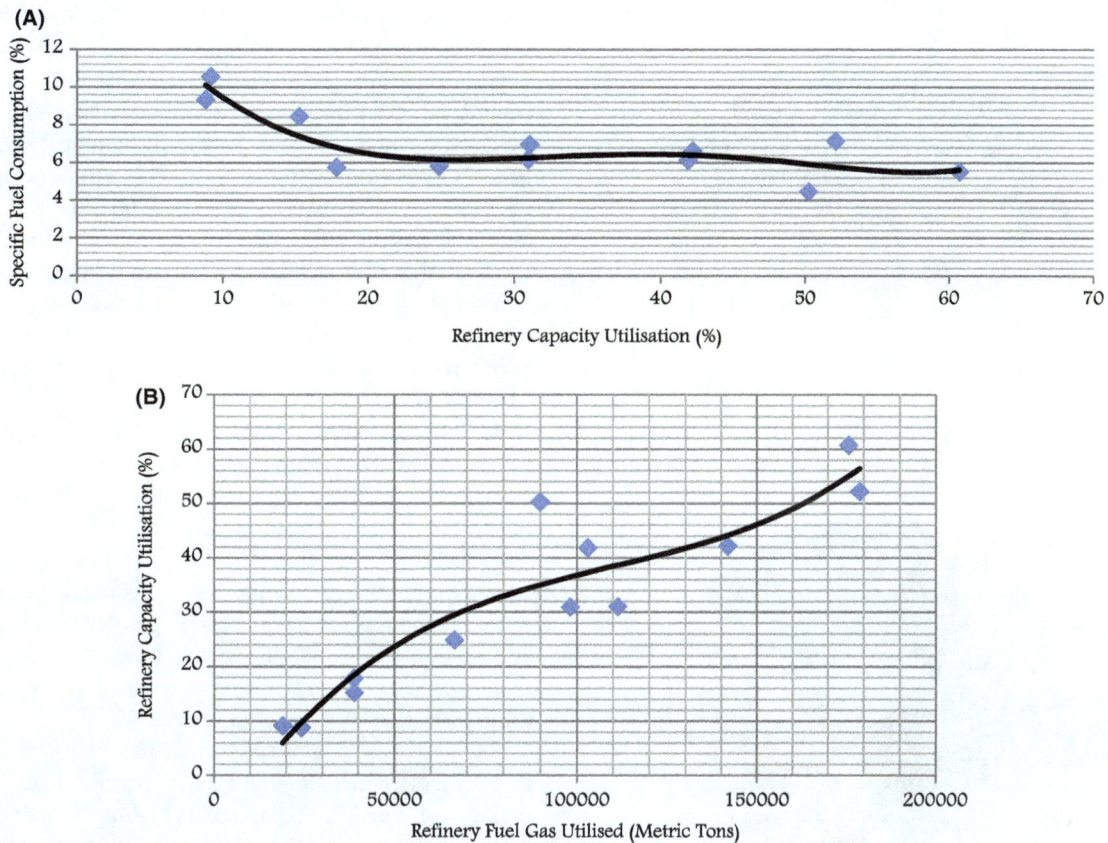

Figure 1. (A) Relationship between specific fuel consumption and refinery capacity utilization. (B) Variation in refinery capacity utilization with refinery fuel gas utilized.

high capacities, however, the specific fuel consumptions are nearly constant (Fig. 1A), but above the design value of 4% as observed by Jesuleye et al. [17]. Here also, the effects of relative proportions of high-carbon and low-carbon fuels on the specific fuel consumption values are not apparent.

Although the "degree of complexity" as a parameter of refinery classification as used by Ocic [34] is not clear, only the years 2009 (9.34%), 2010 (10.57%), and 2011 (8.45%) fall outside the benchmark range (4–8%) for refinery specific fuel consumption specified by him. However, according to Oniwon [35], the international benchmark for capacity utilization is 90%, whereas the international benchmark for fuel consumption as far back as 2008 was 6%. Around the same period of this analysis (1998–2008), Brazilian refineries had an average specific fuel consumption of 6% and a capacity utilization of at least 82% while U.S. refineries had a minimum capacity utilization of 79% around the same period [36].

Besides, energy is totally internally sourced in Nigerian refineries. There is no externally purchased steam or electricity. In the particular case of PHRC, the fuel consumption is thus utilized to:

(a) electricity for general use through steam turbine generators (STGs) and gas turbine generators (GTGs)
(b) raise steam for general plant use
(c) fire heaters for process heat generation

Of the three uses enumerated above, only (b) and (c) may stop entirely when the plants are shut down. For as long as the refinery still opens for business, whether the plants are running or not, electricity would still be utilized. However, process steam and fuel consumption may go down depending on the level of activities in the plant. As processed/refined crude is the only commodity for which the refinery exists, it is customary to "bill" the refined and/or processed crude for all the energy consumed. This explains why the specific energy consumption goes up when the capacity is underutilized.

Fuel-mix trend and its effect on overall fuel energy consumption trend

Apparently, the main fuel utilized in process units fired heaters is largely the refinery fuel gas. This is because the refinery fuel gas consumed is directly proportional to the

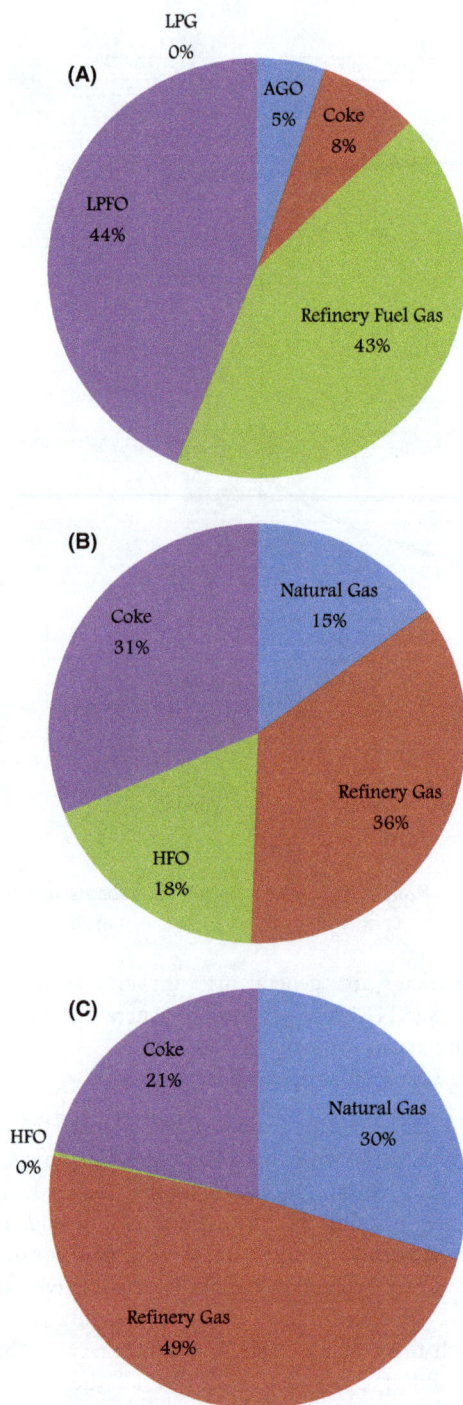

Figure 2. (A) Average PHRC fuel mix from the Year 2000 to the Year 2011. (B) Brazilian refineries fuel mix as at 2008 (Source: Solomon [37] quoted in de Lima and Schaeffer [36]). (C) U.S. refineries fuel mix as at 2008 (Source: Solomon [37] quoted in de Lima and Schaeffer [36]).

capacity utilization, which is a function of crude oil quantity processed (Fig. 1B). On average, throughout the 12-year period under consideration (Fig. 2A), the most

consumed fuel is fuel oil (44%), closely followed by refinery fuel gas (43%). Although liquefied petroleum gas (LPG) has not been used at all, the least utilized fuel has been AGO (5%). An 8% coke utilization suggests that the FCC unit where the fuel is usually combusted has not been in operation for some years. This can be seen vividly in Table 3 where years 2004, 2006–2008, as well as 2010 and 2011 record zero coke consumption values. Hence, leaving aside the peculiar case of coke, high-carbon fuel consumption in the refinery has still been higher (49%) than that of low-carbon fuel (43%), especially when compared with what is obtainable in the same industry in other countries.

In 2008, Brazilian refineries (Fig. 2B) consumed coke: 31%, natural gas: 15%, refinery fuel gas: 36%, and high pour fuel oil: 18%. This means that, aside from coke, high-carbon fuels were only 18% while low-carbon fuels were 51%. Besides, the FCC unit has been quite operational, with 31% petroleum coke consumed. In 2008, US refineries consumed 30% natural gas, 49% refinery gas, and 21% coke (Fig. 2C). This implies, 79% low-carbon fuel consumption and no high-carbon fuel consumption aside from that of petroleum coke.

However, one positive aspect of the PHRC fuel mix is that it uses refinery fuel gas as its process fuel, with the major combustible components being ethane (35%), methane (29%), and hydrogen (12%) as detailed in Table 4. The fuel gas is also free from hydrogen sulfide (H_2S), which is environmentally unfriendly. All these have culminated in a good average LHV of 51.91 MJ/kg.

Fuel substitution option

As a result of the observation in the section Fuel-mix Trend and its Effect on Overall Fuel Energy Consumption Trend above, substituting low-carbon fuels for high-carbon fuels looks promising. The options are LPG and/or natu-

Table 4. PHRC fuel gas properties [16, 38–40].

Fuel component	LHV (MJ/kg)	Gravimetric contribution (kg)
H_2	120.21	1.267
CH_4	50.03	2.9856
C_2H_6	47.51	3.675
C_3H_6	45.661	0.7812
C_3H_8	46.33	0.0704
C_4H_{10}	45.72	0.3422
C_4H_8	45.173	0.028
C_5H_{12}	44.945	0.108
C_6H_{14}	44.1401	0.0774
N_2	–	1.0752

LVH, lower heating value.

ral gas. LPG is a product of the crude oil refining process and thus only available in small quantities. Furthermore, its carbon content is more than that of natural gas, although they have comparable LHVs. On the other hand, natural gas is a direct by-product of crude oil exploration and the quantity that is flared annually (after utilizing a part) is enormous, when compared with the quantity required for refinery fuel.

It is observed in Table 5 that during the years when FCC unit was operational and coke was utilized, change in mean LHV (between the proposed one and the present one) varied from 1.223 MJ/kg in 2002 to 3.256 MJ/kg in 2009. In contrast, whenever coke was not utilized, the change in LHV varied from 2.594 MJ/kg in 2004 to 3.742 MJ/kg in 2010. This implies that the degree of FCC capacity utilization also affects the specific fuel consumption. Hence, there are larger mean LHV gains with natural gas fuel substitution when FCC is not operational and consequently more significant reductions in the specific fuel consumptions. This is not surprising as coke, which is the FCC fuel, is a high-carbon fuel. However, as the LHV still responds favorably to substitution of high-carbon fuels

with natural gas, even when the FCC is in operation, it follows that the fuel substitution is still a way out of the present high specific fuel consumption in the refinery.

Table 5. Impact of coke consumption on the refinery fuel mean lower heating values.

Year	Coke consumption (tons)	Present mean LHV (MJ/kg)	Proposed mean LHV (MJ/kg)
2000	4934	46.083	48.184
2001	53,229	44.781	46.384
2002	82,710	43.970	45.193
2003	27,395	44.298	46.381
2004	0	46.001	48.596
2005	17,790	45.362	47.628
2006	0	44.719	47.827
2007	0	45.356	48.211
2008	0	44.490	47.606
2009	4435	43.035	46.291
2010	0	42.641	46.383
2011	0	43.823	47.129

LVH, lower heating value.

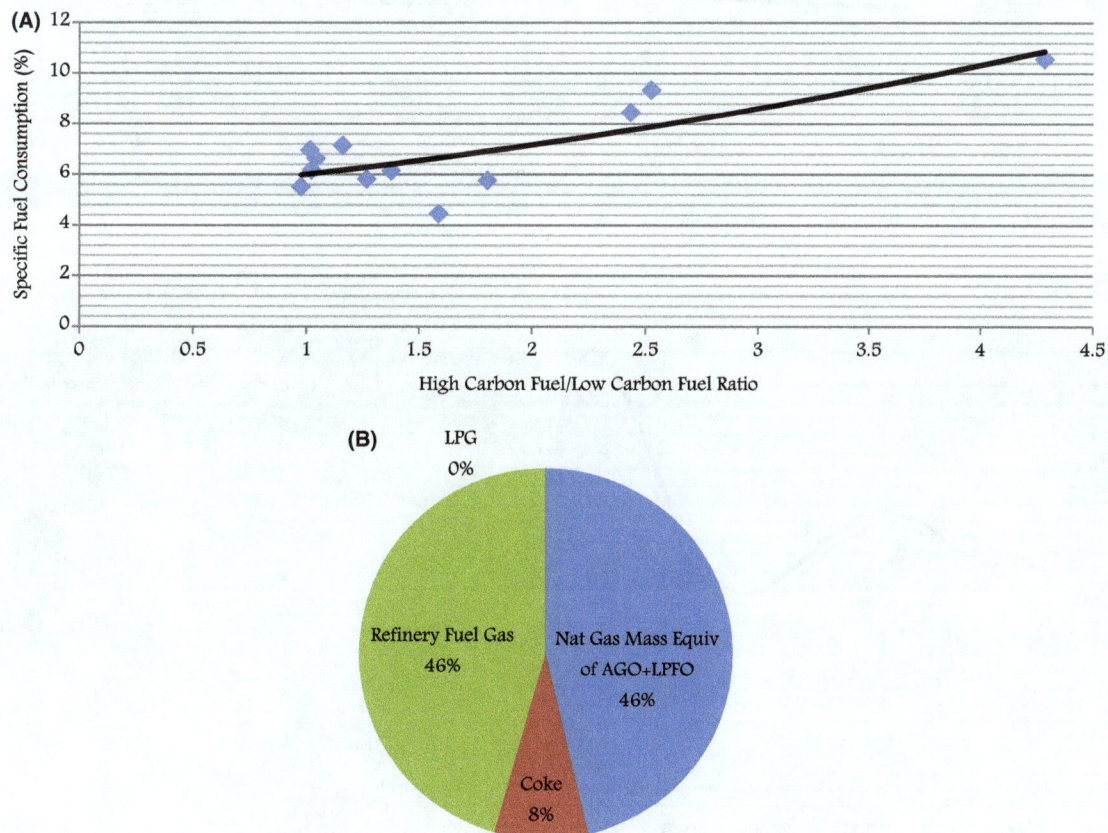

Figure 3. (A) Effect of proportions of high-carbon fuels to low-carbon fuels on specific fuel consumption over the Years. (B) New average fuel mix when AGO and low pour fuel oil (LPFO) are replaced with natural gas.

It should also be noted that natural gas as fuel has an additional advantage of lowest carbon content (being mainly CH_4), giving rise to very low CO_2 emissions. Presently, natural gas is not in use as a refinery fuel in PHRC.

In principle, when natural gas is used as refinery fuel, all other refinery fuels above (apart from coke which is utilized in the FCC unit) can be added to the product streams. This replacement of fuel oil with natural gas has been done successfully in the Iranian refining industry [41]. Besides, the volume of flared natural gas is correspondingly reduced and, being a low-carbon fuel, atmospheric carbon emission is also reduced. Thus, there is

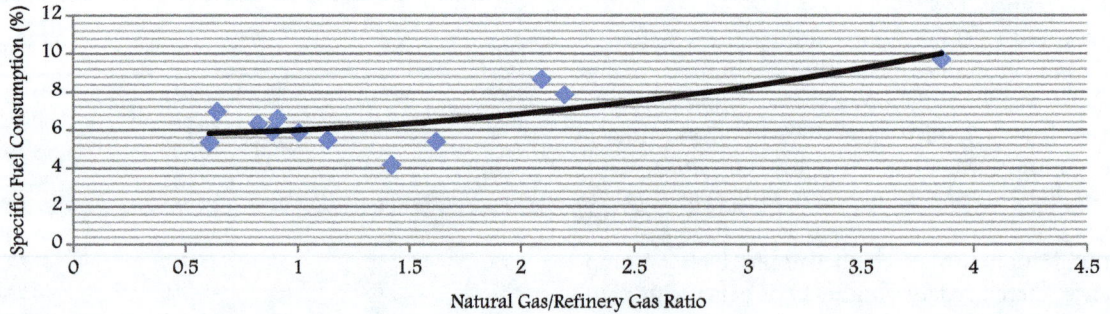

Figure 4. New effect of high-carbon fuel/low-carbon fuel ratio on specific fuel consumption.

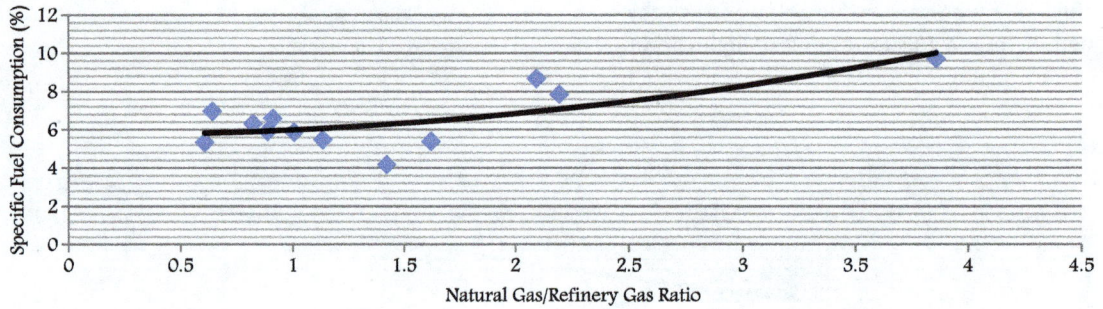

Figure 5. Effect of natural gas/refinery gas ratio on specific fuel consumption.

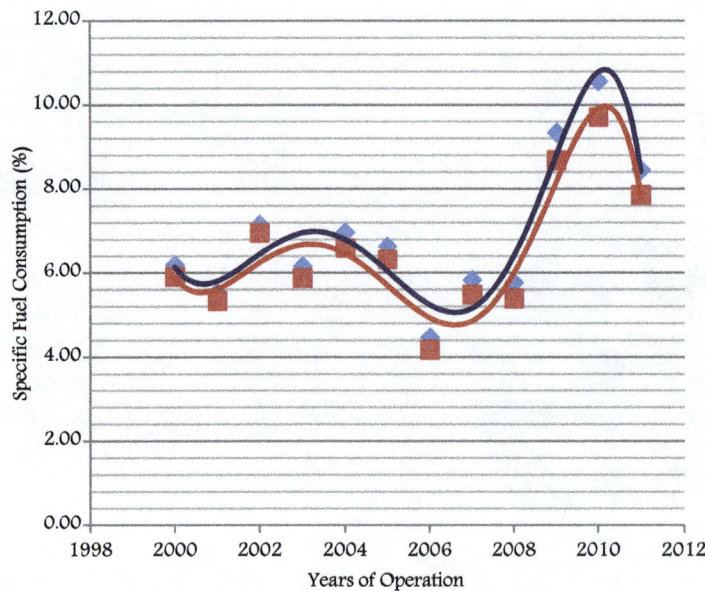

- ■ Proposed Specific Fuel Consumption (%)   ◆ Present Specific Fuel Consumption (%)

Figure 6. Comparison between the present and proposed specific fuel consumptions.

more fuel availability for both domestic and industrial use; there are both financial gains and environmental protection as a result of gas flare reduction.

One additional point worthy of consideration is the fact that the present ratio of high-carbon fuel consumption to low-carbon fuel consumption influences the specific fuel consumption negatively. The specific fuel consumption is directly proportional to the ratio (Fig. 3A). Our proposal is that the present AGO and low pour fuel oil consumption levels are totally replaced with natural gas equivalently. This would yield the following fuel mix: 46% refinery fuel gas, 46% natural gas, and 8% coke. This, in effect, means low-carbon fuels of 92% and coke of 8%. It would also lead to a specific fuel consumption that is averagely unaffected by high-carbon/low-carbon fuel consumption ratios (Fig. 4). However, due to the relatively high LHV nature of PHRC refinery fuel gas (51.91 MJ/kg), compared with that of natural gas (44.95 MJ/kg), the specific fuel consumption is still going to be mildly sensitive to the natural gas/refinery fuel gas ratio, signifying that refinery fuel gas may be more preferred in this case (Fig. 5). Nevertheless, natural gas utilization still has its main advantage in its flare/waste, and consequent environmental degradation reduction.

Finally, Figure 6 indicates that the proposed fuel mix would generally lead to lower specific fuel consumption values, thus saving energy and reducing costs.

Conclusion

A fuel-mix analysis of energy utilization in PHRC has been carried out for the period 2000–2011. The analysis has unraveled capacity underutilization and fuel mixes that can still be improved upon for environmental sustainability and higher utilization efficiency as the major challenges. Substituting high-carbon fuels with low-carbon fuels is the main proposal of this study. Capacity utilization also has to be improved upon in order to curb the present energy waste in the refinery.

Conflict of Interest

None declared.

References

1. EIA. 2009. Nigeria. Country Analysis Briefs. Available at www.eia.doe.gov (accessed 28 April 2012).
2. Energy Commission of Nigeria, ECN. 2003. National Energy Policy. The Presidency, Federal Republic of Nigeria, Abuja.
3. Obih, H. 2001. Fuel distribution and logistics. Sub-Regional Conference on the Phase-out of Leaded Gasoline in Nigeria & Neighbouring Countries. November 15–16, Abuja, Nigeria.
4. Nigerian National Petroleum Corporation, NNPC. 2010. Operators in the Industry. Available at http://www. nnpcgroup.com/operations (accessed 18 May 2010).
5. Okafor, E. E. 2007. Rethinking African development: a critical overview of recent developments in the petroleum sub-sector in Nigeria. J. Soc. Sci. Kamla–Raj. 15: 83–93.
6. Iwayemi, A. 2008. Nigeria's dual energy problems: policy issues and challenges. Pp. 17–21 in Proc. 31st International Conference of the International Association for Energy Economics. Istanbul, Turkey. 18–20 June 2008.
7. Maconachie, R., A. S. Tanko, and M. Zakariya. 2009. Descending the energy ladder? Oil price shocks and domestic fuel choices in Kano, Nigeria. Land Use Policy. 26:1090–1099.
8. Adelekan, I. O., and A. T. Jerome. 2006. Dynamics of household energy consumption in a traditional African City, Ibadan. Environmentalist 26:99–110.
9. Aigbedion, I., and S. E. Iyayi. 2007. Diversifying Nigeria's petroleum industry. Int. J. Physic. Sci. 2:263–270.
10. Petroleum Products Pricing Regulation Agency, PPPRA. 2010. Available at http://www.pppra-nigeria.org/pricingtemplate. asp (accessed 10 June 2010).
11. Bamisaye, O. A., and A. S. Obiyan. 2006. Policy analysis of oil sector in Nigeria. Eur. J. Soc. Sci. 3:37–48.
12. Isah, A. G., M. Alhassan, and M. U. Garba. 2006. Feed quality and its effect on the performance of the fluid catalytic cracking unit (a case study of Nigerian based oil company). Leonardo Electron. J. Pract. Technol. 9:113–120.
13. Akpabio, E. J., and E. J. Ekott. 2012. Integrating delayed coking process into Nigeria's refinery configuration. Indian J. Sci. Technol. 5:2923–2927.
14. Tonnang, H. E. Z., and A. Olatunbosun. 2010. Neural network controller for a crude oil distillation column. ARPN J. Eng. Appl. Sci. 5:74–82.
15. Akpa, J. G., and J. U. Okoroma. 2012. Pinch analysis of heat exchanger networks in the crude distillation unit of Port-Harcourt refinery. J. Emerg. Trends Eng. Appl. Sci. 3:475–484.
16. Uzukwu, P. U., and E. T. Iyagba. 2012. Process analysis of refinery crude charge heater. Cont. J. Eng. Sci. 7:31–51. doi: 10.5707/cjengsci.2012.7.2.31.51
17. Jesuleye, O. A., W. O. Siyanbola, S. A. Sanni, and M. O. Ilori. 2007. Energy demand analysis of Port-Harcourt refinery, Nigeria and its policy implications. Energy Policy. 35:1338–1345.
18. Badmus, I., O. M. Oyewola, and R. O. Fagbenle. 2012. A review of performance appraisals of Nigerian Federal Government-Owned Refineries. Energy Power Eng. 4:47–52. doi: 10.4236/epe.2012.41007
19. FAO Regional Wood Energy Development Programme in Asia/Technology and Development Group, University of Twente, Netherlands. 1997. Energy and environment basics. 2nd ed. RWEDP Report No. 29, Bangkok.

20. Gillet, M, ed. 1996. IPCC Good Practice Guidance and Uncertainty Management in National Greenhouse Gas Inventories. Chap. 2. Energy. IPCC, Montreal.

21. Ogbonnaya, E. A., K. T. Johnson, H. U. Ugwu, and C. U. Orji. 2010. Component model-based condition monitoring of a gas turbine. ARPN J. Eng. Appl. Sci. 5:40–49.

22. Nigerian National Petroleum Corporation, NNPC. 2000. Annual statistical bulletin. Corporate Planning and Development Division, Abuja.

23. Nigerian National Petroleum Corporation, NNPC. 2001. Annual statistical bulletin. Corporate Planning and Development Division, Abuja.

24. Nigerian National Petroleum Corporation, NNPC. 2002. Annual statistical bulletin. Corporate Planning and Development Division, Abuja.

25. Nigerian National Petroleum Corporation, NNPC. 2003. Annual statistical bulletin. Corporate Planning and Development Division, Abuja.

26. Nigerian National Petroleum Corporation, NNPC. 2004. Annual statistical bulletin. Corporate Planning and Development Division, Abuja.

27. Nigerian National Petroleum Corporation, NNPC. 2005. Annual statistical bulletin. Corporate Planning and Development Division, Abuja.

28. Nigerian National Petroleum Corporation, NNPC. 2006. Annual statistical bulletin. Corporate Planning and Development Division, Abuja.

29. Nigerian National Petroleum Corporation, NNPC. 2007. Annual statistical bulletin. Corporate Planning and Development Division, Abuja.

30. Nigerian National Petroleum Corporation, NNPC. 2008. Annual statistical bulletin. Corporate Planning and Development Division, Abuja.

31. Nigerian National Petroleum Corporation, NNPC. 2009. Annual statistical bulletin. Corporate Planning and Development Division, Abuja.

32. Nigerian National Petroleum Corporation, NNPC. 2010. Annual statistical bulletin. Corporate Planning and Development Division, Abuja.

33. Nigerian National Petroleum Corporation, NNPC. 2011. Annual statistical bulletin. Corporate Planning and Development Division, Abuja.

34. Ocic, O. 2005. Oil Refineries in the 21st Century: energy efficient, cost effective, environmentally benign. WILEY-VCH Verlag GmbH & Co. KGaA, Weinheim.

35. Oniwon, A. 2011. Oil and gas in Nigeria's national development: an assessment. A presentation at the National Defence College. Abuja.

36. de Lima, R. S., and R. Schaeffer. 2011. The energy efficiency of crude oil refining in Brazil: a Brazilian refinery plant case. Energy 36:3101–3112. doi: 10.1016/j.energy.2011.02.056

37. Solomon's Fuels Refinery Performance Analysis. 2008. Available at: http://solomononline.com/benchmarking-performance/refining/ (accessed 15 April 2012).

38. Greenhouse gases, Regulated Emissions and Energy use in Transportation. GREET Transportation Fuel Cycle Analysis Model; GREET 1.8b, released 5 September 2008. Argonne National Laboratory, Argonne, IL. Available at http://www.transportation.anl.gov/modeling_simulation/GREET/index.html (accessed 1 February 2011).

39. International Energy Agency, IEA. 2005. Energy Stat. Manual, 9 rue de la Fédération, 75739 Paris Cedex 15. France

40. Engineering Toolbox. http://www.engineeringtoolbox.com/heating-values-fuel-gases-d_823.html (accessed 2 April 2013)

41. Sattari, S., and A. Avami. 2008. Structural, energy and environmental aspects in Iranian oil refineries. 3rd IASME/WSEAS International Conference on Energy and Environment, University of Cambridge, U.K., 23–25 February 2008.

17

Effect of ester compounds on biogas production: beneficial or detrimental?

Heri Yanti[1,2], Rachma Wikandari[3], Ria Millati[4], Claes Niklasson[1] & Mohammad J. Taherzadeh[3]

[1]Department of Chemical and Biological Engineering, Chalmers University of Technology, Gothenburg, Sweden
[2]Department of Chemical Engineering, Universitas Gadjah Mada, Yogyakarta, Indonesia
[3]Swedish Centre of Resource Recovery, University of Borås, Borås, Sweden
[4]Department of Food and Agricultural Product Technology, Universitas Gadjah Mada, Yogyakarta, Indonesia

Keywords
Ester, inhibition, methane, minimum inhibitory concentration

Abstract

Esters are major flavor compounds in fruits, which are produced in high volume. The widespread availability of these compounds in nature attracts interest on their behavior in anaerobic digestion in waste and wastewater treatments. The aim of this work was to study the effects of various esters at different concentrations in anaerobic digestion followed by determination of their minimum inhibitory concentration (MIC), and to study the effect of chain length of functional group and alkyl chain of ester on methane production. Addition of methyl butanoate, ethyl butanoate, ethyl hexanoate, and hexyl acetate at concentration up to $5 \ g \ L^{-1}$ increased methane production, while their higher concentrations inhibited the digestion process. The MIC values for these esters were between 5 and $20 \ g \ L^{-1}$. Except hexyl acetate, the esters at concentration $5 \ g \ L^{-1}$ could act as sole carbon source during digestion. For ethyl esters, increasing number of carbon in functional group decreased methane production. For acetate esters, alkyl chain longer than butyl inhibited methane production. Effect of ester on methane production is concentration-dependent.

Introduction

Exploring and exploiting renewable and green energy is necessary in today's energy life style, not only because of the inevitable depletion of conventional sources of fossil energy but also due to the ecological-environmental effects caused by the conventional energy consumption. Biogas is a clean and renewable form of energy, which has a wide range of applications, such as vehicle fuel, cooking, heating, lighting, and electricity production. Biogas is produced by anaerobic degradation of organic substrates. Anaerobic digestion is one of the oldest processes and the most efficient treatment technologies, widely used for treating industrial wastes, municipal waste, and stabilization of wastewater sludge [1]. However, the fermenting organisms in biogas processes are sensitive to the process conditions and the substrate used. A wide variety of chemical substances have been reported to be inhibitory to the anaerobic digestion processes, resulting in decreasing or stopping the biogas production [2].

One of the chemical substances that could impact biogas production is ester. Ester is one of the most important classes of chemicals produced in high volume [3]. It is widely used in various industries such as solvents, plasticizer for cellulose, nail polish removers, perfumery product, plastic tubing, floor tiles, furniture, automobile upholstery, insect repellents, and as paint additive [3, 4]. Ester can be manufactured via chemical reaction by condensation of alcohols and carboxylic acids. In addition, it can be found in nature as the major volatile compounds in fruits. For instance, esters constitute 78–92%, 25–90%, and 13% of the total volatile mass of apple, strawberry, and raspberry, respectively [5]. Therefore, it is used as a flavor and fragrance compound in food industry. The widespread production and ability of some esters to migrate, make esters to be easily found in the environment [6].

Since anaerobic digestion is mostly used to tread wastewaters, investigation of effect of ester on anaerobic digestion is important. Esters might be beneficial or detrimental to the anaerobic digestion. Some esters are reported to be degraded by *Acetobacterium woodii* and *Eubacterium limosum*, which are involved in anaerobic digestion process [7]. However, some phthalate esters (PAE) could reduce the biogas production at a concentration higher than 60 mg L^{-1} [8]. Besides, some esters are reported having antimicrobial activity against some standard microorganisms. Butyl acetate and hexyl acetate at 4 μL and 8 μL, respectively, were toxic to *Conidia* germination after incubation for 24 h at 22°C [9]. Hexyl acetate at concentration 150 mg L^{-1} has significant inhibitory effect against *E. coli*, *S. enteritidis*, and *L. monocytogenes* that were isolated both in model system and in fresh-sliced apples [10]. To our knowledge, scarce information is available on the effect of esters in methane production.

In this study, some esters that are available in nature as fruit flavor compounds and mainly used in food, cosmetic, and solvent industries were examined. The aim of this study was to investigate the effects of the ester compounds on biogas production, and their minimum inhibitory concentration (MIC). In addition, the effect of chain length of functional group and alkyl chain of ester on biogas production was investigated.

Material and Methods

Microorganisms and chemicals

Inoculum (sludge) with 22 g VS L^{-1} (volatile solid per liter) was obtained from a thermophilic biogas plant (55°C) at Borås Energy and Environment AB (Borås, Sweden). The inoculum was then stored in an incubator at temperature 55°C for 3 days. A synthetic medium solution was prepared from 20 g L^{-1} of nutrient broth (Sigma-Aldrich, St. Louis, MO), 20 g L^{-1} of yeast extract (Merck KgaA, Darmstadt, Germany), and 20 g L^{-1} of glucose (Merck). Nutrient broth contained 10 g L^{-1} beef

extract, 10 g L^{-1} peptone, and 5 g L^{-1} sodium chloride. The ester compounds (Sigma-Aldrich) used in this work were methyl butanoate, ethyl butanoate, ethyl hexanoate, hexyl acetate, methyl acetate, ethyl acetate, and butyl acetate.

The experiments were carried out in two steps. The first step was to investigate the effect of ester compounds on biogas production. In this step, the concentrations of methyl butanoate, ethyl butanoate, ethyl hexanoate, and hexyl acetate were 0.05, 0.5, 5, 10, and 20 g L^{-1} for each compound. The second step was to investigate the effect of chain length of functional group and alkyl chain of ester on biogas production. In this step, the concentration of ethyl hexanoate, ethyl butanoate, methyl acetate, ethyl acetate, butyl acetate, and hexyl acetate was 5 g L^{-1}.

Anaerobic digestion

The method used for anaerobic digestion was adapted from the method described by Hansen et al. [11] and the OECD [12] with minor modifications. The experiments were carried out in a 120-mL glass bottle containing 50 mL of sludge, 1 mL of medium, and 2.5 mL of ester solution or distilled water (for a control). In order to measure methane production from the inoculum, the inoculum was incubated without addition of medium and ester solution in the glass bottle containing 50 mL of sludge and 3.5 mL of distilled water. The bottles were closed tightly and the headspace was filled with gas mixture containing 80% N_2 and 20% CO_2 to achieve anaerobic condition. The bottles were then incubated at 55°C for 28 days. During incubation, the bottles were shaken twice a day using water bath shaker (55°C) at 150 rpm. Gas samples were taken from headspace of the reactors through the septum using a syringe with pressure lock (VICI; Precision Sampling Inc., Baton Rouge, LA). Samples were taken at 3, 6, 9, 12, 15, 20, 25, and 30 days. The initial methane production rate was measured as the mean of methane production per day during the first 10 days.

Analytical methods

Biogas production from the digestion experiments was measured using Gas Chromatography (Varian 450 GC, Palo Alto, CA) with a capillary column (J 6 W scientific GS-Gas Pro, bonded silica based 30 m × 0.32 mm, Agilent Technologies, Santa Clara, CA) and a thermal conductivity detector (Varian, Palo Alto, CA). The conditions for the analysis were injector temperature of 75°C, oven temperature of 100°C, detector temperature of 120°C, and column flow (N_2) 2 mL min^{-1}.

All experiments were carried out in triplicate batch experiment; the result was presented in average ± standard deviation. At the end of digestion, the pH of the sludge was analyzed. For statistical analysis, analysis of variance (ANOVA) with significance level of 0.05 was performed using a statistical package for the social sciences (SPSS version 21, International Business Machine Corporation, Armonk, NY).

Results and Discussion

The widespread application and high-volume production of esters have caused high possibility of ester found in wastewater of many industries. Various esters at different concentrations and different chain lengths of functional group and alkyl chain were added into a batch anaerobic digestion system. The gas production was measured to indicate the inhibitory or enhancement effect of the esters to the system.

Are esters beneficial or detrimental for biogas production?

In order to investigate the effect of the ester on biogas production, methyl butanoate, ethyl butanoate, ethyl hexanoate, and hexyl acetate were added at different concentrations. The inoculum was incubated in the presence of

these esters at five different concentrations that are 0.05, 0.5, 5, 10, and 20 g L^{-1}. Table 1 and Figure 1 present a summary of the effects of esters on methane production.

The results show that after 30 days, addition of ethyl butanoate, ethyl hexanoate, and hexyl acetate at concentrations up to 5 g L^{-1} resulted in increased biogas production (Table 1). Similar result was obtained from methyl butanoate at concentrations up to 10 g L^{-1}. The highest methane productions for ethyl butanoate, ethyl hexanoate, and hexyl acetate were 117.4 $\pm$ 6.6, 105.0 $\pm$ 5.1, and 112.0 $\pm$ 3.8 mL, respectively, in the presence of ester at a concentration of 5 g L^{-1}. These results correspond to 63.5%, 46.2%, and 56% increase in methane production. The highest methane production for methyl butanoate was obtained by adding 10 g L^{-1} was 96.79 $\pm$ 0.94 mL, corresponding to 34.8% increase in methane yield. The effect of the esters on the initial digestion rate is presented in Figure 2. With the exception of hexyl acetate, adding other ester compounds up to 10 g L^{-1} is still beneficial for the digestion in the early process. This higher methane yield in the presence of esters is most probably due to the esters consumption by the anaerobic digesting microorganisms. However, increasing concentration of the esters to more than 5 g L^{-1} had an inhibitory effect. It was even more obvious when the concentration of the esters added was

Table 1. The effects of esters added during methane production.

Ester	Concentration (g L^{-1})	Cumulative methane (mL)	Methane yield (L g^{-1} VS)	Enhancement[1](%)	pH
Inoculum	–	46.7 $\pm$ 11.5	0.3 $\pm$ 0.19	0	7.74
Control (medium + inoculum)	–	71.8 $\pm$ 11.3	1.2 $\pm$ 0.19	0	7.74
Methyl butanoate	0.05	65.1 $\pm$ 2.4	1.08 $\pm$ 0.04	−9.4[2]	7.94
	0.5	82.3 $\pm$ 3.7	1.33 $\pm$ 0.06	15.4	7.86
	5	89 $\pm$ 10	1.48 $\pm$ 0.17	24	7.89
	10	96.8 $\pm$ 0.9	1.61 = 0.02	34.8	7.84
	20	3.3 $\pm$ 0.6	0.05 $\pm$ 0.01	−95.5	5.42
Ethyl butanoate	0.05	68.9 $\pm$ 9	1.15 $\pm$ 0.15	−4.1[2]	7.84
	0.5	85.7 $\pm$ 2.6	1.43 $\pm$ 0.04	19.4	7.76
	5	117.4 $\pm$ 6.6	1.96 $\pm$ 0.11	63.6	7.77
	10	60.8 $\pm$ 7	1.01 $\pm$ 0.12	−15.3	5.8
	20	9.3 $\pm$ 5.4	0.15 $\pm$ 0.09	−87.1	5.31
Ethyl hexanoate	0.05	78.2 $\pm$ 2.8	1.3 $\pm$ 0.05	8.9	7.84
	0.5	95.1 $\pm$ 5.6	1.53 = 0.09	32.4	7.78
	5	105 $\pm$ 5.1	1.75 $\pm$ 0.08	46.2	7.77
	10	74.9 $\pm$ 6.8	1.19 = 0.01	−1[2]	6.65
	20	20.2 $\pm$ 2.9	0.34 $\pm$ 0.05	−71.9	6.61
Hexyl acetate	0.05	69.8 $\pm$ 12	1.16 $\pm$ 0.19	−2.8[2]	7.81
	0.5	92.3 $\pm$ 0.5	1.54 $\pm$ 0.01	28.5	7.83
	5	112 $\pm$ 3.3	1.37 $\pm$ 0.06	56	7.84
	10	4.2 $\pm$ 0.7	0.7 $\pm$ 0.01	−94.2	7
	20	3.9 $\pm$ 1.5	0.06 $\pm$ 0.02	−94.6	6.91

[1]Enhancement = $\frac{\text{CH}_4 \text{ produced by sample containing ester} - \text{CH}_4 \text{ produced by control}}{\text{CH}_4 \text{ produced by control}} \times 100\%$

Negative value shows inhibition.

[2]Not significantly different from the control.

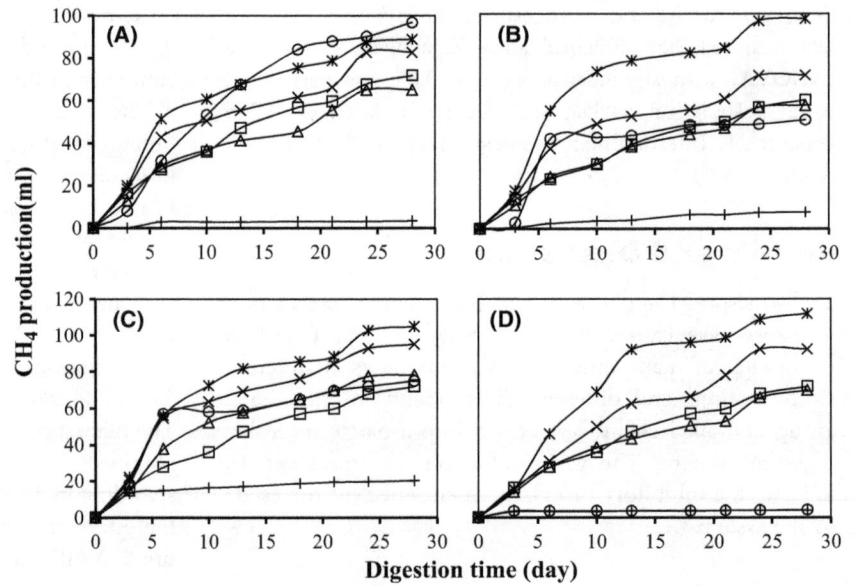

Figure 1. Cumulative methane production with addition of ester compound at different concentrations (+)20 g L^{-1}, (○) 10 g L^{-1}, (*) 5 g L^{-1}, (x) 0.5 g L^{-1}, (Δ) 0.05 g L^{-1}, and (□) control.

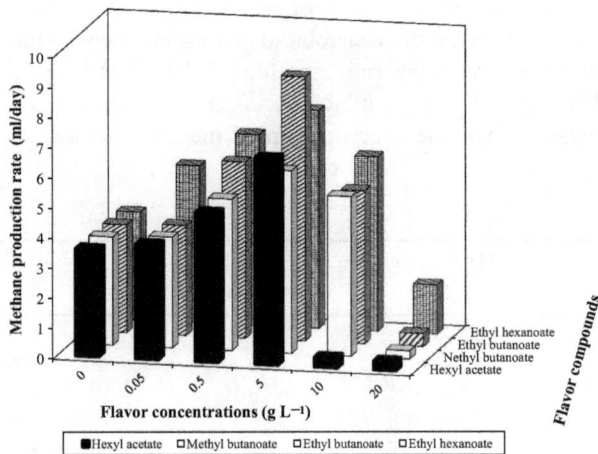

Figure 2. Initial methane production rate with addition of esters at various concentrations.

20 g L^{-1}. At the end of digestion when 20 g L^{-1} was added, the methane yield decreased by 71.9–95.5% (Table 1). Addition of 20 g L^{-1} of ester reduced methane production by 47–90% and showed inhibitory effect since the early stage of digestion (Fig. 2). These results indicate that the effect of esters on anaerobic digestion is concentration-dependent. At concentrations up to 5 g L^{-1}, esters are beneficial, whereas at higher concentrations than 5 g L^{-1}, the esters gave negative effect on biogas production.

In order to evaluate the effect of esters on biogas composition, a methane content ratio was determined with the assumption that only methane and carbon dioxide produced during digestion. The results are presented in

Figure 3. The biogas composition of the control experiment was 63.6% of methane and 36.4% of carbon dioxide. Once esters were added up to 5 g L^{-1}, the methane content ratio of the reactors did not significantly vary compared with that of the control experiment without the esters. However, the methane content ratio declined with the presence of 10 g L^{-1} esters, and the biogas only contained 3–15% of methane in the presence of 20 g L^{-1} of esters. In the case of methyl butanoate, the methane production increased up to 34.8% at a concentration of 10 g L^{-1}, but the methane content ratio reduced from 63.6% to 48.9%.

A MIC is defined as the lowest concentration of ester that will inhibit the activity of microorganisms during anaerobic digestion. Accordingly, the MIC of methyl butanoate was between 10 and 20 g L^{-1}, and for ethyl butanoate, ethyl hexanoate, and hexyl acetate it was 5–10 g L^{-1}. The precise action of antimicrobial activity of ester is not yet clear. One explanation could be due to the changes in permeability of the cell membranes, which cause leakage of cellular components and influence the metabolism of bacteria [13].

At the end of anaerobic digestion, the pH was measured which showed that the pH had decreased from 7.5–8 to 5–6 when ester at concentrations higher than 5 g L^{-1} (for ethyl butanoate and ethyl hexanoate) and 10 g L^{-1}(for methyl butanoate) were added (Table 1). The decrease in pH is caused by accumulation of volatile fatty acids as intermediate products that are produced during acetogenesis reaction. Inhibition could occur when accumulation of volatile fatty acids could no longer be handled by the buffering system of anaerobic digestion [14, 15]. Infantes et al. [16] reported that high energy is

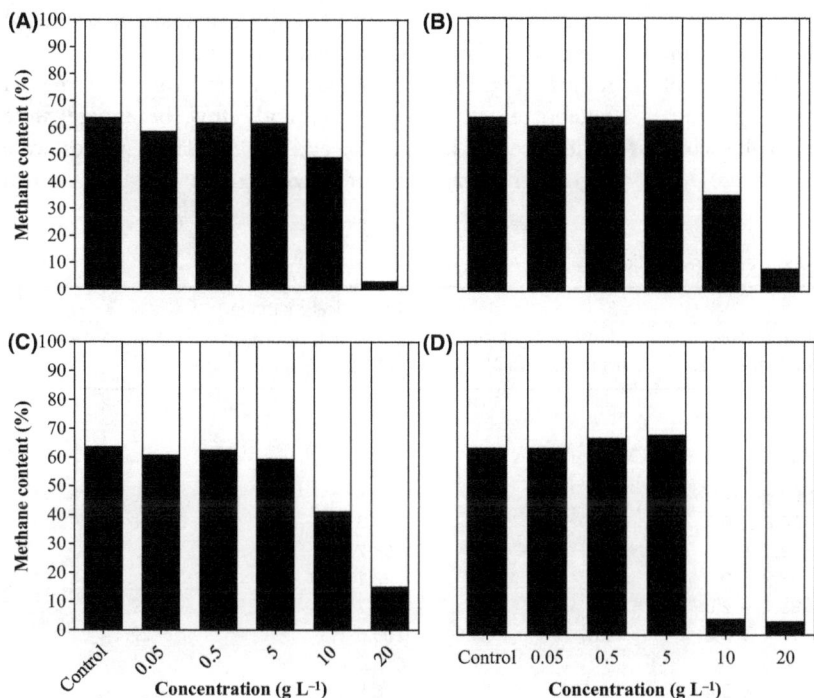

Figure 3. Methane content ratio (■) CH$_4$, (□) CO$_2$ with addition of various ester compounds at different concentrations.

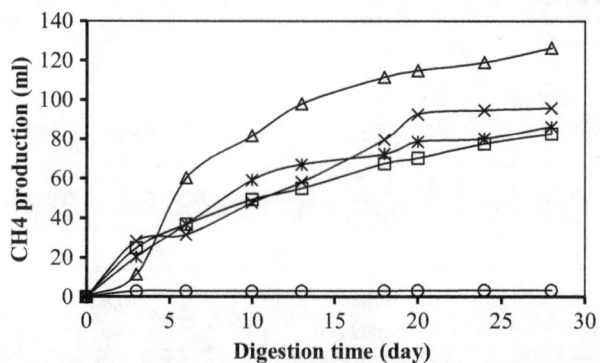

Figure 4. Methane production with addition of various ester compounds (△) methyl butanoate, (x) ethyl butanoate, (*) ethyl hexanoate, (○) hexyl acetate, and (□) control at 5 g L^{-1}

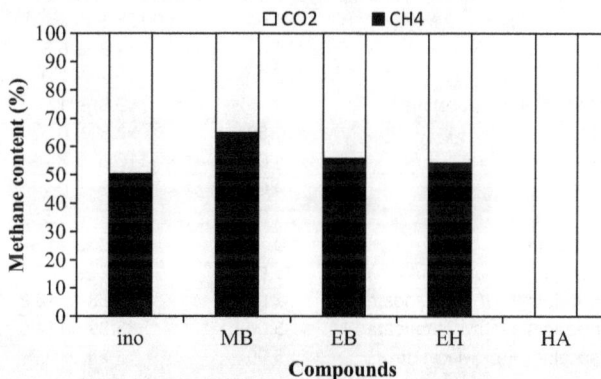

Figure 5. Methane content ratio of inoculum (ino), methyl butanoate (MB), ethyl butanoate (EB), ethyl hexanoate (EH), and hexyl acetate (HA) as sole carbon source compounds at 5 g L^{-1}.

consumed to maintain the intercellular pH at low pH which inhibits the cell metabolism.

Ester consumption

In order to confirm the assumption that esters can act as a carbon source, experiments using 5 g L^{-1} of esters as the sole carbon source were conducted. As control, the anaerobic glass digester was filled with inoculum without an added medium. The results are presented in Figure 4 and show that methyl butanoate, ethyl butanoate, ethyl hexanoate, and hexyl acetate gave higher methane production than did the control. This result is

in accordance with a previous study reporting that methyl esters could be degraded to carboxylic acid and alcohol and would be further converted into methane by *A. woodii* and *E. limosum* [7]. Those microorganisms are found in all four stages of anaerobic biochemical reactions which have ability to degrade methyl ester of acetate, propionate, butanoate, and isobutanoate into carboxylic acid and methanol under anaerobic condition [7, 14]. According to the mechanisms of methyl ester degradation, methyl butanoate will be degraded into methanol and acetic acid. Ethyl butanoate and ethyl hexanoate probably also follow this mechanism.

Hydrolytic bacteria degrade ethyl butanoate and ethyl hexanoate into ethanol and butanoic and hexanoic acid, respectively [15]. The carboxylic acid and alcohol will be further degraded through acetogenesis stage into acetic acid and hydrogen, which will finally be converted into methane [14].

In order to evaluate the biogas composition produced from the esters, the methane content ratio was determined (Fig. 5). The ratios produced from the methyl butanoate, ethyl butanoate, and ethyl hexanoate were varied from 64.7, 55.7%, and 54.1%. These values were slightly higher compared with those of the control

Table 2. Theoretical methane production from ester and ester consumption.

Sample	Ester conc. (g L^{-1})	Cumulative methane (mL)	Methane produced from ester (mL)[1]	Theoretical methane production from ester (mL)[2]	Ester consumption (%)[3]
Control (medium + inoculum)	0	71.8 ± 11.3	–[4]	–	–
Medium+	0.05	65.1 ± 2.4	–	1.9	–
Methyl butanoate	0.50	82.8 ± 3.7	11.0	19.1	57.8
	5.00	89 ± 10	17.2	190.9	9.0
	10.0	96.8 ± 0.9	25.0	381.8	6.5
	20.0	3.3 ± 0.6	–	763.7	–
Medium + ethyl butanoate	0.05	68.9 ± 9	–	2.1	–
	0.50	85.7 ± 2.6	13.9	20.7	67.5
	5.00	117.4 ± 6.6	45.6	206.6	22.1
	10.0	60.8 ± 7	–	413.2	–
	20.0	9.3 ± 5.4	–	826.5	–
Medium + ethyl hexanoate	0.05	78.2 ± 2.8	6.4	2.3	280.5
	0.50	95.1 ± 5.6	23.3	22.9	101.7
	5.00	105 ± 5.1	33.2	228.9	14.5
	10.0	74.9 ± 6.8	–	457.7	–
	20.0	20.2 ± 2.9	–	915.4	–
Medium + hesylacetate	0.05	69.8 ± 12	–	23	–
	0.50	92.3 ± 0.5	20.5	22.9	89.4
	5.00	112 ± 3.8	40.2	228.9	17.6
	10.0	4.2 ± 0.7	–	457.7	–
	20.0	3.9 ± 1.5	–	915.4	–
Inoculum	0	46.7 ± 11.5	–	–	–
Inoculum + methyl butanoate	5.00	126.2 ± 10.28	80.7	190.09	42.5
Inoculum + ethyl butanoate	5.00	95.88 ± 18.8	40.8	228.86	19.8
Inoculum + ethyl butanoate	5.00	86.39 ± 14.5	50.3	206.62	22.0
Inoculum + hexylacetate	5.00	3.29 ± 0.2	–	228.86	–

[1]Methane produced from ester = methane produced by sample methane produced by control.
[2]Theoretical methane production from ester was calculated from following equation:
$CcHhOoNnSs = yH_2O \rightarrow xCH_4 + nNH_3 + sH_2S + (c - x)CO_2$
[3]Ester consumption = $\dfrac{\text{Methane produced from ester}}{\text{Theoretical methane production from ester}} \times 100\%$
[4](–)No production/no consumption.

Table 3. Cumulative methane production with addition of acetate ester and ethyl ester.

Ester	Number of carbon		Cumulative methane (mL)	Methane yield (L g^{-1} VS)
	Alkyl group	Functional group		
Inoculum	–	–	46.7 ± 11.5	0.8 ± 0.19
Ethyl hexanoate	2	6	89.4 ± 14.5	1.4 ± 0.2
Ethyl butanoate	2	4	96.0 ± 8.3	1.6 ± 0.1
Methyl acetate	1	2	111.5 ± 9.8	1.9 ± 0.2
Ethyl acetate	2	2	126.1 ± 10.3	2.1 ± 0.2
Butyl acetate	4	2	111.7 ± 15.7	1.9 ± 0.3
Hexyl acetate	6	2	3.3 ± 0.1	0.1 ± 0

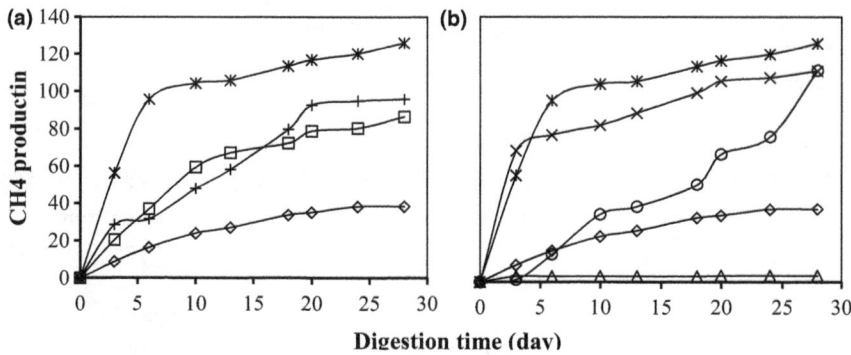

Figure 6. Methane production with addition of various ester compounds (x) Methyl acetate, (*) Ethyl acetate, (o) Butyl acetate, (Δ) Hexyl acetate, (+) Ethyl butanoate, (□) Ethyl hexanoate, and (◊) control at 5 g L^{-1}.

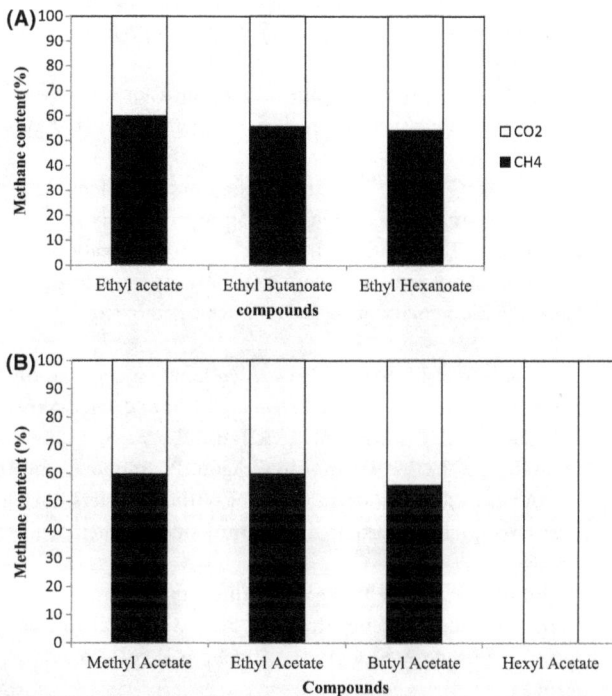

Figure 7. Methane content ratio with addition of ethyl esters (A) and acetate esters (B) at concentration of 5 g L^{-1}.

(50.2%). However, no methane was produced from hexyl acetate.

Esters are mostly organic compounds. As an organic compound, the chemical reaction of degradation of organic matters is given in equation (1).

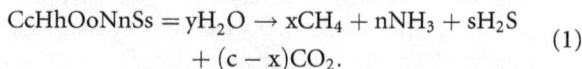

$$CcHhOoNnSs = yH_2O \rightarrow xCH_4 + nNH_3 + sH_2S + (c - x)CO_2. \quad (1)$$

Thus, theoretically, the potential methane production from esters can be calculated. According to the methane obtained from ester, ester consumption can be determined. The effect of concentrations and the presence of medium were investigated and the results are presented in Table 2. The results show that in the presence of a med-

ium, the highest ester consumption was obtained at their concentration of 0.5 g L^{-1}, which was up to 57.8–100%. However, almost no esters were consumed at concentration higher than 10 g L^{-1} for all esters examined. In the absence of medium, ester consumption is in the range of 17.8–42.5% at concentration of 5 g L^{-1}. For ethyl butanoate and ethyl hexanoate, the ester consumption was similar either in the presence or absence of medium, whereas ester consumption of methyl butanoate was higher in the absence of medium. Interestingly, hexyl acetate exhibited a different behavior in terms of medium. It was consumed in the presence of medium while it acted as an inhibitor in the absence of medium, as shown by the significantly lower methane production than that of the inoculum. This might indicate a synergy effect between medium and hexyl acetate, enabling hexyl acetate to be consumed at this concentration.

Effect of chain length of the functional group (alkyl carboxylic chain) and Alkyl Chain of ester

Esters consist of a functional group and alkyl chain. In this study, various carbon lengths of functional group and alkyl chains of esters were added as sole carbon source at concentration of 5 g L^{-1}. The results are presented in Table 3, and Figures 6 and 7. The functional groups (R_1COO-) of the ester were acetate (C2), butanoate (C4), and hexanoate (C6). The alkyls (-R_2) of the ester were methyl (C1), ethyl (C2), butyl (C4), and hexyl (C6).

For the experiments with ethyl ester, an increasing number of carbons in the functional group required longer time to achieve maximum methane production and it decreased the methane production (Fig. 6A). For the experiments with acetate esters, the cumulative methane production was much higher compared to that of the control. When the alkyl was butyl (C4), it exhibited a longer lag phase compared to those of methyl (C1) and ethyl (C2). It was observed that strong inhibition occurred when the alkyl was hexyl (C6) and no methane was produced (Fig. 6B). Moreover, the methane content ratio decreased with an increasing number of carbons both in the functional group and alkyl chain (Fig. 7). The methane content ratio decreased from 59.8% to 54.1% with increasing number of carbon in functional group and it decreased from 59.8% to zero with increasing number of carbons in alkyl chain of esters. A correlation between increasing length of the ester side-chain and decreasing biodegradability as well as increasing the toxicity has been confirmed by Gavala et al. [8] Similarly, Merkl et al. [17] found that the derivative of phenolic acids, butyl ester is approximately three times more toxic than methyl ester. The hydrophobicity of ester increases with increasing chain length of ester. Esters have both a hydrophilic head and hydrophobic tail, which resemble to bipolar membrane of the bacterial cell wall. This similarity might be the reason the hydrophobic tail of ester targets the microorganisms cell membranes by penetrating and disrupting normal function of cellular membranes, thus resulting in the leakage of cells or even killing the cells [18].

Conclusion

Addition of methyl butanoate, ethyl butanoate, ethyl hexanoate, and hexyl acetate at concentrations up to 5 g L^{-1} increased methane production. Addition of esters to anaerobic digestion at concentrations higher than 5 g L^{-1} decreased methane production. The MIC for methyl butanoate was between 10 and 20 g L^{-1}, while for ethyl butanoate, ethyl hexanoate, and hexyl acetate the MIC was between 5 and 10 g L^{-1}. Methyl butanoate, ethyl butanoate, and ethyl hexanoate at a concentration of 5 g L^{-1} could function as sole carbon source for the anaerobic digesting bacteria. For ethyl ester, the methane production decreased with increase in the number of carbons in the functional group. For acetate esters, with an alkyl chain longer than butyl, the number of carbons in the functional group started to give an inhibitory effect for methane production.

Acknowledgment

We express our gratitude to Directorate General of Higher Education, Ministry of National Education of Republic of Indonesia through Competitive Research Grant of International joint Research for International Publication (grant no LPPM-UGM/696/LIT/2013), the Swedish International Development Agency (SIDA), the Swedish Gas Centre for financial support of this study, and to Borås Energy and Environment AB (Borås, Sweden) for the inoculum.

Conflict of Interest

None declared.

References

1. Yadvika Santosh, T. R. Sreekrishnan, S. Kohli, and V. Rana. 2004. Enhancement of biogas production from solid substrates using different techniques – a review. Bioresour. Technol. 95:1–10.

2. Kroeker, E. J., D. D. Schulte, A. B. Sparling, and H. M. Lapp. 1979. Anaerobic treatment process stability. J. Water Pollut. Control Fed. 51:718–727.

3. Papa, E., F. Battaini, and P. Gramatica. 2005. Ranking of aquatic toxicity of esters modelled by QSAR. Chemosphere 58:559–570.

4. Kirk, R. E., and D. F. Othmer. 1988. Encyclopedia of chemical technology, 4th ed. Vol. 21. John Wiley and Sons Inc., New York, NY.

5. Berger, R. G. 2007. Flavours and fragrances – chemistry, bioprocessing and sustainability. Springer, Germany.

6. Parkerton, T. F., and W. J. Konkel. 2000. Application of quantitative structure-activity relationships for assessing the aquatic toxicity of phthalate esters. Ecotoxicol. Environ. Saf. 45:61–78.

7. Liu, S., and J. M. Suflita. 1994. Anaerobic biodegradation of methyl esters by Acetobacterium woodii and Eubacterium limosum. J. Ind. Microbiol. 13:321–327.

8. Gavala, H. N., F. Alatriste-Mondragon, R. Iranpour, and B. K. Ahring. 2003. Biodegradation of phthalate esters during the mesophilic anaerobic digestion of sludge. Chemosphere 52:673–682.

9. Filonow, A. B. 1999. Yeasts reduce the stimulatory effect of acetate esters from apple on the germination of Botrytis cinerea Conidia. J. Chem. Ecol. 25:1555–1565.

10. Lanciotti, R., N. Belletti, F. Patrignani, A. Gianotti, F. Gardini and M. E. Guerzoni. 2003. Application of hexanal, (E)-2-hexenal, and hexyl acetate to improve the safety of fresh-sliced apples. J. Agric. Food Chem. 51:2958–2963.

11. Hansen, T. L., J. E. Schmidt, I. Angelidaki, E. Marca, J. I. C. Jansen, H. Mosbaek, et al. 2004. Method for determination of methane potentials of solid organic waste. Waste Manage. 24:393–400.

12. Organisation for Economic Cooperation and Development (OECD). 2007. Determination of the inhibition of the activity of anaerobic bacteria - reduction of gas production from anaerobically digesting sewage sludge, O.f.E.C.-o.a.D. in Test No. 224. OECD Publishing, Paris, France. ISBN: 9264067337, 9789264067332.

13. Deng, W. L., T. R. Hamilton-Kemp, M. T. Nielsen, R. A. Andersen, G. B. Collins and D. F. Hildebrand.1993. Effect of six-carbon aldehydes and alcohols on bacterial proliferation. J. Agric. Food Chem. 41:506–510.

14. Deublein, D., and A. Steinhauser. 2008. Biogas from waste and renewable resources. Wiley-VCH, Germany.

15. Gerardi, M. H. 2003. The microbiology of anaerobic digesters. John Wiley & Sons Inc, Canada.

16. Infantes, D., A. González Del Campo, J. Villaseñor, and J. Fernandez. 2011. Influence of pH, temperature and volatile fatty acids on hydrogen production by acidogenic fermentation. Int. J. Hydrogen Energy 36:15595–15601.

17. Merkl, R., I. Hrádková, V. Filip, and J. Šmi drkal2010. Antimicrobial and antioxidant properties of phenolic acids alkyl esters. Czech J. Food Sci. 28:275–279.

18. Huang, C. B., B. George, and J. L. Ebersole. 2010. Antimicrobial activity of n-6, n-7 and n-9 fatty acids and their esters for oral microorganisms. Arch. Oral Biol. 55:555–560.

18

Engineering evaluation of direct methane to methanol conversion

Arno de Klerk

Department of Chemical and Materials Engineering, University of Alberta, Edmonton, Alberta, T6G 2V4, Canada

Keywords

Gas-to-liquids, methane, methanol, natural gas, partial oxidation

Abstract

Investigations into direct methane to methanol conversion are justified based on the avoidance of synthesis gas generation, which accounts for around 60% of the capital cost of synthesis gas to methanol conversion. A significant body of information already exists on the process chemistry, but little has been reported on the engineering of such a process. An engineering evaluation of the process was performed and the potential of this process as a platform technology for small-scale gas-to-liquids (GTL) applications was evaluated. It was found that direct methane to methanol conversion had 35% carbon efficiency and 28% thermal efficiency, which were about half of the process efficiencies of indirect methanol synthesis using synthesis gas. The poor process efficiency was mainly a consequence of the irreversible loss of carbon to CO_x during conversion. The direct methane to methanol process also required an air separation unit, which eroded the stated benefit of avoiding a synthesis gas generation step in the process. The utility footprint was typical of GTL processes, with large gas compression duties and cooling duties. Overall, the engineering evaluation indicated there was no benefit to employ direct methane to methanol conversion instead of indirect methanol synthesis (the industry standard), and there was no specific benefit of direct methane to methanol conversion, irrespective of scale, for GTL applications.

Introduction

Methanol is a large volume commodity chemical. Industrially methanol is produced mainly by indirect liquefaction of natural gas [1]. The indirect liquefaction process involves three steps (Fig. 1). The first step is reforming of natural gas to synthesis gas, which is then in the second step converted into methanol. Synthesis gas conversion to methanol is conducted over a Cu-ZnO-based catalyst and methanol is obtained with high selectivity. The conversion of synthesis gas to methanol is equilibrium limited [2]. In the third step, the crude methanol is recovered from the unconverted synthesis gas and purified, while the unconverted synthesis gas is recycled. The most expensive step in process is synthesis gas generation, which accounts for 60% of the capital cost of the process [3].

Methanol can also be produced by the direct partial oxidation of methane to methanol. There is a considerable body of literature on the process chemistry of direct

methane to methanol conversion [4]. It is one of a class of processes that involves the reaction of methane with an oxidant and direct oxidative methane conversion is often discussed collectively [5, 6, 7, 8]. Studies that consider the engineering aspects of processes for direct oxidative methane conversion are less abundant [9, 10].

On paper the direct conversion of methane to methanol seems like a good idea. The trade-off between conversion and product selectivity is anticipated and can in principle be overcome by the recycling of unconverted methane. Examples of industrial processes for the direct oxidation of methane to methanal (formaldehyde) can be found [4, 11, 12], but there are no examples of direct oxidation of methane to methanol, which technically requires only a change in reactor operation. Yet, the industrial methane to methanal conversion cited was of such low capacity (<1 t/day) that it would only be referred to as large pilot-scale in current terms. To quote Arutyunov [4]:

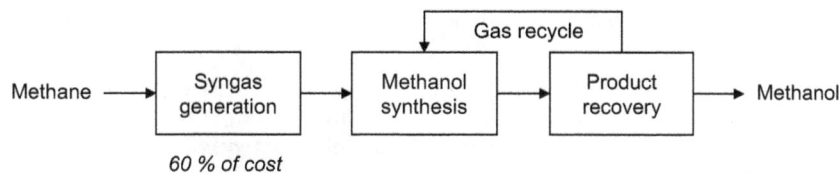

Figure 1. Simplified block flow diagram of indirect methanol synthesis.

... partial oxidation of dry natural gas, which requires more stringent conditions and gives lower yield of the target products [than heavier hydrocarbon gases], has never been mastered by the industry.

What is it then that prevents direct methane to methanol conversion to be applied?

The objective of this work was to perform an engineering evaluation of the direct methane to methanol conversion process. Of specific interest was to evaluate the potential of this process to be employed as a platform technology for small-scale gas-to-liquids (GTL) applications.

Process Chemistry

The first challenge in the direct conversion of methane is overcoming the stability of the C–H bonds in methane. Strategies to activate C–H in methane are discussed in literature, for example, [13], but activation of the C–H bond is only part of the challenge. Free radical partial oxidation is a quite successful strategy to activate C–H bonds in methane. Despite many claims in literature of the beneficial effects of employing catalysts, catalysts do not appear to offer meaningful yield advantages [4, 14]. It appears that the role of catalysis is mainly to reduce the severity of process conditions. The partial oxidation of methane to methanol is therefore more simply conducted as a gas phase free radical process, which avoids gas cleaning to remove potential catalyst poisons. The free radical partial oxidation of methane to methanol is not deleteriously affected by the presence of hydrogen sulfide or other contaminants that may be present in natural sources of methane, such as natural gas.

The other part of the direct methane conversion challenge is to prevent further reaction of the products from partial methane oxidation. The homolytic bond dissociation energy of the C–H in methane (439 kJ · mol^{-1}) is higher than that in methanol (402 kJ · mol^{-1}), methanal (369 kJ · mol^{-1}), or in methanoic acid (402 kJ · mol^{-1}) [15]. Whatever C–H bond activation strategy, it will work better on the products of methane conversion than on methane itself. (An analogous problem presents itself for microbial conversion, where it is necessary to inhibit methanol conversion in methane to methanol processes, because methanol is more easily metabolized) [16].

If heavier hydrocarbons are present, such as ethane and propane, the reaction proceeds more readily, since it is easier to oxidize heavier hydrocarbons.

The main carbon-containing products that are produced during the partial oxidation of methane at temperatures below 450°C are methanol (eq. 1), methanal (eq. 2), methanoic acid (eq. 3), carbon monoxide (eq. 4), and carbon dioxide (eq. 5):

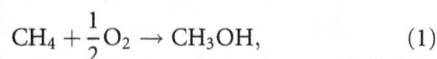

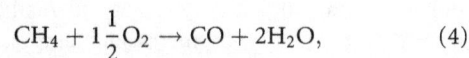

$$CH_4 + \frac{1}{2}O_2 \rightarrow CH_3OH, \tag{1}$$

$$CH_4 + O_2 \rightarrow CH_2O + H_2O, \tag{2}$$

$$CH_4 + 1\frac{1}{2}O_2 \rightarrow HCOOH + H_2O, \tag{3}$$

$$CH_4 + 1\frac{1}{2}O_2 \rightarrow CO + 2H_2O, \tag{4}$$

$$CH_4 + 2O_2 \rightarrow CO_2 + 2H_2O. \tag{5}$$

The reaction engineering strategy that is employed to maximize the methanol selectivity during gas phase oxidation of methane is to limit the single-pass conversion. This is achieved by limiting the concentration of the oxidant. When the oxidant is the limiting reagent, successive oxidation reactions are inherently limited and the selectivity to primary oxidation products is increased. Some general observations about the operating conditions and process chemistry can be made: [4, 17, 18]

(1) Typical operating temperatures for noncatalytic partial oxidation of methane to methanol are in the range 370–470°C. It is impractical to perform the reaction at temperatures below 350°C, because the induction time is long and the methane oxidation rate is low. At higher temperatures, there is an increase in oxidative coupling reactions to form heavier hydrocarbons, such as ethane and ethene. No literature to the contrary was found and there is high confidence in the operating temperature range indicated.

(2) Methanol and other desirable products are intermediate oxidation products. Oxidation of methanol and the other oxygenates is easier than oxidation of methane. This is an example of an "A→B→C" reaction where "B" is desired product. With increasing O$_2$ feed concentration the selectivity to methanol decreases, irrespective of the temperature. Since the O$_2$ concentration in the feed limits the conversion, this is a typical conversion-selectivity

trade-off. In order to minimize combustion reactions (eqs. 4 and 5), it was claimed that the oxidant concentration must not exceed 2.8% and should ideally be in the range 2.0–2.8% [17]. There is also a restriction on the methane to O_2 ratio, with methanol selectivity being quickly eroded as the methane to O_2 ratio decreases below 30:1.

(3) Operating at higher pressures is beneficial to methanol selectivity and the overall yield of oxygenates (eqs. 1–3). This is understandable from Le Chatelier's principle, because methanol synthesis causes molar contraction, whereas the converse is true for oxidation to CO_x (eqs. 4–5). At pressures below 8 MPa, the oxygenate yield decreases monotonously with pressure, but at pressures above 8 MPa the reaction becomes fairly insensitive to pressure. The suggested minimum practical pressure of operation is, therefore, around 8 MPa. Less was published on pressure effects and the threshold pressure value is a best estimate. There is moderate confidence in the minimum operating pressure threshold.

(4) In the temperature range 370–450°C at typical operating conditions described above, the selectivity to methanol is around 40%, the selectivity to methanal increases from 5% to 13% with increasing temperature and the selectivity to carboxylic acids increased from 0.5% to 0.7% with increasing temperature. There is only moderate to low confidence in the selectivity values, which represent a best estimate based on literature.

(5) Cofeeding free radical initiators can initiate the oxidation reaction and eliminate the induction time when employing O_2 as oxidant. This strategy was employed for formaldehyde production by direct partial oxidation of methane, where a small quantity of NO was cofed [4, 11]. Direct methane to methanol conversion making use of N_2O with a catalyst to conduct the reaction at lower temperature was reported [19], and low temperature methane to methanol synthesis could also be achieved by NO catalyst activation [20].

As an alternative to oxygen as oxidant, hydrogen peroxide (H_2O_2) in combination with a catalyst enables low temperature oxidation of methane to methanol [21]. Other examples can also be found in the literature. Many oxidants are capable of oxidation at the carbon atom [22, 23]. The selection of an appropriate oxidant is an economic decision, since oxidant consumption is stoichiometric with respect to the oxidation products. Air or O_2 separated from air is often the most economical choice for oxidation of a total feed stream, as is the case in methane to methanol conversion.

The reaction network is complex, which is typical of any partial oxidation reaction involving a free radical mechanism. The reactions that were listed (eqs. 1–5) are not a reaction network. A far more complex interrelationship exists between the various species. Not all reactions in such a reaction network require oxygen to be present. Thermal decomposition and disproportionation reactions can proceed without oxygen. For example, the thermal decomposition of methanal to produce CO, H_2 and methanol (eqs. 6–7) [24].

$$CH_2O \rightarrow CO + H_2 \qquad (6)$$

$$2CH_2O \rightarrow CH_3OH + CO \qquad (7)$$

There is no lack of information on the chemistry, but constructing a credible reaction network and obtaining kinetic data for the individual reactions are more challenging tasks. This information was conspicuously absent from discussions on the process chemistry of direct methane to methanol conversion.

Process Description

Process flow diagrams for partial oxidation of hydrocarbons with recycle, can be found in the literature [4, 9, 11, 12]. For the purpose of an engineering evaluation, it is necessary to develop the key elements of the process more fully, so that the utility requirements can be determined. The utility footprint of GTL processes that employ synthesis gas is substantial [25]. Although the direct methane to methanol conversion process does not involve synthesis gas generation, many of the process elements leading to a high utility footprint are still present, such as oxidant compression, gas recycling, and high temperature operation.

The process flow diagram of direct methane to methanol conversion is shown in Figure 2. The equipment table is given in Table 1.

There are two raw material streams, natural gas and the oxidant. These input streams are compressed and preheated separately to avoid premature oxidation. Oxidation must be controlled to obtain good selectivity to

Table 1. List of equipment shown in the process flow diagram in Figure 2.

Number	Equipment
C-01	Natural gas feed compressor
C-02	Oxidant feed compressor
C-03	Recycle gas compressor
E-01	Interstage cooler for compressor C-01
E-02	Interstage cooler for compressor C-02
E-03	Oxidant heater (doubles as start-up heater)
E-04	Feed-product heat exchanger
E-05	Product coolers
R-01	Partial oxidation reactor
V-01	High-pressure phase separator vessel

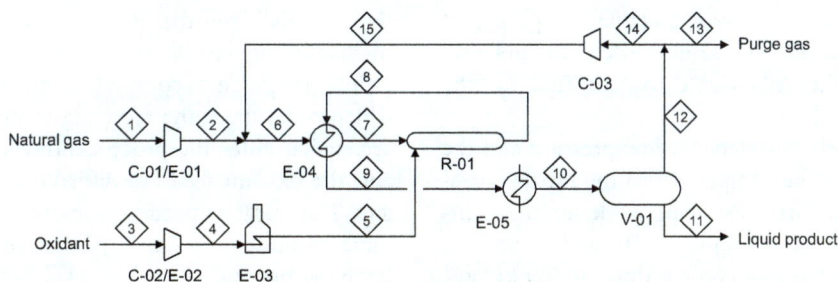

Figure 2. Process flow diagram of direct methane to methanol conversion.

methanol and much of the technology know-how is related to the design of the partial oxidation reactor. In the partial oxidation reactor, the methane is partially oxidized to produce methanol, with the oxidant being the limiting reagent. The product gas is cooled and the methanol is recovered by condensation. The recovered crude methanol and side-products can be separated and purified by conventional distillation. Product work-up is not unique to the direct methane to methanol synthesis process and it is not shown or considered in this evaluation. Part of the unconverted natural gas is recycled and part is purged. The purge of some unconverted natural gas is necessary to limit the build-up of inert material in the process.

The main elements of the design and design decisions associated with each will be discussed.

Natural gas feed

The hydrocarbon feed to the process is natural gas and not pure methane. In most instances, the natural gas will be available at an elevated pressure. The gas pressure will depend on the source of the natural gas. If it is associated gas, which is coproduced with crude oil, the gas may only be available at low pressure, for example, 0.2 MPa, while connected natural gas in a distribution network will typically be available at higher pressure, for example, 8 MPa [26]. The extent of compression and intercooling required for the process depends on the feed conditions and the calculated relationship is shown (Fig. 3). The ideal shaft work is shown, from which the actual shaft work and utility requirements can be calculated. The cooling duty was calculated based on the intercooling requirements for a realistic compressor design. Cooling duty becomes zero at an inlet pressure of ~1.2 MPa, when no intercooling is required for the compressor design.

The composition of natural gas is origin dependent and the main associated gases are other hydrocarbons, carbon dioxide (CO_2), nitrogen (N_2), and hydrogen sulfide (H_2S) [27]. Of these, nitrogen is the most difficult to separate from the unconverted natural gas, because it

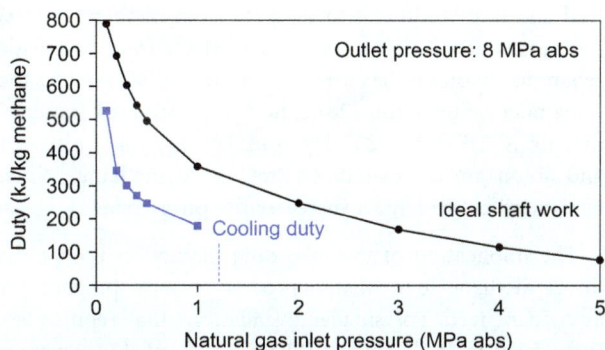

Figure 3. Ideal shaft work and cooling duty for natural gas compression to 8 MPa.

requires cryogenic distillation, or some form of pressure swing adsorption. Thus, of all the natural gas components nitrogen will have the most impact on the purge rate. For the purpose of this evaluation, a generic natural gas composition was employed (Table 2). This gas composition does not include H_2S or Hg, which are likely to be present in some sources of natural gas [27].

Oxidant feed

It is in principle possible to use air, oxygen-enriched air, or oxygen as oxidant feed for the direct methane to meth-

Table 2. Generic natural gas composition.

Compound	Natural gas
Composition (mol fraction)	
Methane (CH_4)	0.950
Ethane (C_2H_6)	0.005
Nitrogen (N_2)	0.030
Carbon dioxide (CO_2)	0.015
Average molecular mass (g · mol^{-1})	16.89
Normal gas density (kg·m^{-3})[1]	0.754
Lower heating value (MJ·m^{-3})	34.3
Higher heating value (MJ·m^{-3})	38.0

[1]Normal conditions: 0°C (273.15 K) and 1 atm (101.325 kPa).

anol conversion process. The cost and the complexity of the process increase in the same order. In practice, purified oxygen is the only viable oxidant feed for the process:

(1) The oxidant feed is available at low pressure and the oxidant feed must also be compressed to the process pressure, around 8 MPa. The ideal shaft work for the compression of pure O_2 is 400 kJ·kg^{-1} O_2 and a realistic compressor design requires a cooling duty of 300 kJ·kg^{-1} O_2 for interstage compressor cooling. For air, these duties change to 1.9 and 1.4 MJ·kg^{-1} O_2, respectively.

(2) Any inert material introduced as part of the oxidant feed and that is difficult to separate from methane, affects the purge rate of the process. Purified O_2 from cryogenic separation typically has a purity of around 99.5% O_2, the remainder being argon [28]. The composition of standard dry air is 21% O_2, 78% N_2, and 1% Ar. Both nitrogen and argon are difficult to separate from methane. When air is used as oxidant, it increases the purge rate.

The implication of the preceding discussion is that the direct methane to methanol process requires purified O_2 as oxidant feed. For smaller installations that require less than 5000 m^3·h^{-1} normal (2 kg·sec^{-1}) of O_2, vacuum pressure swing adsorption (VPSA) can be considered, but the purity of the oxygen is only in the range 90–93% O_2 [28]. For larger capacities or to reduce the impact of the inert gas in the oxidant feed, cryogenic air separation is required. This is a utility unit and it is not reflected in either Figures 1 or 2, but it clearly will have an impact on the size and cost of the process.

Liquid product

The oxygenate products, methanol, methanal, and methanoic acid (eqs. 1–3), are recovered as a crude liquid aqueous product mixture by condensation under pressure. In addition to these products, some of the CO_2 and H_2S are removed by dissolution under pressure in the liquid phase [29, 30]. The temperature at which phase separation is performed affects the relative amounts of products and soluble gases that are removed in the liquid phase. A small amount of the inert gases are also removed, but it is insufficient to affect the purge rate. For example, the solubility of N_2 in water is about 0.1 mol % at 7 MPa [31].

Purge gas product

It was pointed out that the inert gas content of the natural gas feed and the oxidant feed both affect the purge rate. This is a generic chemical engineering problem that is encountered whenever a gas recycle is employed. The problem stems from the difficulty in separating the reactive material from the inert material. In direct methane to methanol conversion, the process chemistry dictates that a low single-pass conversion of methane is required, which necessitates the recycling of methane. Nitrogen and argon that enter the process through the natural gas feed and the oxidant feed are difficult to separate from methane. The purge is needed to ensure material balance. The same amount of N_2 and Ar that enter the process must leave the process.

In practice, there are three design decisions that affect the purge rate. First, the feed composition can be selected or manipulated to ensure that the least amount of inert material enters the process. Second, a separation technology can be installed to selectively remove inert material from the recycle gas. This is unlikely to be a practical option for this process. The separation of methane from N_2 and Ar requires cryogenic distillation with prior CO_2 removal to avoid solid CO_2 plugging of the cryogenic section [28]. Third, the concentration of the inert material in the recycle gas can be manipulated.

The impact of the inert composition of the oxidant feed and the inert concentration in the recycle gas on the purge rate of methane was calculated and it is illustrated by Figure 4. In the calculations, the natural gas feed composition in Table 2 and 30:1 methane to O_2 feed ratio were employed. For 99.5% O_2 from cryogenic separation the least amount of methane is lost through the purge. For air, which has only 21% O_2, the methane loss through the purge is substantial, even at a high inert concentration in the recycle gas.

The calculated methane loss through the purge gas (Fig. 4) is also affected by the inert content in the natural gas feed. The effect of the inert content in the natural gas feed decreases in significance as its contribution to the overall inert content entering the process decreases. For example, if the N_2 content in the natural gas is 1% instead of 3% (Table 2), the methane loss with a 99.5%

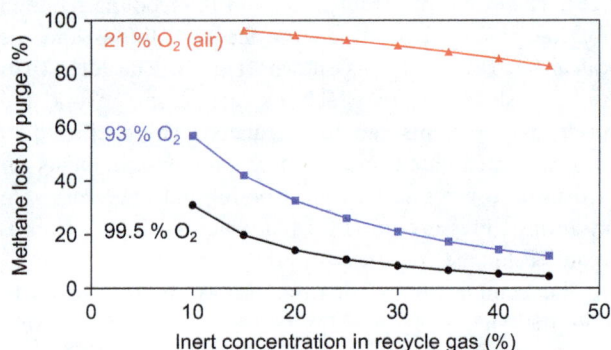

Figure 4. Impact of O_2 purity in oxidant feed and inert concentration in the recycle gas on the methane loss through the purge gas.

O_2 feed is 60% smaller than that indicated in Figure 4, but with a 93% O_2 feed the methane loss is decreased by only 20%. The methane loss is determined by the total amount of inert material that enters the process.

The methane in the purge gas can be employed as fuel gas in the process. Since the volumetric heating value of the purge gas decreases with increasing inert concentration, there is a trade-off between the fuel gas needs, fuel gas quality and the design of the gas recycle.

Gas recycle

The size of the gas recycle determines the sizing of equipment in the gas loop. With reference to Figure 2, the equipment in the gas loop is the partial oxidation reactor (R-01), feed-product heat exchanger (E-04), product coolers (E-05), phase separator vessel (V-01), and recycle gas compressor (C-03).

The gas recycle is particularly large in comparison to the fresh feed streams due to the low single-pass conversion. For an ideal reactor and separator with no inert materials in the feed, the ratio of the recycle flow rate to the fresh feed flow rate can be calculated in terms of the single-pass conversion, x (eq. 8):

$$\text{Recycle ratio} = (1 - x)/x. \qquad (8)$$

In the direct methane to methanol process, the single-pass conversion is limited to around 3% by limiting the oxidant in the feed to the partial oxidation reactor. A recycle to fresh feed ratio of ~30:1 is anticipated from equation 8, the exact ratio depending on the efficiency of separation and the impact of the purge rate. The recycle to fresh feed ratio increases as the design concentration of inert material in the gas recycle is increased. The recycle to fresh feed ratio also increases as the inert content in the feed materials decrease, because the system becomes more efficient and less methane is lost through the purge (i.e., not recycled). The calculated impact of O_2 purity on the inert concentration in the recycle is shown in Figure 5. A lower recycle ratio does not necessarily imply that in absolute terms the flow rate through the partial oxidation reactor is less, because purge losses are compensated for by an increased fresh feed flow rate.

Partial oxidation reactor

Product selectivity depends on temperature, pressure, O_2 concentration, and residence time. The proposals for reactor designs and improvements in reactor designs are based partly on an appreciation of the reaction variables that need to be controlled and on empirical observations. It was pointed out that a well-founded reaction network with kinetic data is conspicuously absent from the literature

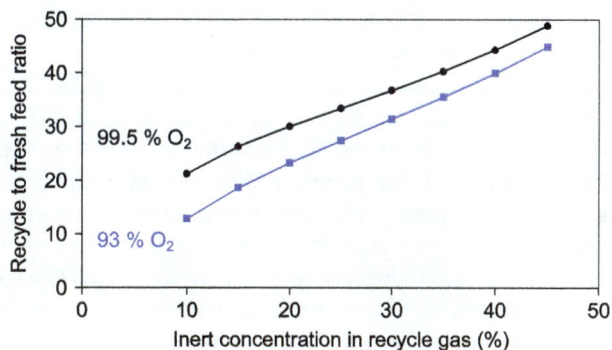

Figure 5. Impact of O_2 purity in oxidant feed and inert concentration in the recycle gas on the recycle to fresh feed ratio of the process.

dealing with direct methane to methanol conversion. To add to this uncertainty, reaction rate and selectivity are affected by the construction material of the reactor [4, 18].

Some general observations about reactor design can be made that agree with experimental observations and reactor designs suggested for this process.

(1) Temperature will affect the reaction rate of oxidation, but it will also affect the contribution of thermal decomposition. Since partial oxidation reactions are exothermic (Table 3), it is preferable to limit the temperature, in particular to avoid loss of methanal (eqs. 6–7) even though some methanal is converted to methanol by thermal decomposition. Heat management is important.

(2) Control of the residence time is not independent of temperature. There are three aspects to consider. First, thermal decomposition reactions can take place independently of the other species present and there will be a threshold temperature below which these reactions become negligible. Control of the residence time, independent of oxygen consumption, is therefore important when the temperature is high enough for thermal decomposition reactions to take place. Second, yield of an intermediate product in a simple reaction is determined by the rate constant of the formation of the intermediate product (k_1) in relation to the rate constant for the consumption of the intermediate product (k_2). Since these reactions typically have different activation energies, the ratio of k_1/k_2 is dependent on the temperature. In a complex reaction network, the time–temperature relationship becomes

Table 3. Standard heats of reaction (ΔH_r) for methane oxidation reactions.

Product	Equation	ΔH_r (kJ · mol^{-1})
Methanol	1	−127
Methanal (formaldehyde)	2	−276
Methanoic acid (formic acid)	3	−546
Carbon monoxide	4	−519
Carbon dioxide	5	−802

important, because it is the global optimum that must be achieved, not the optimum for a single product. Third, there is an induction period in free radical oxidation that can be shortened or eliminated by the use of an initiator. During the induction period, oxidation is negligible. The length of the induction period is dependent on the temperature and it is actually the time–temperature profile once O_2 comes into contact with the methane that determines the length of the induction period. There is of course also an implicit risk for the reactor design if the absence of initiators in the reactor feed cannot be guaranteed.

(3) Oxygen availability determines product selectivity. The methane to O_2 ratio threshold assumes a well-mixed system. What this implies is that after mixing the methane and O_2, the mixture must be homogenized before the end of the induction period.

The detailed reactor design is very important. Although the reactor is not very complex, the design of the reactor is very complex. Scale-up without a fundamental description of the reaction network and the flow dynamics is likely to involve trail-and-error. In this respect, history serves as lesson. The oxygenate yield that was predicted Gutehoffnungschütte-process for direct methane to methanal conversion was 35%, but on scale-up the actual yield was only around 10% [4].

Process Evaluation

Material balance

A material balance for the process (Table 4) was calculated based on the natural gas composition in Table 2

and a natural gas feed rate of 1 kg · sec^{-1}. The feed basis is equivalent to a volumetric natural gas flow rate at standard conditions of 115,000 m^3 per day (4 million scf per day). With reference to Figure 2, other important assumptions and design decisions that were made in the preparation of the material balance are:

(1) The natural gas feed (stream 1) is available at 25°C and 2.0 MPa absolute. In practice, this value will be dependent on the nature of the natural gas resource. For example, natural gas from deep reservoirs will have a much higher pressure, whereas associated natural gas available after crude oil depressurization may be available at lower pressure [26].

(2) The recycle (stream 15) to fresh feed (stream 1) ratio was fixed at 30:1 on a volumetric basis. It is not claimed that this is the optimum, but it provides a credible trade-off between recycling, inert content in the recycle and purging rate. In principle this number can be increased, but not by much, otherwise it invalidates the constraints on reactor operation for the product selectivity values employed.

(3) The oxidant feed (stream 3) is 99.5% O_2 from a cryogenic air separation unit. The only inert introduced in this way is 0.5% Ar. The oxidant is available at 25°C and 0.1 MPa absolute.

(4) The oxidant feed rate was fixed at 2.5% O_2 concentration on a volumetric basis in the combined reactor feed. This value resulted in a methane to O_2 ratio just over 30:1.

(5) The reactor inlet conditions are 8 MPa and 370°C.

(6) The overall pressure drop between the feed supply and product recovery is 500 kPa.

(7) Recoveries for the oxygenate products in the liquid phase is 99.9%. This is based on high-pressure recovery at 25°C by phase separation (V-01).

Table 4. Material balance for direct methane to methanol conversion process shown in Figure 2.

	Flow rate (kg · sec^{-1})[1]						
Compounds	Gas feed #1	Oxidant feed #3	Reactor feed #7	Reactor product #8	Liquid product #11	Purge gas #13	Recycle #15
CH_4	0.902	–	25.485	24.892	–	0.309	24.583
C_2H_6	0.009	–	0.009	0.000	–	0.000	0.000
O_2	–	1.514	1.551	0.038	–	0.000	0.037
N_2	0.050	–	3.999	3.999	–	0.050	3.949
Ar	–	0.009	0.764	0.764	–	0.009	0.755
CO	–	–	0.409	0.414	–	0.005	0.409
CO_2	0.039	–	3.035	3.792	0.758	0.038	2.996
H_2O	–	–	0.000	0.716	0.716	0.000	0.000
Alcohols	–	–	0.000	0.480	0.479	0.000	0.000
Aldehydes	–	–	0.000	0.146	0.146	0.000	0.000
Carboxylic acids	–	–	0.000	0.012	0.012	0.000	0.000
Σ (kg · sec^{-1})	1.000	1.523	35.252	35.252	2.111	0.412	32.729
Molar flow (mol · sec^{-1})	59.2	47.5	1882.8	1882.8	77.0	22.4	1776.0

[1]Values that are 0.000 represent flow rates that are <0.0005 kg · sec^{-1}.

(8) The CO_2 removal due to dissolution in the aqueous product is 20% based on partial pressure of CO_2 of the system [30]. Although some of the other gases also dissolve to some extent in the liquid product, these contributions were neglected.

(9) Conversion of O_2 in the reactor (R-01) is complete.

(10) Conversion of CO present in the recycle gas to the reactor is complete. The ratio of $CO:CO_2$ produced by partial oxidation of methane was assumed to be 85:15 based on typical values from literature [4, 18].

(11) The product selectivities are based on what has been claimed in literature [4, 17]. Both material and atom balance closure was ensured. Water production was calculated under the assumption that no H_2 will be produced. The molar ratio of water to methanol reported was 2.4 [4], and was reasonably close to the value of 2.7 obtained from material and atom balance closure.

(12) Ethane in the natural gas feed is converted into ethanol, ethanal (acetaldehyde), and ethanoic acid (acetic acid). Conversion of ethane is assumed to be complete, because it is easier to convert heavier hydrocarbons than methane.

The natural gas feed and oxidant feed flow rates are of the same order of magnitude, but even combined, these flow rates are an order of magnitude smaller than the recycle (Table 4). The low single-pass methane conversion and the need for a large methane recycle dominates the material flow in the process. The material balance supports the qualitative predictions that were made using simplified calculations (Fig. 5). Likewise, the methane loss through the purge is substantial, as qualitatively predicted using simplified calculations (Fig. 4).

Energy and utility use

The main energy and utility consumers in the direct methane to methanol process are the compressors, heat exchangers, and air separation unit. Additional energy and utilities are required during product purification and treatment of waste streams, but as indicated before, these units were excluded from the scope of the evaluation.

The energy and utility consumption for the process is given in Table 5. The values for utility use are influenced by the process design and equipment selection.

Due to the potential application of this process for small-scale GTL, reciprocating compressors, rather than axial compressors might be a more realistic selection [32]. A compression efficiency of 0.9 and a mechanical efficiency of 0.9 were employed for shaft work calculations. The use of natural gas powered internal combustion engines is more practical for a small-scale GTL application since the infrastructure for delivering the required amount of electric power might not be available [33]. The power consumption was converted into a natural gas consumption for internal combustion engine drives, with a drive efficiency of 0.37 [34]. As an alternative, the purge gas can be employed instead of natural gas, albeit with some loss in efficiency. The purge gas is 86 vol % methane with a lower heating value (LHV) of 31.0 MJ · m^{-1} and higher heating value (HHV) of 34.4 MJ · m^{-3}.

The heat exchanger network was not optimized and it is likely that a slightly lower utility use can be achieved by better heat integration. One fired heater is necessary for start-up, even though a steady-state optimization will show that it is not necessary. Product cooling (E-05) is

Table 5. Energy and utility use in direct methane to methanol conversion process shown in Figure 2 and material balance in Table 4.

Equipment or unit	Duty (MW)[1]	Power (MW)[2]	Utilities (kg · sec^{-1})[3]		
			Natural gas	Cooling water	Steam[4]
Natural gas feed compressor (C-01)	0.29	0.79	0.017	–	–
Oxidant feed compressor (C-02)	0.75	2.03	0.045	–	–
Recycle gas compressor (C-03)	0.35	0.94	0.021	–	–
Interstage cooler for C-01 (E-01)	–	–	–	–	–
Interstage cooler for C-02 (E-02)	0.46	–	–	5.4	–
Oxidant heater (E-03)	0.35	0.41	0.009	–	–
Feed-product heat exchanger (E-04)	31.02	–	–	–	–
Product coolers (E-05)	7.20	–	–	86.0	–
Partial oxidation reactor (R-01)	7.89	–	–	–	-3.7
Air separation unit	0.88	2.38	0.052	4.1	–
Σ			0.144	95.5	-3.7

[1]Shaft work for compressors, heat flow for heat exchangers.
[2]Depends on compressor drive-type or heat source employed for heating.
[3]Positive is consumption, negative is production.
[4]Low pressure steam, 150°C, 400 kPa gauge.

shown as a single exchanger, but in practice it will be multiple exchangers, which was considered in the calculations. The assumption was made that all cooling duty is supplied by cooling water, although it is likely that in practice air coolers will be used for some of that cooling duty. It was further assumed that cooling water is available at 20°C and that 40°C was an acceptable cooling water return temperature. For small-scale GTL application, a closed cooling water system is likely to be more practical, albeit more expensive, since sufficient fresh water might not be available to allow evaporative cooling [33].

Although product purification was outside of the scope of the evaluation, some steam will be required for distillation. Low pressure steam, 0.5 MPa absolute, was produced where practical.

The work required by the air separation unit to produce 99.5% O_2 is 0.35 kWh per normal m^3 of unpressurized oxygen [28]. This is equivalent to 0.88 MW $\cdot$ kg^{-1} O_2. The air separation unit also requires cooling water for the initial cooling of the outlet of the primary air compressor [28]. Using the same utility assumptions as for the rest of this evaluation, the required cooling water flow is 4.1 kg $\cdot$ kg^{-1} O_2.

It is best to view the utility requirements in Table 5 as indicators of the utility footprint. In an optimized design the absolute numbers will differ, but the numbers will be of the same order of magnitude. The process requires much cooling capacity and power for gas compression. Compared to the natural gas feed, an additional 14% natural gas is required for heating and power. The cooling water flow rate is 95 kg $\cdot$ kg^{-1} natural gas, that is, almost two orders of magnitude more on a mass flow basis.

A large utility footprint is a common feature of GTL facilities and one of the challenges in the development of a credible process configuration for small-scale GTL applications [25, 35].

Process efficiency

There are various metrics that can be used to assess the efficiency of processes. Two of the most commonly employed are carbon efficiency and thermal efficiency. The carbon efficiency is defined as the fraction of the carbon in the feed to the process that is contained in useful products. The thermal efficiency is defined as the fraction of the LHV of the feed to the process that is retained by the useful products and it is more often employed when the products are fuels. In both instances, it is important to be clear on the system boundary. These values can be calculated for the process streams only, or they can be calculated for the overall process. When the process is not energy intensive there will be little difference in the calcu-

lated efficiencies, but not for energy intensive processes. For example, the carbon efficiency of indirect methanol synthesis based on the process streams only is better than 95% [3], but for the overall process it is 65–68% [9]. Somewhat lower efficiency is likely if the indirect methanol synthesis process is scaled down.

The overall process efficiency was assessed by comparing the thermal efficiency and the carbon efficiency of the direct methane to methanol conversion with indirect methanol synthesis (Table 6) [9]. In the calculation of the overall efficiency of the direct methane to methanol process it was assumed that the purge gas can be used as fuel gas to supply all of the energy of compression and heating requirements (Table 5), so that no additional natural gas is consumed for these purposes.

Strictly speaking the comparison in Table 6 is not on the same basis, because the efficiency values for indirect methanol synthesis include carbon and energy losses from product purification, and it is at larger scale. However, the direct methane to methanol process has excess heating value available as steam and from the use of the purge gas as fuel gas. However, the difference between the two processes is so large that the difference in scale and battery limit does not change the conclusion.

It is clear that the overall process efficiency of indirect methanol synthesis is much higher than that of direct methane to methanol conversion. The main reason for this is reaction selectivity. During direct methane to methanol synthesis, the production of CO and CO_2 (eqs. 4–5) is undesirable and these reactions erode carbon efficiency. Conversely, during indirect methanol synthesis, CO and CO_2 are potential product-forming reactants. The single-pass conversion during indirect methanol synthesis is equilibrium limited and not limited by side-reactions. In fact, the selectivity to methanol even at near equilibrium conversion is of the order of 99% [2]. Thus, almost no carbon that enters the gas loop during indirect methanol synthesis is lost due to irreversible side-reactions, whereas the irreversible carbon loss due to side-reactions during direct methane to methanol synthesis is 40–50% (46% based on the assumptions used in this work). Based on the calculated efficiencies (Table 6), direct partial oxidation of methane to methanol must

Table 6. Comparison of overall process efficiency for direct and indirect methanol synthesis.

Description	Overall process efficiency (%)	
	Carbon	Thermal
Direct methane to methanol	35	28
Indirect methanol synthesis	65–68	51–54

achieve a combined methanol and methanal selectivity of the order 90% to become competitive with indirect methane to methanol synthesis.

Discussion

The objective of the present investigation was to perform an engineering evaluation of the process and evaluate the potential of this process for small-scale GTL applications.

There is reasonable agreement between the heat transfer and compression duties reported by Kuo [9], and that calculated during the present investigation. These duties cannot be avoided even though more optimized heat integration would change the numeric values somewhat. However, the overall outcome of the evaluation was more pessimistic than the evaluation by Kuo. This is primarily due to a difference in process chemistry, with Kuo's analysis being based on the optimistic methanol selectivity of 90% and 15% single-pass methane conversion. These optimistic values were based on the work by Gesser, Hunter, and coworkers [36]. The bulk of the literature suggests that such methanol selectivities are not obtained [4]. There are also fundamental reasons why such optimistic data are questionable, notably the implicit adiabatic temperature increase and the relative ease of further oxidation of methanol compared to oxidation of methane.

One of the major disappointments from this study was that a credible engineering design for the direct methane to methanol process required high purity O_2. The anticipated cost and complexity advantage over indirect methane to methanol synthesis could, therefore, not be realized, because an air separation unit was required. Hence, for small-scale GTL applications, direct methane to methanol had no specific advantage over indirect methane to methanol. When this is combined with the low methanol selectivity and low single-pass methane conversion, there is little technical incentive to justify further interest in direct oxidation of methane as a potential process for methanol synthesis. Furthermore, based on the fundamentals of gas recycling and the fundamentals of free radical oxidation, it is doubtful that direct methane to methanol conversion can ever be competitive with indirect methane to methanol conversion.

Conclusions

The objective of the present investigation was to perform an engineering evaluation of the direct methane to methanol conversion process and evaluate the potential of this process for small-scale GTL applications. The most noteworthy observations from the engineering evaluation were:
(1) An air separation unit was required to provide high purity O_2 as oxidant. The process cannot be operated with air as the oxidant feed. The need for an air separation unit diluted the benefit of avoiding synthesis gas generation as a process step (Fig. 1). If one considers synthesis gas generation by autothermal reforming and its associated air separation unit, only part of the complexity and cost is avoided by direct methane to methanol conversion.
(2) Natural gas is not pure methane and oxygen from air separation is not pure O_2. The introduction of some inert gases is unavoidable and this affects the design. The recycle ratio, capacity of the gas loop, and purge gas loss all depend on the amount of inert gases introduced in the feed.
(3) The utility footprint of the process is significant, with compressor and cooling duties dominating utility requirements. This is typical of GTL processes. The large utility footprint of GTL processes is one of the key challenges for small-scale GTL applications and in this respect direct methane to methanol conversion offers little benefit.
(4) Overall the carbon efficiency is 35% and the thermal efficiency is 28%, which is about half of that of indirect methanol synthesis (65–68% and 51–54%, respectively).
(5) Low oxygenate selectivity during direct partial oxidation of methane is the largest contributor to the loss of process efficiency compared to indirect synthesis of methanol. The process chemistry is burdened by some fundamental constraints. First, the reactivity of methanol and methanal (desirable products) to further oxidation is higher than that of methane. Second, there are thermal reactions that can proceed in the absence of the oxidant to decrease oxygenate selectivity, so that limiting oxidant availability on its own is insufficient to achieve better selectivity. Third, CO and CO_2 are fatal products and these products represent an irreversible carbon loss.

Conflict of Interest

None declared.

References

1. Tijm, P. J. A., F. J. Waller, and D. M. Brown. 2001. Methanol technology developments for the new millennium. Appl. Catal. A 221:275–282.
2. Lee, S. 1990. Methanol synthesis technology. CRC Press, Boca Raton.
3. Rostrup-Nielsen, J., and L. J. Christiansen. 2011. Concepts in syngas manufacture. Imperial College Press, London.
4. Arutyunov, V. 2014. Direct methane to methanol. Elsevier, Amsterdam.
5. De Klerk, A., and V. Prasad. 2012. Methane for transportation fuel and chemical production. Pp. 327–384 in T. M. Letcher and J. L. Scott, eds. Materials for a sustainable future. Royal Society of Chemistry, Cambridge, UK.

6. Alvarez-Galvan, M. C., N. Mota, M. Ojeda, S. Rojas, R. M. Navarro, and J. L. G. Fierro. 2011. Direct methane conversion routes to chemicals and fuels. Catal. Today 171:15–23.

7. Holmen, A. 2009. Direct conversion of methane to fuels and chemicals. Catal. Today 142:2–8.

8. Lunsford, J. H. 2000. Catalytic conversion of methane to more useful chemicals and fuels: a challenge for the 21st century. Catal. Today 63:165–174.

9. Kuo, J. C. W. 1992. Engineering evaluation of direct methane conversion processes. Pp. 483–526 in E. E. Wolf, ed. Methane conversion by oxidative processes. Van Nostrand Reinhold, New York.

10. Kuo, J. C. W., C. T. Kresge, and R. E. Palermo. 1989. Evaluation of direct methane conversion to higher hydrocarbons and oxygenates. Catal. Today 4:463–470.

11. Asinger, F. 1968. Paraffins chemistry and technology. Pergamon Press, Oxford.

12. Sittig, M. 1962. Combining oxygen and hydrocarbons for profit. Gulf Publishing Co., Houston.

13. Hubert, A. J. 1983. Methane. Pp. 245–261 in W. Keim, ed. Catalysis in C1 chemistry. D. Riedel Publishing Company, Dordrecht.

14. Gesser, H. D., and N. R. Hunter. 1992. The direct conversion of methane to methanol (DMTM). Pp. 403–425 in E. E. Wolf, ed. Methane conversion by oxidative processes. Van Nostrand Reinhold, New York.

15. Blanksby, S. J., and G. B. Ellison. 2003. Bond dissociation energies of organic molecules. Acc. Chem. Res. 36:255–263.

16. Han, J.-S., C.-M. Ahn, B. Mahanty, and C.-G. Kim. 2013. Partial oxidative conversion of methane to methanol through selective inhibition of methanol dehydrogenase in methanotrophic consortium from landfill cover soil. Appl. Biochem. Biotechnol. 171:1487–1499.

17. Pawlak, N. A., V. I. Vedeneev, and R. W. Carr. Method and apparatus for producing methanol. Patent US 8293186, 2012.

18. Shtern, V. Y. 1964. The gas phase oxidation of hydrocarbons. Pergamon Press, New York.

19. Starokon, E. V., M. V. Parfenov, S. S. Arzumanov, L. V. Pirutko, A. G. Stepanov, and G. I. Panov. 2013. Oxidation of methane to methanol on the surface of FeZSM-5 zeolite. J. Catal. 300:47–54.

20. Sheppard, T., C. D. Hamill, A. Goguet, D. W. Rooney, and J. M. Thompson. 2014. A low temperature, isothermal gas-phase system for conversion of methane to methanol over Cu-ZSM-5. Chem. Comm. 50:11053–11055.

21. Hammond, C., N. Dimitratos, R. L. Jenkins, J. A. Lopez-Sanchez, S. A. Kondrat, A. Hasbi, et al. 2013.

Elucidation and evolution of the active component within Cu/Fe/ZSM-5 for catalytic methane oxidation: from synthesis to catalysis. ACS Catal. 3:689–699.

22. Chinn, L. J. 1971. Selection of oxidants in synthesis. Oxidation at the carbon atom; Marcel Dekker, New York.

23. Hudlický, M. 1990. Oxidations in organic chemistry (ACS Monograph Ser. 186). American Chemical Society, Washington DC.

24. Steacie, E. W. R. 1954. Atomic and free radical reactions. 2 ed. (ACS Monograph Ser. 125), Reinhold Publishing Corporation, New York, NY.

25. Zennaro, R. 2013. Fischer-Tropsch process economics. Pp. 149–169 in P. M. Maitlis and A. De Klerk, eds. Greener Fischer-Tropsch processes for fuels and feedstocks. Wiley-VCH, Weinheim.

26. Szilas, A. P. 1986. Production and transport of oil and gas. 2 ed. Elsevier, Amsterdam.

27. Peebles, M. W. H. 1992. Natural gas fundamentals. Shell, London.

28. Häring, H. -W. ed. 2008. Industrial gases processing. Wiley-VCH, Weinheim.

29. Carroll, J. J., and A. E. Mather. 1989. The solubility of hydrogen sulphide in water from 0 to 90°C and pressures to 1 MPa. Geochim. Cosmochim. Acta 53:1163–1170.

30. Mason, D. M., and R. Kao. 1980. Correlation of vapor-liquid equilibria of aqueous condensates from coal processing. ACS Symp. Ser. 133:107–138.

31. Chapoy, A., A. H. Mohammadi, B. Tohidi, and D. Richon. 2004. Gas solubility measurement and modeling for the nitrogen + water system from 274.18 K to 363.02 K. J. Chem. Eng. Data 49:1110–1115.

32. Bloch, H. P. 2006. A practical guide to compressor technology. 2 ed. Wiley-Interscience, Hoboken, NJ.

33. De Klerk, A. 2014a. Consider technology implications for small-scale Fischer-Tropsch GTL. Gas Process July/August:41–48.

34. Ulrich, G. D., and P. T. Vasudevan. 2004. Chemical engineering process design and economics. A practical guide. 2 ed. Process Publishing, Durham, NH.

35. De Klerk, A. 2014b. Small-scale gas-to-liquids using Fischer-Tropsch synthesis: opportunity or myth?. Prepr. Pap. -Am. Chem. Soc. Div. Energy Fuels 59:823–824.

36. Gesser, H. D., N. R. Hunter, and C. B. Prakash. 1985. The direct conversion of methane to methanol by controlled oxidation. Chem. Rev. 85:235–244.

Permissions

List of Contributors

Y.-H Percival Zhang
Biological Systems Engineering Department, Virginia Tech, 304 Seitz Hall, Blacksburg, Virginia, 24061
Institute for Critical Technology and Applied Science (ICTAS), Virginia Tech, Blacksburg, Virginia, 24061
Gate Fuels Inc., 2200 Kraft Drive, Suites 1200B, Blacksburg, Virginia, 24060
Cell-Free Bioinnovations Inc., Blacksburg, Virginia, 24060

Samad Jafarmadar
Mechanical Engineering Department, University of Urmia, Urmia, West Azerbaijan 57561-15311, Iran

Alborz Zehni
Mechanical Engineering Department, University of Sahand, Sahand, Iran

Jibrail Kansedo
School of Chemical Engineering, Universiti Sains Malaysia Engineering Campus, Seri Ampangan, 14300, Nibong Tebal, Pulau Pinang, Malaysia

Keat Teong Lee
School of Chemical Engineering, Universiti Sains Malaysia Engineering Campus, Seri Ampangan, 14300, Nibong Tebal, Pulau Pinang, Malaysia

Lin Yang
Agricultural Economics, Purdue University, 403 West State St., West Lafayette, Indiana 47907

Wallace E Tyner
Agricultural Economics, Purdue University, 403 West State St., West Lafayette, Indiana 47907

Kemal Sarica
Agricultural Economics, Purdue University, 403 West State St., West Lafayette, Indiana 47907

Grisel Corro
Instituto de Ciencias, Benemerita Universidad Autonoma de Puebla, 4 sur 104, 72000 Puebla, Mexico

Umapada Pal
Instituto de Física, Benemerita Universidad Autonoma de Puebla, Apdo. Postal J-48, 72570 Puebla, Mexico

Surinam Cebada
Instituto de Ciencias, Benemerita Universidad Autonoma de Puebla, 4 sur 104, 72000 Puebla, Mexico

Sonil Nanda
Lassonde School of Engineering, York University, Ontario, Canada

Ajay K. Dalai
Department of Chemical and Biological Engineering, University of Saskatchewan, Saskatchewan, Canada

Janusz A. Kozinski
Lassonde School of Engineering, York University, Ontario, Canada

Matias Eriksson
Umea° University, SE-901 87 Umea° , Sweden
NorFraKalk AS, Kometveien 1, 7650 Verdal, Norway
Nordkalk Oy Ab, Skra¨bbo¨ leva¨ gen 18, 21600 Pargas, Finland

Bodil Ho¨ kfors
Umea° University, SE-901 87 Umea° , Sweden
Cementa AB, A° rstaa¨ ngsva¨ gen 25, SE-11743 Stockholm, Sweden

Rainer Backman
Umea° University, SE-901 87 Umea° , Sweden

Robert W. Howarth
Department of Ecology & Evolutionary Biology, Cornell University, Ithaca, New York 14853

Nickson Langat
Kenyatta University, P.O. Box 43844-00100, Nairobi, Kenya

Thomas Thoruwa
Pwani University, P.O. Box 195-80108, Kilifi, Kenya

John Wanyoko
Tea Research Foundation of Kenya, P.O. Box 820-20200, Kericho, Kenya

Jeremiah Kiplagat
Kenyatta University, P.O. Box 43844-00100, Nairobi, Kenya

Brian Plourde
University of St. Thomas, St. Paul, Minnesota, 55105-1079

John Abraham
University of St. Thomas, St. Paul, Minnesota, 55105-1079

Joaquin Sanchez
Laboratorio Nacional de Fusion, CIEMAT, 28040 Madrid, Spain

Motoi Sekine
Chiba University, Matsudo, Chiba, Japan

Yukiharu Ogawa
Chiba University, Matsudo, Chiba, Japan

Nobuhiro Matsuoka
Chiba University, Matsudo, Chiba, Japan

Yoshiya Izumi
Biomaterial in Tokyo Co., Ltd., Kashiwa, Chiba, Japan

Morgan L. Thomas
Lassonde School of Engineering, York University, 4700 Keele Street, Toronto, Ontario, Canada M3J 1P3
Department of Chemical and Biological Engineering, University of Saskatchewan, 57 Campus Drive, Saskatoon, Saskatchewan, Canada S7N 5A9

Ian S. Butler
Department of Chemistry, McGill University, 801 Sherbrooke Street West, Montreal, Quebec, Canada H3A 2K6

Janusz A. Kozinski
Lassonde School of Engineering, York University, 4700 Keele Street, Toronto, Ontario, Canada M3J 1P3
Department of Chemical and Biological Engineering, University of Saskatchewan, 57 Campus Drive, Saskatoon, Saskatchewan, Canada S7N 5A9

Darryl D. Siemer
Nuclear Energy Department, Idaho State University, 921 S. 8th Ave Mail Stop 8060 Pocatello, ID 83209-8060, USA

Lingjun Zhu
State Key Laboratory of Clean Energy Utilization, Zhejiang University, Hangzhou 310027, China

Shi Yin
State Key Laboratory of Clean Energy Utilization, Zhejiang University, Hangzhou 310027, China

Qianqian Yin
State Key Laboratory of Clean Energy Utilization, Zhejiang University, Hangzhou 310027, China

Haixia Wang
State Key Laboratory of Clean Energy Utilization, Zhejiang University, Hangzhou 310027, China

Shurong Wang
State Key Laboratory of Clean Energy Utilization, Zhejiang University, Hangzhou 310027, China

Lorenzo Albanese
Istituto di Biometeorologia, CNR, via Caproni 8, 50145 Firenze, Italy

Rosaria Ciriminna
Istituto per lo Studio dei Materiali Nanostrutturati, CNR, via U. La Malfa 153, 90146 Palermo, Italy

Francesco Meneguzzo
Istituto di Biometeorologia, CNR, via Caproni 8, 50145 Firenze, Italy

Mario Pagliaro
Istituto per lo Studio dei Materiali Nanostrutturati, CNR, via U. La Malfa 153, 90146 Palermo, Italy

Ismaila Badmus
Mechanical Engineering Department, University of Ibadan, Nigeria

Richard Olayiwola Fagbenle
Mechanical Engineering Department, University of Ibadan, Nigeria
Mechanical Engineering Department, Obafemi Awolowo University, Ile-Ife, Nigeria

Olanrewaju Miracle Oyewola
Mechanical Engineering Department, University of Ibadan, Nigeria

Heri Yanti
Department of Chemical and Biological Engineering, Chalmers University of Technology, Gothenburg, Sweden
Department of Chemical Engineering, Universitas Gadjah Mada, Yogyakarta, Indonesia

Rachma Wikandari
Swedish Centre of Resource Recovery, University of Boras, Boras, Sweden

Ria Millati
Department of Food and Agricultural Product Technology, Universitas Gadjah Mada, Yogyakarta, Indonesia

Claes Niklasson
Department of Chemical and Biological Engineering, Chalmers University of Technology, Gothenburg, Sweden

Mohammad J. Taherzadeh
Swedish Centre of Resource Recovery, University of Boras, Boras, Sweden

Arno de Klerk
Department of Chemical and Materials Engineering, University of Alberta, Edmonton, Alberta, T6G 2V4, Canada